Fault Diagnosis, Prognosis, and Reliability for Electrical Machines and Drives

Fault Diagnosis, Prognosis, and Reliability for Electrical Machines and Drives

Elias G. Strangas
Michigan State University
East Lansing, Michigan

Guy Clerc
University of Lyon
Villeurbanne, France

Hubert Razik
University of Lyon
Villeurbanne, France

Abdenour Soualhi
University of Lyon
Saint Etienne, France

WILEY

This edition first published 2022

Published by John Wiley & Sons, Inc., Hoboken, New Jersey.
Published simultaneously in Canada.

Library of Congress Cataloging-in-Publication Data
A catalogue record for this book is available from the Library of Congress

Hardback ISBN: 9781119722755; ePub ISBN: 9781119722786; ePDF ISBN: 9781119722809;
oBook ISBN: 9781119722823.

Cover image: © Sandipkumar Patel/Getty
Cover design by Wiley

Set in 9.5/12.5pt STIXTwoText by Integra Software Services Pvt. Ltd, Pondicherry, India

10 9 8 7 6 5 4 3 2 1

MIX
Paper from responsible sources
FSC® C013604

From the authors to their families

Contents

Contributors

Elias Strangas
Department of Electrical and Computer Engineering, Michigan State University, East Lansing, Michigan, USA

Guy Clerc
Univ Lyon, University Claude Bernard Lyon 1, Ecole Centrale de Lyon, INSA Lyon, Ampère Laboratory, CNRS UMR5005, F-69622, Villeurbanne, France

Hubert Razik
Univ Lyon, University Claude Bernard Lyon 1, Ecole Centrale de Lyon, INSA Lyon, Ampère Laboratory, CNRS UMR5005, F-69622, Villeurbanne, France

Abdenour Soualhi
Univ Lyon, University of Jean Monnet, Laboratory LASPI EA-3059, F-42100 Saint Etienne, France

Acknowledgments

The authors would like to thank their doctoral students and collegues for their contributions to research in the area of failure diagnosis and prognosis in the field of the electrical engineering and their contribution to the writing of this book:

- Anmol AGARWAL
- Andrew BABEL
- Mohamed BEN MARZOUG
- Romain BREUNEVAL
- Jorge CINTRON-RIVERA
- Alexandre EID
- Samuel EKE
- Nathan EVANS
- Shanelle FOSTER
- Reemon HADDAD
- Malorie HOLOGNE
- Abdesselam LEBAROUD
- John NEELY
- Rodney SINGLETON
- Diego VELAZCO
- Sajjad ZAIDI
- Wesley ZANARDELLI
- Yang ZETONG
- Shaopo HUANG
- Selin AVIYENTE

A. E. G. H.

Acronyms

α, β	Direct and quadrature axis of a fixed frame
α	Firing delay angle
β	Weibull parameter
λ	Failure rate
λ	Flux linkages
ω	Speed of the rotor for two-pole equivalent machine, electrical rotor speed
ω_m	Mechanical rotor speed, $\omega_s = \omega/p$
ω_r	Angular speed of rotor variables, for an equivalent 2-pole machine
ω_s	Speed of the stator variables for an equivalent two-pole machine
Φ	Flux in Wb
Φ	Transition matrix
B_R	Remanence
H_c	Coercivity
$L_1 0$	The number of revolutions or hours that 90% of a group of apparently identical components will complete or exceed before failure
L_R	Rotor inductance
L_S	Stator inductance
p	Number of pole pairs
s	Slip, $s = (\omega_s - \omega)/\omega_s$
T	Torque in Nm
AC	Alternative current
ANN	Artificial Neural Network
BN	Baysian Network
CBM	Condition-Based Maintenance
CDF	Cumulative Distribution Function
d, q, 0	Direct, quadrature and homopolar axis of a synchronous reference frame

DAG	Directed Acyclic Graph
DC	Direct current
DF	Dissipation factor
EKF	Extended Kalman Filter
EMF	Electromotive Force
FEA	Finite element analysis
FFT	Fast Fourier Transform
FS	Fault signature
FT	Fault tree
FTA	Fault tree analysis
FTC	Fault-tolerant control
GaN	Gallium Arsenide
HHT	Hilbert-Huang Transform
HSCT	High-sensitivity current transformer
HSMM	Hidden Semi-Markov Model
IGBT	Insulated-gate bipolar transistor
ISI	Insulation health indicator
ITSC	Inter-turn short circuit
MBF	Fuzzy Membership Functions
MC	Markov Chain
MC	Monte Carlo
MFC	Metallized film capacitor
MMF	Magnetomotive force, $\mathcal{F}$
MOSFET	Metal–oxide–semiconductor field-effect transistor
MRAS	Model Reference Adaptive System
MTTF	Mean time to failure
MTTR	Mean time to repair
NN	Neural network
PF	Particle filter
PI	Polarization index
PMAC	Permanent magnet AC
PWM	Pulse width modulation
R	Reliability function
RAC	Reliability associated costs
RBD	Reliability block diagram
RFC	Rotor field-oriented control
RMSD	Root-mean-square deviation
RNN	Recurrent neural network
RUL	Remaining useful life
Si	Silicon

SiC	Silicon carbide
STFT	Short-time Fourier transform
SVM	Support vector machine
TF	Time-Frequency
WT	Wavelet transform

Introduction

The progress in electrification of manufacturing processes, transportation, commercial, and residential applications is accelerating exponentially. This movement is supported by an increasing acceptance and use of electrical drives, which have progressed in terms of cost, size, efficiency, and performance. This progress enabled the use of drives in current and new applications that benefit from these characteristics. This resulted in lower environmental pollution, and applications requiring higher flexibility, such as electric and hybrid vehicles, more electric airplanes and electric ships, new energy sources, industrial controls, consumer electronics, health devices, etc.

Electrical drives that are of interest in this book are of widely varying sizes, from large wind generators and ship propulsion systems, down to miniature ones used in medical devices. The drives invariably use an electrical machine, power electronics, controllers, sensors, and occasionally batteries to operate. This ongoing expansion has resulted from progress on many fronts:

- Electrical machines: new materials, advanced manufacturing processes, and accurate and fast design tools
- Power electronics: new switches with higher energy density, switching characteristics, and efficiency, advanced topologies, continuously improving manufacturing
- Advanced control methods and increasingly powerful computation, both at the operational and at the design level
- Increased collected knowledge from extensive experience

Mechanical actuators, motion systems, generators, and controls have always relied on Condition Based Maintenance (CBM) to ensure minimal interruptions of service and safety of people, processes, and equipment. Replacing mechanical actuators and controls with electromechanical ones has resulted in higher overall efficiencies, cost savings, and improved performance. It was then natural to improve CBM through the use of health monitoring. It was made possible, with the sensors and signal processing power which are integrated in electrical drives to enable incipient fault identification and diagnosis and to some degree failure prognosis. Toward that end, operating variables have to be monitored, evaluated, and acted upon. To realize these goals, a supervisory system is needed that resides in one or more CPUs. The implementation of this system possibly needs additional sensors installed on some components and systems to filter and process their outputs, a number of internal or external redundant components, subsystems, or systems, a central or distributed controller with algorithms to identify a fault and its severity, predict its development, and take and then implement automatically. What remained an open question and a challenge, at least for early such systems of diagnosis and prognosis, has been whether they result in improvement of reliability. This question and some skepticism, to a large degree, stem from the increased complexity of drive systems, both in the number of parts and the algorithms used, and to a lesser degree from the distrust of the new. Due to the complexities of drives, the number of failure modes has increased, and the ability of the operator to manage them directly has decreased.

The answer to these issues is the development of drive systems that have demonstrably higher reliability while remaining affordable. This means, at best, long, predictable, uninterrupted, as-designed operation, but also the ability to identify any deviation of the performance at present or in the future, and minimally take action to manage it. Besides detection and identification of incipient faults which can result in failure, and diagnosis of the fault severity, it means accurate determination of the expected time before failure, so that appropriate action should be taken. Such action may include continuing normal operation as designed, modification of control, employment of redundancies, or modification of the required performance or mission. To accomplish the tasks associated with diagnosis, prognosis, and reliability enhancement, two components are required beyond the hardware discussed earlier:

- Models of the drive and its components, based on analytical, physics-based models, on collected data from extensive testing of similar components and drives, and even data collected during the drive operation.
- Algorithms based on a large body of diverse work that includes, among others, signal processing, statistical and Bayesian work, and artificial intelligence tools.

Reliability of a system has been a characteristic that is determined during the design stage, from the reliability of the components, their methods of interaction, and when possible, from past experience of similar systems with the same components. Once decided, the system is launched, and in the classic concept, not modified except through maintenance. Instead, including in the drive at the design stage, a diagnosis and prognosis module integrated with redundancies and decision mechanisms is expected to improve reliability.

Widespread use of fault diagnosis and failure prognosis and enhanced reliability through these have been elusive goals for many years. Despite extensive research results, applications had been limited to niche markets. It is only recently that related research work has translated and resulted in broader adoption in industry. Thousands of scientific papers and many books have preceded this one, discussing in detail the narrow subjects associated with electrical drives, and more broadly, with electromechanical systems. It is the hope of the authors that this book will be a contribution and of use to students and practicing engineers, scientists, and researchers in furthering the application of the rapidly expanding and maturing research results in the interest of improving the safety and the environment.

The book integrates the results of research efforts, framed by the interests and research of the authors, and consists of a discussion of basic tools, applications and to specifics: power electronics, capacitors, batteries, sensors, electrical machines, and closes with the integration of fault diagnosis and failure prognosis to the enhancement of the reliability of electrical drives.

The first chapter deals with the components involved in the diagnosis and prognosis of breakdowns. It presents the different sensors, the signals obtained, their conditioning, and the different methods using the information extracted from these signals. Thus, time, frequency domain, or time frequency representation could be used to build relevant information. Classifiers, neural networks, rain flow, etc., estimate the present or future state of the system from this information.

The second chapter deals with the different components of the electrical drives from the supply to the electrical actuator: static switches, capacitors, asynchronous, and synchronous motors. A presentation of their physics and their different faults is introduced before detailing some applications in the diagnosis and prognosis domain.

The last chapter addresses the use of fault diagnosis and failure prognosis discussed in the previous chapters in the improvement of the reliability of electrical drives. Going beyond the classical definitions of reliability and its estimation at the design stage, it presents how failure prognosis and decision systems enhance reliability. It proposes additional measures of reliability based on the cost of fault management and of failure prognosis, as well as of the necessary modification of the mission profile.

1

Basic Methods and Tools

1.1 General Approach

This first part is devoted to a synthesis of the different methods and concepts used in the various fault diagnosis and failure prognosis techniques (Figure 1.1).

Whatever the approach, these methods require informative data. Therefore we dedicate the first section to the signals that can be extracted from the drive system (Section 1.2). Thus, the different technology of sensors will be presented (Subsection 1.2.1). From these signals are then extracted time, frequency, statistical (Subsection 1.2.2), or model (Subsection 1.2.3) parameters. They could come also from analytical redundancies (Subsection 1.2.4).

However some parameters may be redundant and the problem may be overdetermined. Therefore, a compromise must be found between a minimum number of parameters and sufficient information (Section 1.3). Thus, two main techniques for reducing the size of the pattern vector will be presented: the principal component analysis (Subsection 1.3.1) and the intercorrelation, which measures the dependency of the parameters (Subsection 1.3.2). The quality of the information is measured with the entropy (Subsection 1.3.3). Then, the main techniques for selecting the best parameters in the case of supervised clustering (Subsection 1.3.4) and unsupervised clustering (Subsection 1.3.5) will be detailed.

Another important problem is the determination of the best number of classes (Subsection 1.3.6), especially in the case of unlabeled data. Some criteria are given in the Subsection 1.3.7 to evaluate the quality of the classification.

Diagnosis is mainly based on classifiers (Section 1.4). They allow automatic classification of the unknown states into healthy class or faulty ones. Some generalities are presented in Subsection 1.4.1. Then the supervised classifiers (Subsection 1.4.2) and unsupervised ones (Subsection 1.4.3) are detailed.

Fault Diagnosis, Prognosis, and Reliability for Electrical Machines and Drives. First Edition.
Elias G. Strangas, Guy Clerc, Hubert Razik, and Abdenour Soualhi.

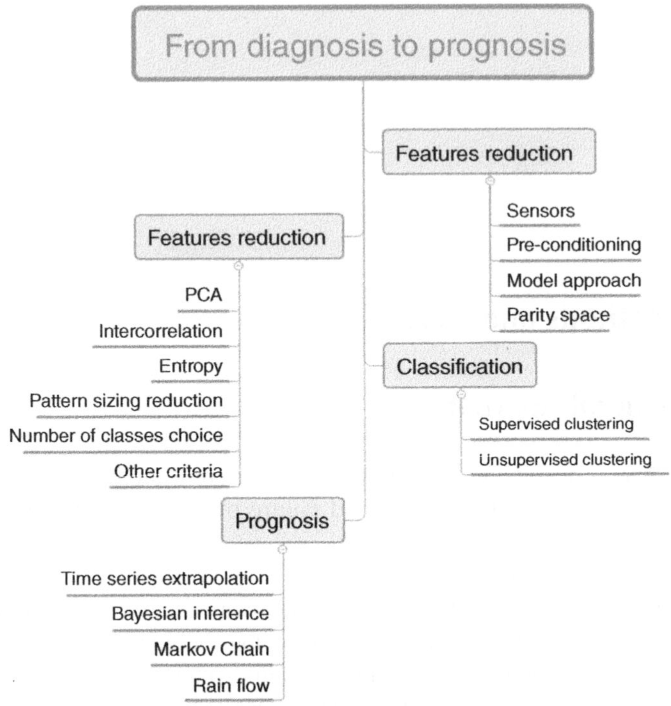

Figure 1.1 General overview of Diagnosis and Prognosis methods

Finally, the methods of failure prognosis are discussed (Section 1.5). First, the general process is described in Subsection 1.5.1. Some techniques are detailed:

- the extrapolation of time series (Subsection 1.5.2) that can be used to forecast the pattern vector,
- the Bayesian inference (Subsection 1.5.2),
- the Markov chain (Subsection 1.5.3) and the hidden Markov chain, which allow the diagnosis to be addressed in terms of stochastic process,
- the rain flow (Subsection 1.5.6) that joins the reliability approaches.

1.2 Feature Extraction: Signal and Preconditioning

1.2.1 Raw Signals: What Kind of Signals and Sensors?

Diagnostic tools need sensors, which are adapted to the faults that are being investigated. The choice of these sensors depends on the nature of the faults, the desired bandwidth, the influence signals but also, of course, the cost and

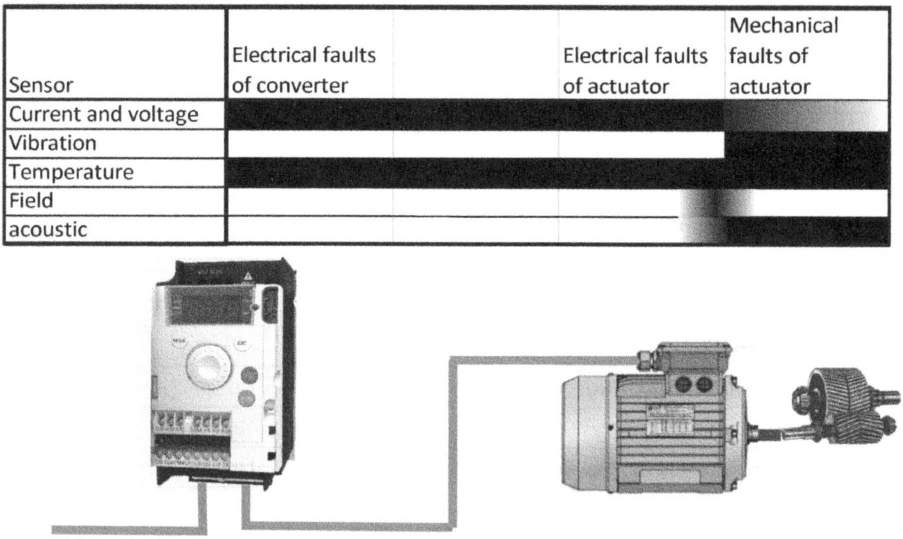

Figure 1.2 Application fields for sensors

the complexity of the interface. The reader can refer to Webster [1998] and Asch [2017] for a more in-depth study of the different types of sensors.

Current sensors and voltage ones are mainly used in motor current signature analysis (MCSA) applications to detect electrical faults in inverter and actuators. They can also give indications of bearing and gear defects.

Vibration and noise sensors are used for the mechanical part of the drive. They also give indications of abnormal torque, eccentricity, and thus the magnetic behavior of the actuator.

Temperature sensors are used in the detection of converter and actuator faults but with a bandwidth of a few hundred Hz maximum.

Axial field sensors are used to detect some defects linked to a magnetic asymmetry (eccentricity, magnetomotive force asymmetry ...).

1.2.1.1 Current Sensors

Current sensors (and voltage sensors) are used for the control, protection, and supervision of electrical drives. Their choice depends on the bandwidth, the influence quantities, the necessity or not of a galvanic isolation, the possibility to measure DC currents and their price. The following tables give the main characteristics of the different technology (Clerc [1989]) and (Asch [2017]).

Table 1.1 gives the characteristics of the shunts (Clerc [1989]).

Table 1.1 Current sensors: shunts 1/8

Characteristics	Classical shunt	Coaxial shunt
Schematic diagram	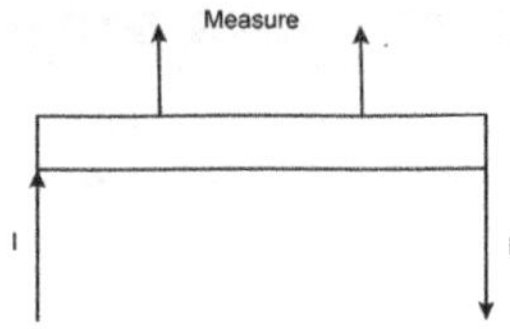	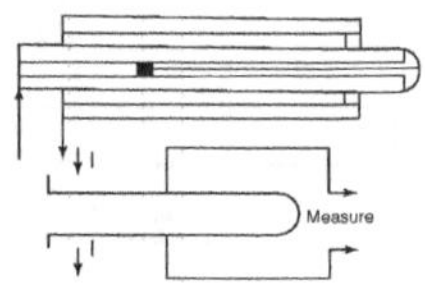
Transfer function	$U = R_{shunt}.I$	$U = R_{shunt}.I$
Galvanic insulation	No	No
DC measurement	Yes	Yes
Frequency bandwidth	Few hundreds of Hz	Up to 1MHz
Output level	Very low: Few tens of mV	Very low: Few tens of mV
Interface	Differential amplifier with high common mode rejection	Differential amplifier with high common mode rejection
Losses	Joule losses: $R_{shunt} \cdot I^2$	Joule losses: $R_{shunt} \cdot I^2$
Influence quantities	-Frequency: Skin effect and interaction with inductances and capacitances of the structure. -Temperature. -Electromagnetic interference.	-Skin effect for $f > 1MHz$ -Interaction of inductances and capacities of the structure, -For the U-shaped structure, edge effects increase the influence of the eddy currents -Temperature.
Limitations	-Sensitivity to interferences for low I. -Long-term overloads degrade the structure by overheating.	Intensity is limited by: -the heating -the electrodynamic stresses that can distort too thin walls.
Remarks	- Major drawback: low output level - Advantage: its price. with inductances and capacitances of the structure. -Temperature: Depends on the material being used. -Electromagnetic interference.	It is more used in prototypes than in industrial series. Major drawback: low output level and its price for high current.

Table 1.2 gives the characteristics of sensors based on temperature and current comparator working at zero flux (Clerc [1989]).

Table 1.2 Current sensors: Temperature-based sensor and current comparator 2/8

Characteristics	Temperature sensor	Current comparator
Schematic diagram	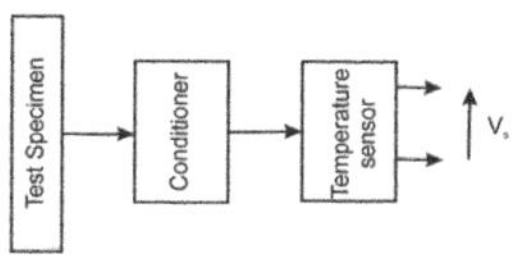	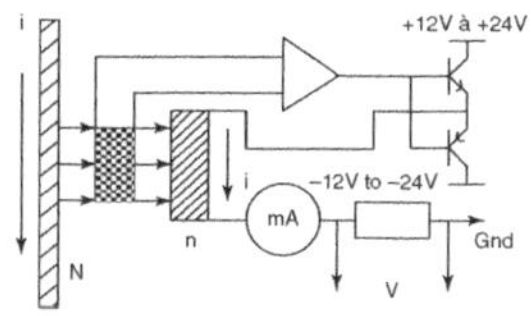
Transfer function	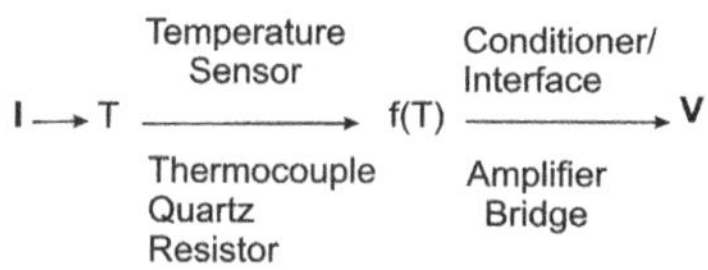	$N.I = n.i$
Galvanic insulation	Yes	Yes
DC measurement	Yes	Yes
Frequency bandwith	The response time depends on the test specimen $> 1s$.	A few tens of kHz Response times $< 1\mu s$
Output level	Low but takes advantage of galvanic isolation. Before amplifier: a few mV	Current A few hundred mA
Interface	For thermocouple: amplifier. For resistance: Bridge and amplifier or potentiometer circuit.	Supply: A few W The interface is described on the schematic diagram.
Losses	Depends on the temperature sensor and the test specimen. Very low with thermocouple.	Average
Influence quantities	Ambient temperature. Supply voltage for the conditioner. Humidity for the 8 resistive sensors.	Temperature. Stray fluxes caused by neighboring conductors.
Limitations	Response time. Internal thermal coupling.	Rather poor reliability. Withstands overloads.

(Continued)

Table 1.2 *(Continued)*

Characteristics	Temperature sensor	Current comparator
Remarks	Temperature sensor: Resistance, Thermocouple, Quartz oscillator, Diode. Test Specimen: Resistance or semiconductor.	Source of errors -Magnetic circuit remanence -Offset current -Magnetic interference -Temperature drift

Table 1.3 gives the characteristics of a current comparator with a detection coil and of a current transformer.

Table 1.3 Current sensors: Current comparator with detection coil and Current transformer 3/8

Characteristics	Current comparator with detection coil	Current transformer
Schematic diagram	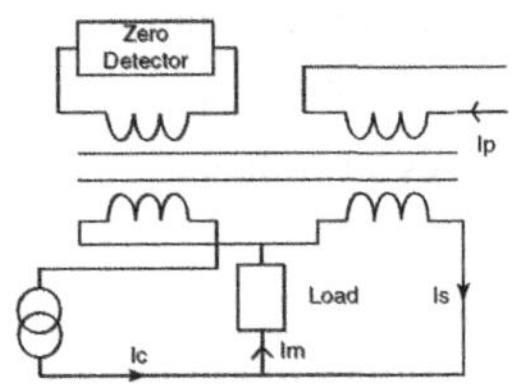	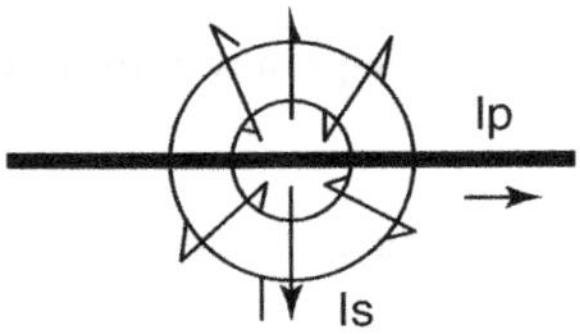
Transfer function	$N_1.I_p = N_2.I_m$	$N_p.I_p - N_s.I_s = I_m \approx 0$
Galvanic insulation	Yes	Yes
DC measurement	No	No
Frequency bandwith	A few kHz	Limited to few kHz
Output level	Current	Current
Interface	An interface is given in the remarks. The adjustment of G and C can be controlled by the signal of the zero detector. An auxiliary source must provide the compensation current.	TI must have a low input impedance. Current/voltage transformer, must have a high input impedance

(Continued)

Table 1.3 *(Continued)*

Characteristics	Current comparator with detection coil	Current transformer
Losses	Part of the measurement energy is taken from the main circuit.	Eddy current and Joule losses. Measurement energy pick-up on the primary.
Influence quantities	They have minimal impact: - interference fluxes caused by neighboring conductors. -frequency (parasitic capacitances).	Stray fluxes caused by neighboring conductors. Load: (which modifies the TI characteristics)
Limitations	Parasitic fields, parasitic capacitances between windings, and between windings and conductive screens.	Overload saturation. Self-heating can cause deterioration of the insulation.
Remarks	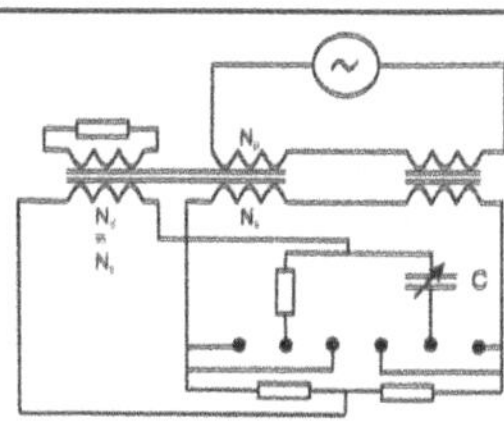	Sources of errors: -magnetizing current -leakage inductances and winding resistances. Saturation can protect measuring devices by limiting I_S

Table 1.4 gives the characteristics of the transducer and of the Rogowski coil. These later give an image of the current gradient.

Table 1.4 Current sensors: Transducer and Rogowski coil

Characteristics	Transducer	Rogowski Coil
Schematic diagram	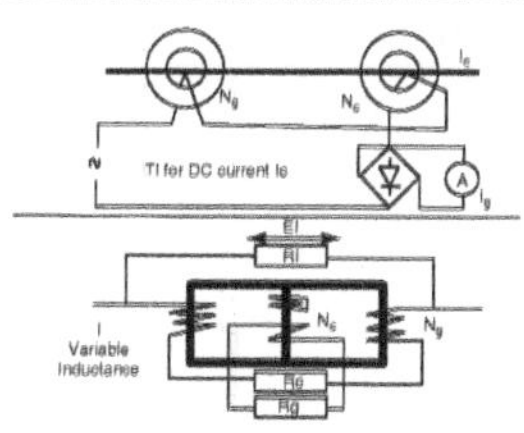	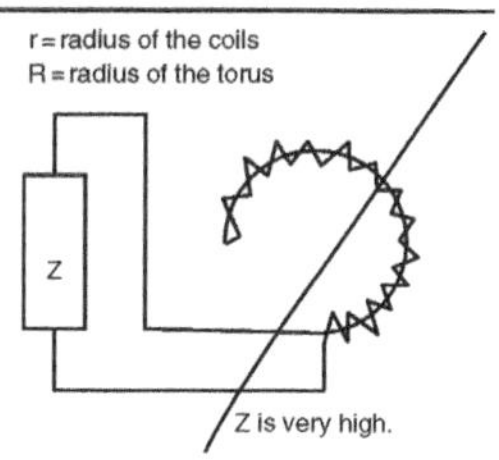

(Continued)

Table 1.4 *(Continued)*

Characteristics	Transducer	Rogowski Coil
Transfer function	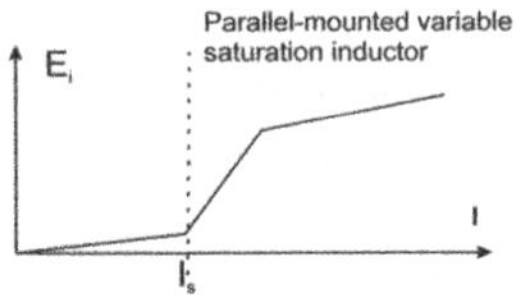 $I_s = I.\frac{N_c}{N_g}.(1 + \frac{R_g}{R_i})$ TI for DC current: $N_c.I_c = N_g.I_g$	$E = -\mu_0 \frac{N}{l}.\frac{dI}{dt}.S$ with $\frac{N}{l}$ linear density of turns
Galvanic insulation	Yes	Yes
DC measurement	Yes	No
Frequency bandwith	Limited to a few kHz	Bandwidth from 0.1Hz to 100kHz. Propagation time: $10\mu s$
Output level	Current	Voltage
Interface	Power Supply: Few tens of W. It is recommended to use a supply signal of a few kHz.	Measurement interface with high input impedance. Differential amplifier
Losses	Important	Very low
Influence quantities	Stray fluxes caused by the neighboring conductors.	- Frequency, - Sensitivity to external magnetic fields very attenuated by the feed-back cable.
Limitations	Saturation on overload	- Minimum measurable voltage. - Breakdown voltage of windings. - Accuracy requirements
Remarks	Drawback: High cost and High consumption.	-Excellent linearity. -Not suitable for low $\frac{dI}{dt}$

Table 1.5 gives the characteristics of the internal inductive flow sensor and of the coreless full shield probe.

Table 1.5 Current sensors: Internal inductive flow sensor and coreless full shield probe 5/8

Characteristics	**Internal inductive flow sensor**	**Coreless full shield probe**
Schematic diagram	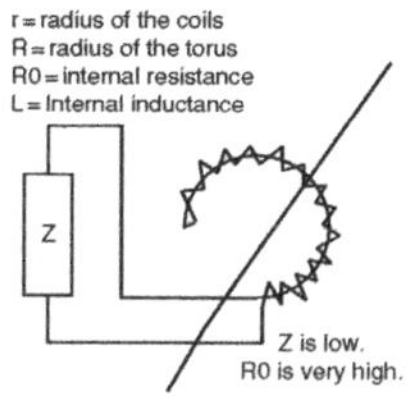	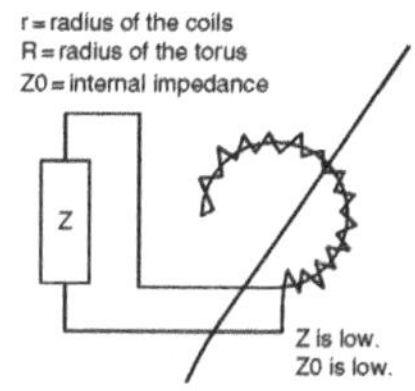
Transfer function	$i = -\frac{L}{n.R_0}.\frac{dI}{dt}$ with $\frac{L\omega}{R_0} << 1$ and n number of turns	$I_s = \frac{I}{n}$
Galvanic insulation	Yes	Yes
DC measurement	No	No
Frequency bandwith	Similar to the Rogowski coil.	Similar to the Rogowski coil.
Output level	Current	Current
Interface	Low input impedance measuring device.	Low input impedance measuring device.
Losses	Very low.	Very low.
Influence quantities	-Frequency. -Sensitivity to external magnetic fields very attenuated by the feedback cable. -Temperature	Sensitivity to external magnetic fields very attenuated by the feedback cable.
Limitations	R_0 must be kept independent of -the temperature and -the frequency.	They depends on -mechanical resistance, -the heating, -the skin effect.
Remarks	Not suitable for low $\frac{dI}{dt}$	

Table 1.6 gives the characteristics of the Hall effect sensor and of the Magnetoresistive probe.

Table 1.6 Current sensors: Hall effect sensor and magnetoresistive probe 6/8

Characteristics	Hall effect sensor	Magnetoresistive probe
Schematic diagram	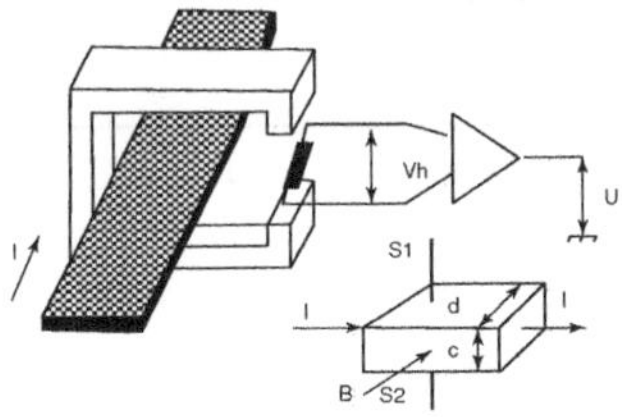	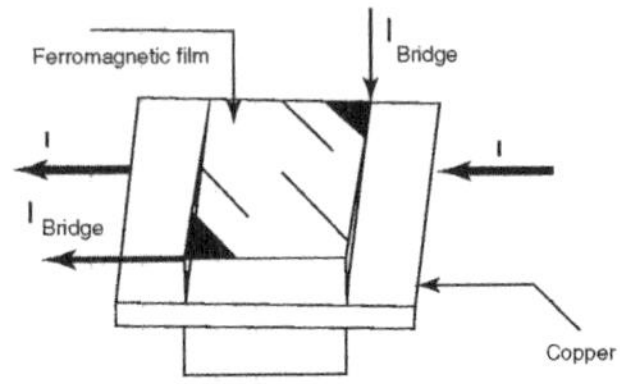
Transfer function	$V_h = K.I$ and $B = k.I$ with $K = \frac{R_h.k.i}{d}$ and $R_h = \frac{1}{N.e}$ and $e = 1.610^{-19}C$	R increases with H_s with a good polarization, the relation is linear.
Galvanic insulation	Yes	Yes
DC measurement	Yes	Yes
Frequency bandwith	Few kHz	Sensor on Fe substrate: 1MHz Sensor on epoxy substrate: 10MHz Sensor with interface : $< 10Mhz$
Output level	Sensitivity: $\approx 2.5V/T$ Output : few Volts for $I = 1000A$	For $\Delta B = \pm 1T, \quad 5 < \frac{R}{R_0} < 15$
Interface	Supply: Few mW Interfaces: -differential amplifier -Summators for serial connection of sensors.	Wheatstone bridge and its supply.
Losses	Very low	Very low
Influence quantities	External magnetic field Temperature	Temperature Supply voltage Amplifier Gain External magnetic field
Limitations	Aging Saturation on overload Limited bandwidth	For low fields, noise in $1/f$ Used to measure fields from a few nT to tens of nT.
Remarks	Sources of errors: -offset of the hall effect generator -magnetic circuit hysteresis	We can supply the bridge with an alternative source to improve the behavior.

Table 1.7 gives the characteristics of the current mirror (especially for MOS-FET transistor) and of the optical sensor.

Table 1.7 Current sensors: Current mirror and optical sensor 7/8

Characteristics	Current mirror	Optical sensor
Schematic diagram	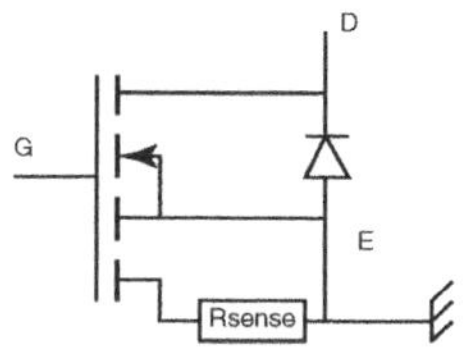	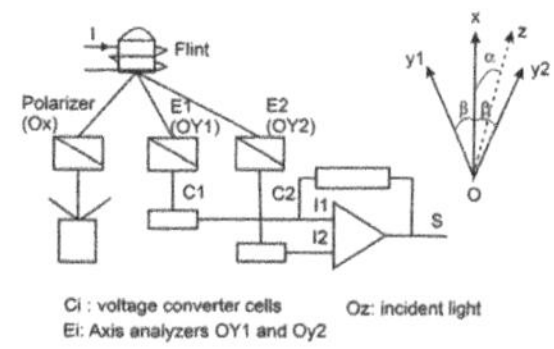
Transfer function	$\frac{I_D}{I_{sense}} = \frac{N.R_{DS_{On}} + R_{sense}}{R_{DS_{On}}}$ with $R_{D_{On}}$ the on state resistance for all the output cells and N the ratio of the number of output cells to the number of measurement cells	$\alpha = p.I.H$ $i_1 = 2A_1 \cos^2(\beta - \alpha)$ $i_2 = 2A_2 \cos^2(\beta + \alpha)$ For $A_1 = A_2 = A$ $\Delta i = A \sin(2\alpha) \sin(2\beta)$ with A_1, A_2, and p coefficients.
Galvanic insulation	No	Yes
DC measurement	Yes	Yes
Frequency bandwith	Bandwidth around 1Mz	Around ten kHz
Output level	Current	The output is the polarization angle of rotation. Typically $p = 10^{-5} rad/A$ for a silicon fiber
Interface	Interface must have a low input impedance	The polarization angle is converted into current or voltage
Losses	Low	Almost no loss.
Influence quantities	The transfer function depends on the junction temperature, the collector current and the frequency	-Temperature (but a little) -p depends on the wavelength

(Continued)

Table 1.7 *(Continued)*

Characteristics	Current mirror	Optical sensor
Limitations	Nonlinearity in case of overload Low accuracy : ≈ 5%	Depends on the sensor. Intensity from mA to 100kA with a resolution of $1.10^{-4}A$ (SAGNAC interferometer)
Remarks	Low price sensor	High voltage applications. Notations: α: rotation of the polarization plan l: length subjected to the field p: Verdet constant

Table 1.8 gives the characteristics of the laser diodes and presents the use of the characteristics of the power devices.

Table 1.8 Current sensors: Laser diodes and device characteristics 8/8

Characteristics	Laser diodes	Device characteristics
Schematic diagram	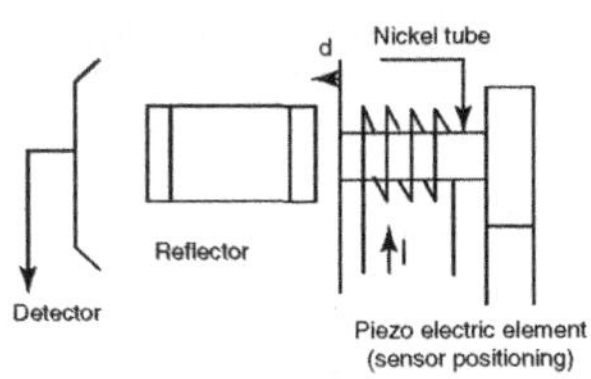	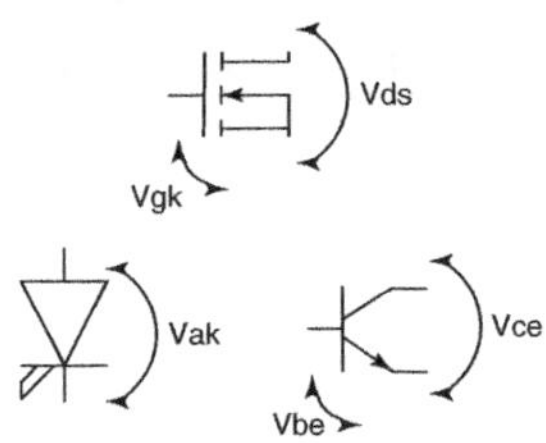
Transfer function	ΔI ↓ ΔH ↓ Δl of the nickel tube ↓ Δl of the reflector ↓ ΔP emitted by the laser diode	Transistors: $V_{CE}(I_c)$ and $V_{BE}(I_c)$ MOS: $V_{ds} = R_{on}.I_d$ Thyristors: $V_{AK}(I_c)$

(Continued)

Table 1.8 *(Continued)*

Characteristics	Laser diodes	Device characteristics
Galvanic insulation	Yes	No
	Yes	Yes
Frequency bandwith	In the range of some kHz	Response time: $\approx 1\mu s$
Output level	Voltage or current	Voltage
Interface	Photodetector output signal processing.	The interfaces must correct the thermal drift.
Losses	Very low	Almost zero
Influence quantities	Laser wavelength Initial position of the reflector Temperature	Temperature Drive Nonlinearity of the transfer function
Limitations	Minimum field strengh: $10^{-10}T$. From $10^{-5}A$ to 10A	Essentially set by the accuracy of the triggering threshold.
Remarks	Improvement of detection conditions: -For variable fields, superimpose a continuous field -Precise positioning of the reflector	Advantages: -Low price -Fast response time Drawback: influence quantities

1.2.1.2 Vibration Measurement and Accelerometers

The diagnosis of mechanical failures or excessive eccentricity of electrical machines often uses vibration measurements (see Table 1.9).

Three type of sensors are used to realize a vibration analysis:

- displacement in mm,
- speed in mm/s,
- acceleration in mm/s^2 or in g ($g = 9.81 m.s^{-2}$).

In this paragraph, we will focus on one of the most used vibration sensors: the acceleration sensors.

An accelerometer is made of a damped mass (called proof mass) and a spring. The acceleration is given by the proof mass movement. The system is damped in order to decrease the measurement disturbances (see Figure 1.3).

Table 1.9 Vibration sensors

Sensors	Bandwidth	Application
Displacement	low frequency: $f \leq 200Hz$	Balourd, misalignment coupling wear, loosening
Speed	medium frequency: $f \leq 1kHz$	Balourd, misalignment coupling wear, loosening faulty meshing highly degraded bearing
Acceleration	high frequency: $f \leq 20kHz$	Faulty meshing, regular degradation or highly degraded bearing, lack of lubrication

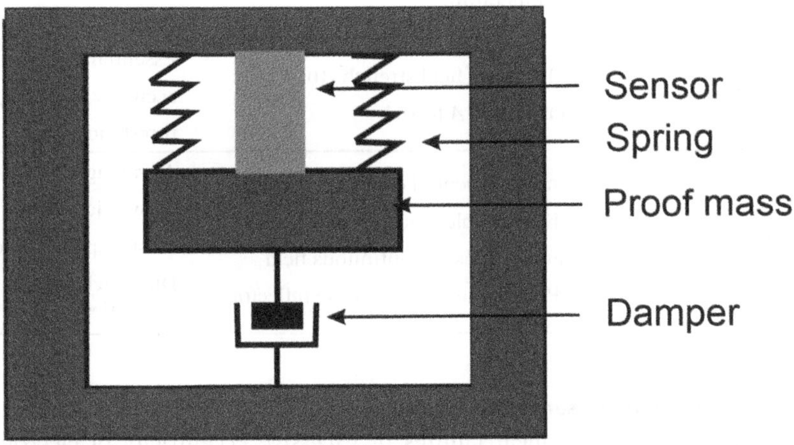

Figure 1.3 Accelerometer

The measurement of the displacement between the proof mass and the frame can be done with piezoelectric, piezoresistive, or capacitive senors. Piezoelectric accelerometer can measure acceleration up to 100g ($g = 9.81m.s^{-2}$), a shock resistance up to 1,000,000g and a bandwidth greater than 50kHz.

There are single accelerometers that measure in only one direction and multi-axis accelerometers that can measure simultaneously in three directions.

1.2.1.3 Temperature Sensors

Temperature is a key feature for diagnosis and and prognosis of electrical drives. But its dynamics, in relation to the thermal constants of the circuit and the temperature sensors, is significantly slower than that of the electrical and mechanical quantities. Nevertheless, the temperature can be a component of the pattern vector or an influencing parameter. It is used for both motor (Mohammed et al. [2019]) and converter diagnosis (van der Broeck et al. [2018]).

The main thermal sensors are:

- Thermistor;
- Thermocouple;
- Resistance temperature detector (RTD);
- Fiber optic thermal sensor;
- Thermo sensitive parameter (TSEP). They will be fully described in Subsection 2.6.6.2.

Table 1.10 gives an overview of the temperature sensors.

Table 1.10 Temperature sensors

Thermistor 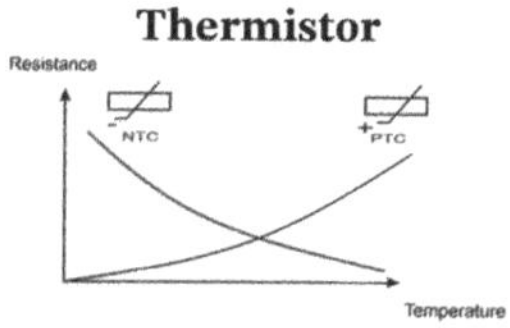	Thermistors use the variation of resistance as a function of temperature. CTN (semiconducting material) has a Negative Temperature Coefficient, CTP (barium titanate) has a Positive Temperature Coefficient, Temperature Range: -100 to 350°C Response time: 0.1 to 10s Accuracy: 0.05 to 1.5°C
Thermocouple 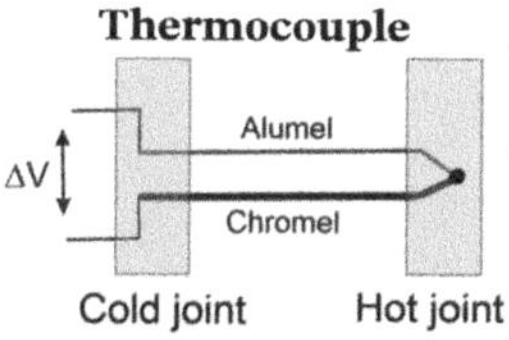	Thermocouples use the Seebeck effect: Two wires of different metals, welded at one of their ends, making a circuit without current flow, when they are submitted to a temperature gradient can exhibit a gradient of voltage at the end of the open loop. Temperature Range: -200 to 2500°C Response time: 0.10 to 10s Accuracy: 0.5 to 5°C
Resistance temperature detector (RTD) **(RTD PT100 and PT1000)** 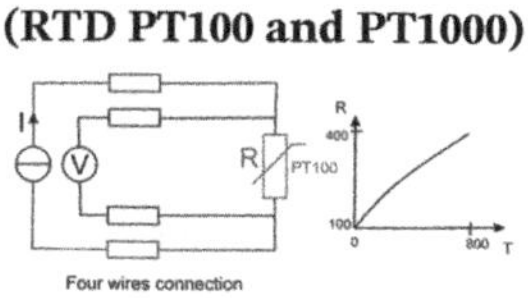	RTD are often made of a length of Platinum wire wrapped around a ceramic or glass core. They are based on the increasing of the resistance with the temperature. Temperature Range: -200 to 650°C Response time: 1 to 10s Accuracy: 0.1 to 1°C

(Continued)

Table 1.10 *(Continued)*

Fiber optic thermal sensor 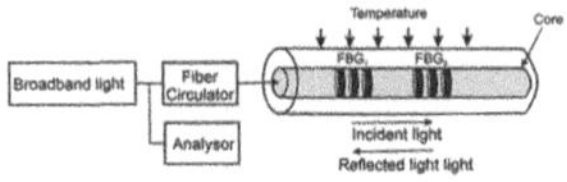	They are mostly based on (Roriz et al. [2020] and Malekzadeh et al. [2014])): - Fiber Bragg Grating (using the selective photo-induced modulation of the optical fiber core refractive index). - Fabry-Pérot interferometer. Temperature Range: Typically -100 to 400°C Response time: 5ms Accuracy: Typically $\pm 0.5°C$
Thermo sensitive parameter	Junction temperature is estimated from some electrical characteristics of the device such as V_{CE}, ... Temperature Range: limited by the junction temperature Response time: depends on the interface - a few ms Accuracy: $> 1°C$. Sensitive to different parameters.

1.2.1.4 Field Sensors

These sensors are mainly made of a measurement coils distributed in the machine body (non intrusive case) or placed inside the machine (intrusive case) (Ewert [2017]).

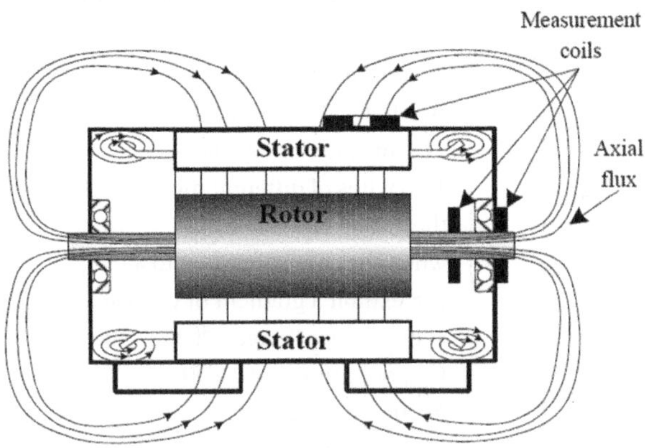

Figure 1.4 Axial flux distribution in an induction motor and possible distribution of measurements coils (Ewert [2017]). Courtesy of IEEE

1.2.1.5 Acoustic Sensors

Many machine breakdowns can be detected acoustically since machines consist of moving parts inducing friction and noise.

Pressure and speed can be used to measure an audible signal. But most sensors use sound pressure. Table 1.11 gives the main technologies (Wikipedia contributors [2020a]).

The choice of the acoustic sensors depends on:

Table 1.11 Acoustic sensors

Condenser microphone 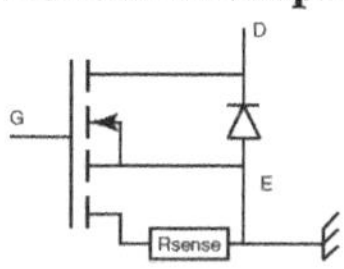 	Electrostatic microphone : a thin metallic membrane is placed near a solid plate and acts as a variable capacitor. Bandwidth: few 10kHz
Dynamic microphone 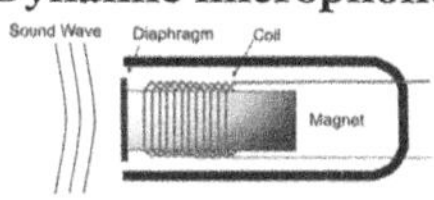 	Moving-coil microphone: The diaphragm drives a coil moving in a magnetic field and induce a voltage. Bandwidth: few 10kHz
Ribbon microphones 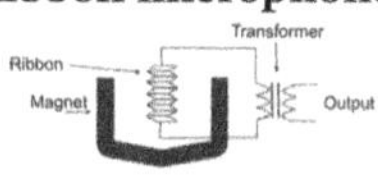 	A thin metal ribbon is suspended in a magnetic field. and moves under the effect of a sound wave, creating an induced voltage. Bandwidth: few 10kHz
Piezoelectric microphone 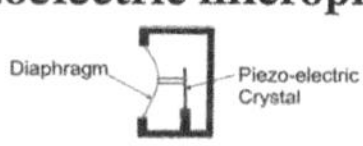 	They use piezoelectric materials which have the ability to produce a voltage when subjected to pressure Bandwidth: few tens kHz
Fiber optic microphones	They are based in a device in which the sound waves modulate the light beam guided in optical fibers. They are use in harsh environment. Bandwidth: few kHz
Micro Electrical-Mechanical System (MEMS) microphones	A pressure-sensitive diaphragm is mounted directly into a silicon wafer using MEMS processing techniques. Bandwidth: up to 1MHz

- the polar patterns: they are the locus of points with the same signal level output from the microphone for a given sound pressure level generated from that point;

- the bandwidth;
- the output impedance.

These sensors can be used, among other things, to detect gear defects (Zurita et al. [2013]), bearing faults (Xue et al. [2017]), stator winding delamination (Zhu et al. [2010]), etc. They have bandwidth up to 1MHz.

1.2.1.6 Other Sensors

Voltage sensors can be used to monitor the state of capacitors and batteries. But, they are also necessary to evaluate the transient power for some diagnosis

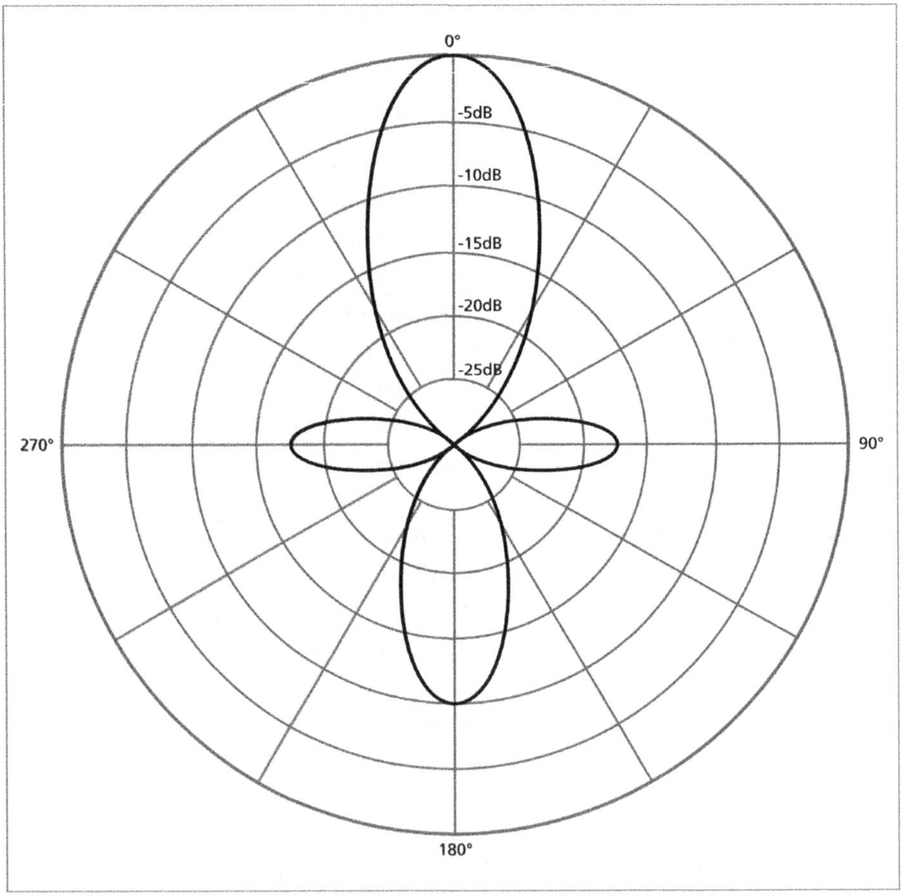

Figure 1.5 Lobar polar pattern

methods of semiconductor devices or actuators. In the latter case, particular attention must then be paid to synchronization with the current channels.

Voltage meters could be (Webster [1998]):

- electromagnetic they are based on the interaction between currents and magnetic fields ($F = N.B.l.I$ with F the force, l the length of conductor cutting the magnetic field, B and N the number of turns of the coil). Some of them only measure DC voltage (with moving coil or magnet);
- electrodynamic: they are based on the interaction between currents ($F = k.I_1.I_2$);
- electrostatic: they are based on electrostatic interactions between the conductors (for a moving plate displaced by ds, $F = U^2\frac{dC}{ds}$ with C the capacitance between conductors);
- thermal: they are based indirectly on thermal effects;
- inductive: they are based on magnetic induction (Voltage Transformers) and only measure AC voltage;
- electronic: they are based on digital to analog interaction and signal processing.

Speed sensors are often integrated in the control. They can enter in the evaluation of the mechanical torque from the absorbed power. However, the inertia strongly smoothes the signal, which makes them unsuitable for fast defect detection. Indicators based on speed are more relevant for defects whose vibratory impact appears in low and medium frequencies.

Two main technologies are used:

- **Tachometer generator**: It could be a DC current Tachometer generator. The DC no-load e.m.f. is proportional to the speed (see Figure 1.6).

  It could be also a synchronous AC tachometer generator where the AC no-load electromotive force (e.m.f.) is proportional to the speed.
- **Pulse angular speed tachometers**: A train of pulses whose frequency is proportional to the speed is generated. It could be a variable reluctance sensor, a magnet sensor with Hall effect receiver, or an optic sensor.

Position could be used in order to find the orientation of the frame where raw signals are projected (see paragraph Symmetric component, Park Component 1.2.2.2). It is also used in diagnostic methods related to cyclostionarity in particular for bearing defects, etc. (Wu et al. [2009]).

There are two types of position sensors:

- The incremental encoder: at least two square waves are generated in quadrature A and B, whose frequency is proportional to the speed. The direction

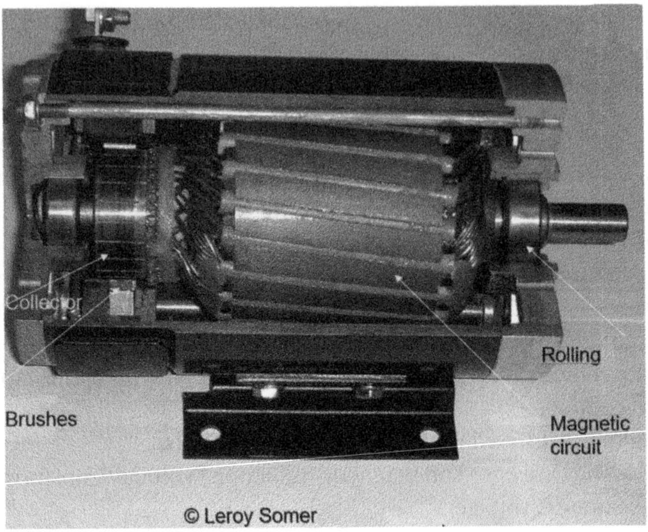

Figure 1.6 Direct current generator @LeroySomer

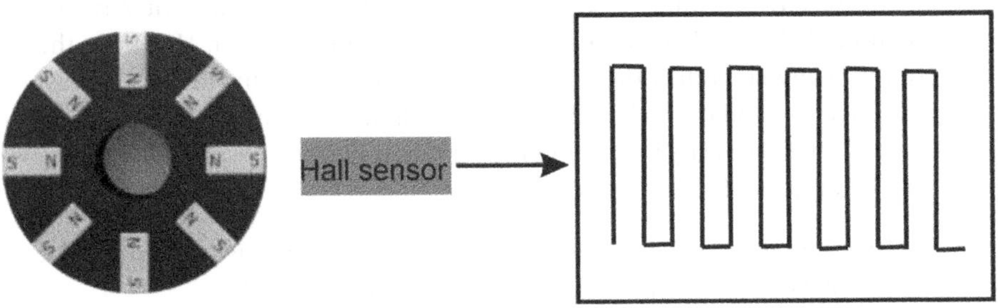

Figure 1.7 Pulse angular speed tachometer

of rotation is indicated by the sign of the phase angle A and B (Figure 1.8). Sometimes, rotary incremental encoders have an index output that provides a pulse at a given angle to synchronize.

- The absolute encoder: They can provide a numeric output (Figure 1.10). They could be mechanic, magnetic (with magneto-resistor or Hall effect sensor), optical, or mechanical sensors. The output signal can be coded in binary or in Grey where two adjacent codes differ by only one bit position, thus reducing coding errors.

It could be also analog (Figure 1.10). A synchro-resolver provides two ±90° shifted sinusoidal signals. Speed and position are obtained by demodulation.

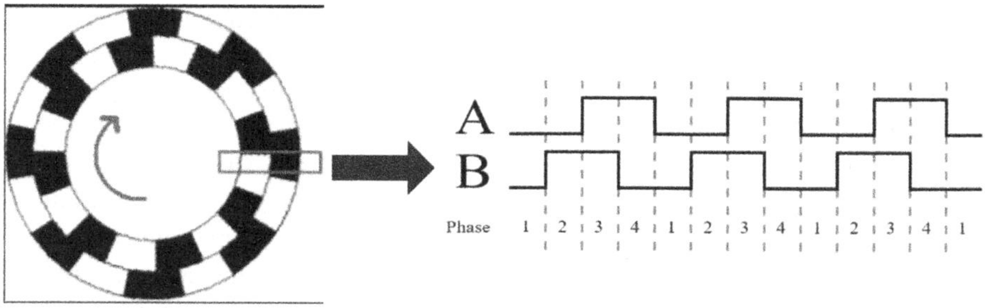

Figure 1.8 Incremental encoder

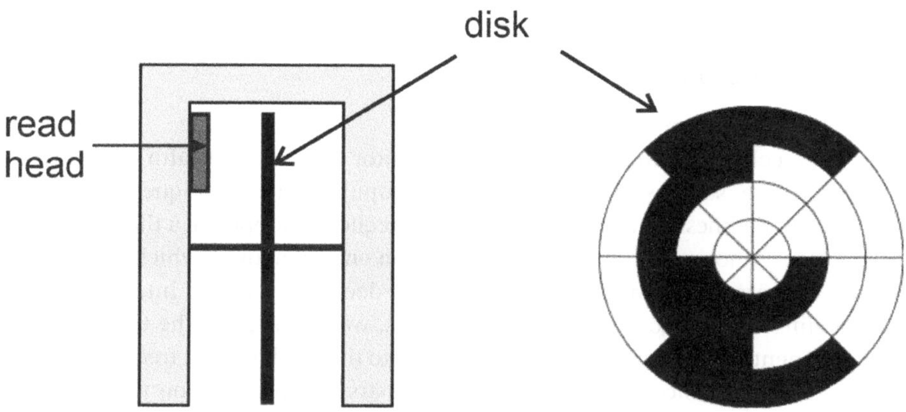

Figure 1.9 Numeric Grey absolute encoder

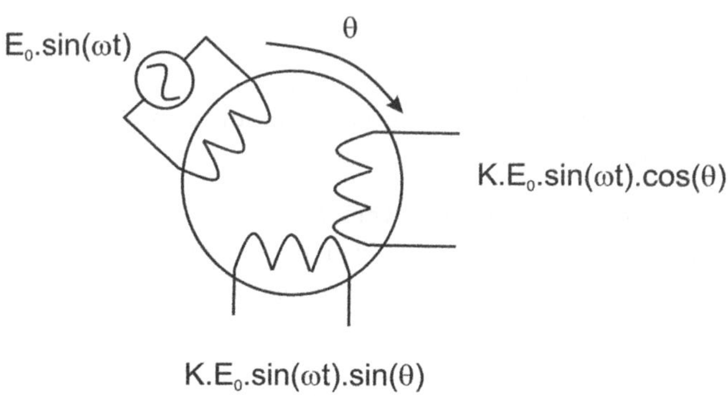

Figure 1.10 Synchro-resolver

Position can be also obtained by virtual sensors such as extended observers (see Subsection 1.2.3.2), or estimated from the e.m.f. of synchronous machines. But in this latter case, the rotor must turn in order to generate an e.m.f. (Zheng et al. [2007] and Comanescu [2012]).

The torque can be used to diagnose electric unbalance, short circuits, and mechanical failures. Torque meters are essentially used for research applications. They can reach bandwidth of a few kHz. However, torque can be computed by extend observer with measured stator currents and estimated flux (Liu et al. [2006] and Wang Chenchen and Li Yongdong [2008]). Torque can be measured from the shear forces experienced on a test body. Their measurement by a stress sensor is transmitted from the rotor to the stator in high frequency by inductive coupling.

1.2.2 Preconditioning

1.2.2.1 Signal Features in the Time Domain

Several techniques are used in order to monitor the state of health of electrical machine, such as the MCSA, which is popular. This technique allows to show up the lines close to the fundamental frequency supplying a three-phase machine. Among all techniques, this one is an online method, which is attractive, because the monitoring of the process doesn't require to interrupt the operating system. To illustrate our meanings, we will present the theoretical components in a spectrum a lot of lines close to the fundamental frequency f_s.

As a consequence, in the case of broken bars for an induction motor, the components frequencies close to f_s are:

$$f_{brb} = \left[\frac{k}{p}(1-s) \pm s\right] f_s \tag{1.1}$$

where $k/p = 1, 5, 7, 11, \ldots$

Knowing the fault frequencies in the current spectrum and their evolution in time, which depend on the fault severity, we can easily estimate the state of health of the three-phase induction machine.

1.2.2.2 Symmetric Component, Park Component

Any three-phase voltage or current can be written as the sum of a forward sequence with a backward and a homopolar sequence. So, the forward sequence is written as follows:

$$[V_F(t)] = \begin{cases} v_{a_F}(t) &=& V_F\sqrt{2}\cos(\omega_e t) \\ v_{b_F}(t) &=& V_F\sqrt{2}\cos(\omega_e t - \frac{2}{3}\pi) \\ v_{c_F}(t) &=& V_F\sqrt{2}\cos(\omega_e t - \frac{4}{3}\pi) \end{cases} \tag{1.2}$$

where ω_e is the electrical pulsation, and V_F is the rms value of the voltage for the forward sequence. For the backward sequence, voltages can be written as follows:

$$[V_B(t)] = \begin{cases} v_{a_B}(t) & = & V_B\sqrt{2}\cos(\omega_e t) \\ v_{b_B}(t) & = & V_B\sqrt{2}\cos(\omega_e t + \frac{2}{3}\pi) \\ v_{c_B}(t) & = & V_B\sqrt{2}\cos(\omega_e t + \frac{4}{3}\pi) \end{cases} \tag{1.3}$$

And for the homopolar sequence, one can write as follows:

$$[V_O(t)] = \begin{cases} v_{a_O}(t) & = & V_O\sqrt{2}\cos(\omega_e t) \\ v_{b_O}(t) & = & V_O\sqrt{2}\cos(\omega_e t) \\ v_{c_O}(t) & = & V_O\sqrt{2}\cos(\omega_e t) \end{cases} \tag{1.4}$$

Let us remember that

$$\begin{cases} \cos p.\cos q & = & \frac{1}{2}(\cos(p-q)+\cos(p+q)) \\ \sin p.\sin q & = & \frac{1}{2}(\cos(p-q)-\cos(p+q)) \\ \sin p.\cos q & = & \frac{1}{2}(\sin(p+q)+\sin(p-q)) \\ \sin p + \sin q & = & 2\sin\left(\frac{p+q}{2}\right).\cos\left(\frac{p-q}{2}\right) \end{cases} \tag{1.5}$$

Let us introduce the not orthonormal Clarke's transformation. The main idea is to transform a three-phase system into a two-phase system. This transformation is based on the fact that a three-phase system is naturally composed of a system shifted by a multiple of $\frac{2}{3}\pi$. Thus, in conclusion, the matrix transformation is written as follows:

$$[C] = \frac{2}{3}.\begin{bmatrix} \cos 0 & \cos\frac{2}{3}\pi & \cos\frac{4}{3}\pi \\ \sin 0 & \sin\frac{2}{3}\pi & \sin\frac{4}{3}\pi \\ \frac{1}{2} & \frac{1}{2} & \frac{1}{2} \end{bmatrix} = \frac{2}{3}.\begin{bmatrix} 1 & -\frac{1}{2} & -\frac{1}{2} \\ 0 & \frac{\sqrt{3}}{2} & -\frac{\sqrt{3}}{2} \\ \frac{1}{2} & \frac{1}{2} & \frac{1}{2} \end{bmatrix} \tag{1.6}$$

If we apply the Clark's transformation to the forward voltage sequence, we obtain:

$$V_{\alpha\beta O} = [C].V_F(t) = \frac{2}{3}.\begin{bmatrix} 1 & -\frac{1}{2} & -\frac{1}{2} \\ 0 & \frac{\sqrt{3}}{2} & -\frac{\sqrt{3}}{2} \\ \frac{1}{2} & \frac{1}{2} & \frac{1}{2} \end{bmatrix} . \begin{bmatrix} V_F\sqrt{2}\cos(\omega_e t) \\ V_F\sqrt{2}\cos(\omega_e t - \frac{2}{3}\pi) \\ V_F\sqrt{2}\cos(\omega_e t - \frac{4}{3}\pi) \end{bmatrix}$$

$$= \begin{bmatrix} V_F\sqrt{2}\cos(\omega_e t) \\ V_F\sqrt{2}\sin(\omega_e t) \\ 0 \end{bmatrix} \tag{1.7}$$

If we apply now the Clark's transformation to the backward voltage sequence, we obtain:

$$V_{\alpha\beta O} = [C].V_B(t) = \frac{2}{3}.\begin{bmatrix} 1 & -\frac{1}{2} & -\frac{1}{2} \\ 0 & \frac{\sqrt{3}}{2} & -\frac{\sqrt{3}}{2} \\ \frac{1}{2} & \frac{1}{2} & \frac{1}{2} \end{bmatrix} . \begin{bmatrix} V_B\sqrt{2}\cos(\omega_e t) \\ V_B\sqrt{2}\cos(\omega_e t + \frac{2}{3}\pi) \\ V_B\sqrt{2}\cos(\omega_e t + \frac{4}{3}\pi) \end{bmatrix}$$

$$= \begin{bmatrix} V_B\sqrt{2}\cos(\omega_e t) \\ -V_B\sqrt{2}\sin(\omega_e t) \\ 0 \end{bmatrix} \tag{1.8}$$

In conclusion, any three-phase system or any unbalanced three-phase system can be expressed as the sum a forward sequence with a backward sequence.

1.2.2.3 Symmetric Component, Park Component

As a three-phase system is composed of a two-phase system, one can express the Park's vector. One can notice that it is an abuse of language because the transformation don't use a rotating frame. Consequently, the expression of the Park's vector components (i_α and i_β) using the Concordia's tranformation, which is a normalized orthonormal matrix, in a reference frame related to the stator is as follows:

$$i_\alpha(t) = \sqrt{\frac{2}{3}}\left(i_A(t) - \frac{1}{2}i_B(t) - \frac{1}{2}i_C(t)\right) \tag{1.9}$$

$$i_\beta(t) = \frac{1}{\sqrt{2}}\left(i_B(t) - i_C(t)\right), \tag{1.10}$$

where $i_A(t)$, $i_B(t)$, and $i_C(t)$ represent the currents line of the induction motor.

$$[T] = \sqrt{\frac{2}{3}}\begin{bmatrix} 1 & -\frac{1}{2} & -\frac{1}{2} \\ 0 & \frac{\sqrt{3}}{2} & -\frac{\sqrt{3}}{2} \\ \frac{1}{\sqrt{2}} & \frac{1}{\sqrt{2}} & \frac{1}{\sqrt{2}} \end{bmatrix} \tag{1.11}$$

Under normal conditions, the motor currents are a forward-sequence system. Therefore, the Park's vector can be written as follows:

$$i_\alpha(t) = \sqrt{3}I_f \sin \omega_e t \tag{1.12}$$

$$i_\beta(t) = \sqrt{3}I_f \cos \omega_e t \tag{1.13}$$

where

I_f is the rms value of the current forward sequence,
ω_e is the angular supply frequency ($2\pi f_s$).

The Park's current vector can be written in a complex form as follows:

$$\vec{i_s}(t) = i_\alpha(t) + j i_\beta(t) = \sqrt{3}I_f e^{j\omega_e t} \tag{1.14}$$

where $j^2 = -1$.

Consequently, using the Park's extended vector, the current for the forward sequence becomes:

$$\begin{aligned} \vec{i_{sf}}(t) = \quad & \sqrt{3}\left[I_f e^{j\left[\omega_e t - \varphi_f\right]}\right. \\ & + I_{lf} e^{j\left[(1-2s)\omega_e t - \varphi_{lf}\right]} \\ & \left. + I_{rf} e^{j\left[(1+2s)\omega_e t - \varphi_{rf}\right]}\right] \end{aligned} \tag{1.15}$$

And by the same way, we can deduce the current for the backward sequence:

$$\begin{aligned}\overrightarrow{i_{sb}}(t) = \quad & \sqrt{3}\,[I_b e^{-j[\omega_e t-\varphi_b]} \\ & +I_{lb} e^{-j[(1-2s)\omega_e t-\varphi_{lb}]} \\ & +I_{rb} e^{-j[(1+2s)\omega_e t-\varphi_{rb}]}]\end{aligned} \tag{1.16}$$

The main advantage of the use of Park's extended vector is to analyze and quantify the contribution of a forward and backward power supply on the stator line current. Moreover, the estimation of all line amplitudes in the spectrum allows to better deduce the quality of the power supply and the state of health of the induction machine.

1.2.2.4 Signal Features in the Frequency Domain

The Fourier transform (Poularikas [2000]) $F\{s(t)\}$ for an aperiodic signal $s(t)$ is given by:

$$S(\omega) = F\{s(t)\} = \int_{-\infty}^{\infty} s(t)e^{-j\omega t}dt \tag{1.17}$$

Thus, for a classical signal as a cosinus signal ($\cos \omega_o t$), the spectral components are:

$$S(\omega) = [\delta(\omega - \omega_o) + \delta(\omega + \omega_o)] \tag{1.18}$$

where δ is the Dirac function. But for the complex signal as the function $e^{j\omega_o t}$, the spectral component is:

$$S(\omega) = [\delta(\omega - \omega_o)] \tag{1.19}$$

Consequently, this approach is recommended for the analysis of the signal in the forward sequence (1.15) as the backward sequence (1.16). All components will appear in the spectral analysis.

(1) Fourier transform and windowing

These methods require the availability of all samples in order to evaluate the spectrum. Moreover, they need a number N of samples equal to a power of two. Consequently, the spectral resolution which depends on $N = 2^r$, is given by the relation $\Delta f = Fe/N$, where Fe is the sampling frequency. The acquisition time of the signal is consequently equal to $T = 1/\Delta f$.

Considering the measured signal $x(p)$ at the sampling time $t = p.Te$, which is characterized by a finite duration sequence of length N and knowing all the previous values $x((p-n)Te)$ $(n = 0, \dots, N-1)$ and starting with the discrete

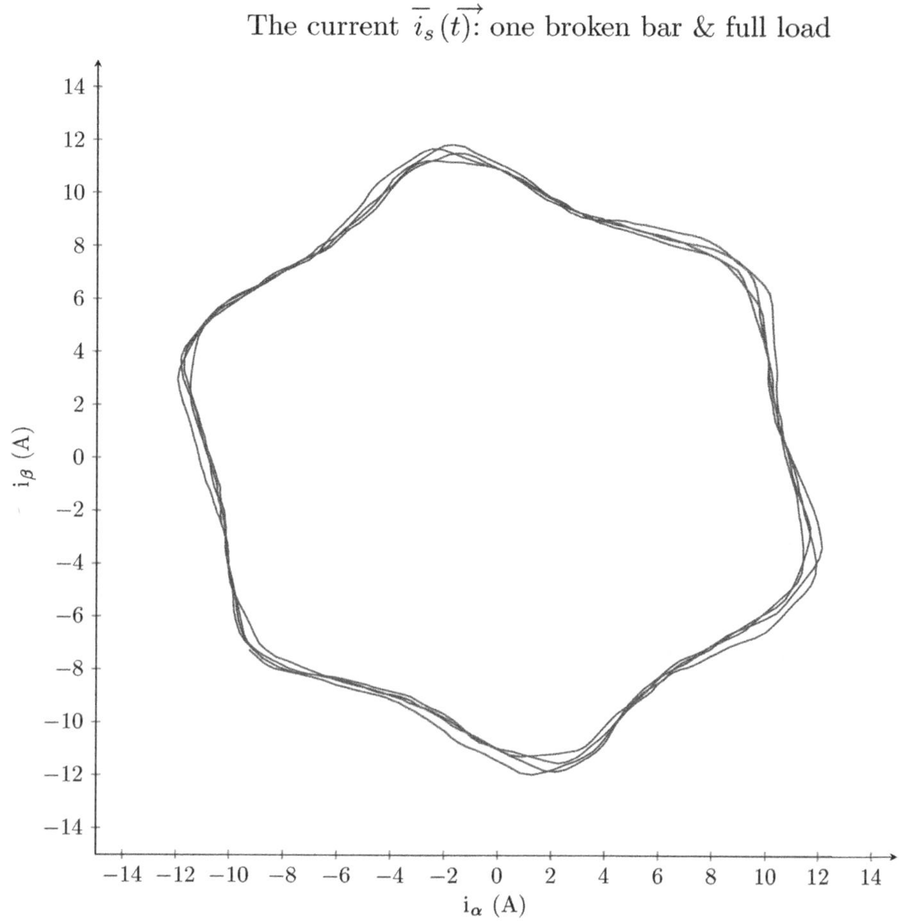

Figure 1.11 The trajectory of the line current: the motor operates under one broken bar and at full load

Fourier transform (DFT) at the time $t = p.Te$, denoted $F_p[k]$, we can evaluate the spectrum thanks to the relation:

$$F_p[k] = \frac{1}{N}\sum_{n=0}^{N-1} x(p-n)W_N^{nk};$$

where $W_N^{nk} = e^{-\jmath 2\pi nk/N}$ with $n, k = 0, 1, 2, \ldots, N-1$. If the signal is complex, then:

$$F_p[k] = \frac{1}{N}\sum_{n=0}^{N-1} \left[\Re\{x(p-n)\} + \jmath\, Im\{x(p-n)\}\right] W_N^{nk}$$

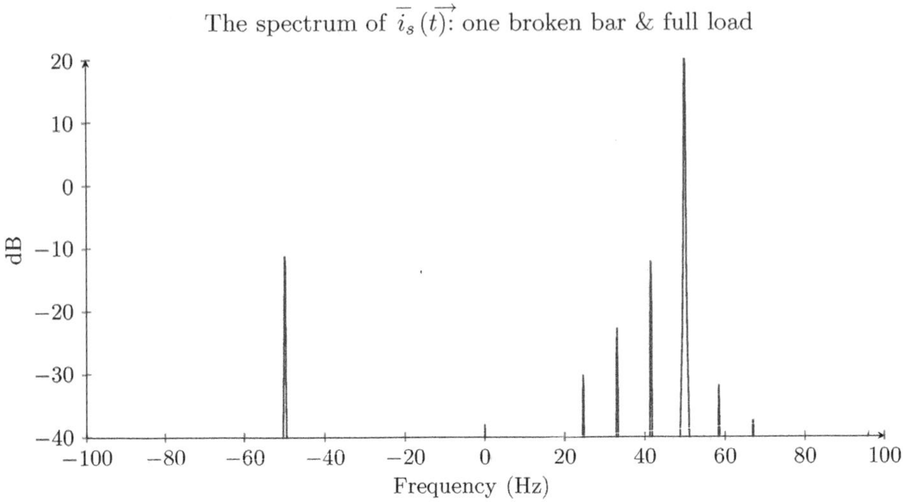

Figure 1.12 The spectrum of the line current: the motor operates under one broken bar and at full load

where $\mathfrak{R}$ and $\mathfrak{I}$ denote the real and the imaginary parts, respectively.

The Hann window: $w(n) = 0.5\left(1 - \cos\left(2\pi\frac{n}{N}\right)\right)$ with $0 \leq n \leq N$.

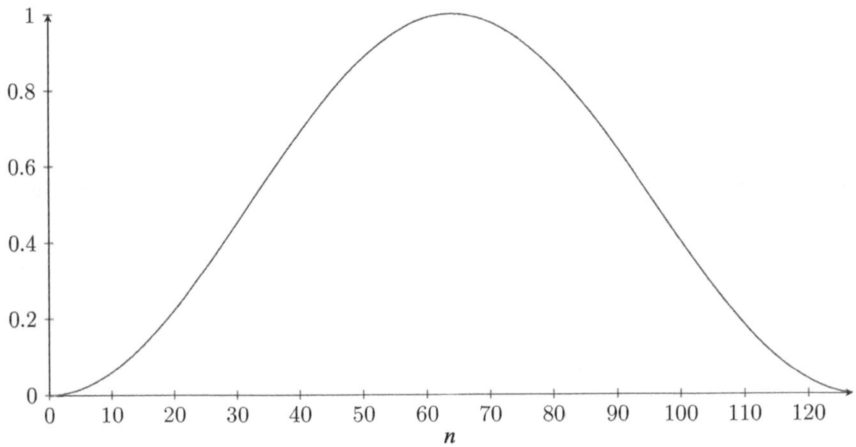

Figure 1.13 The Hann window

The Hamming window: $w(n) = 0.54 - 0.46\cos\left(2\pi\frac{n}{N}\right)$ with $0 \leq n \leq N$.

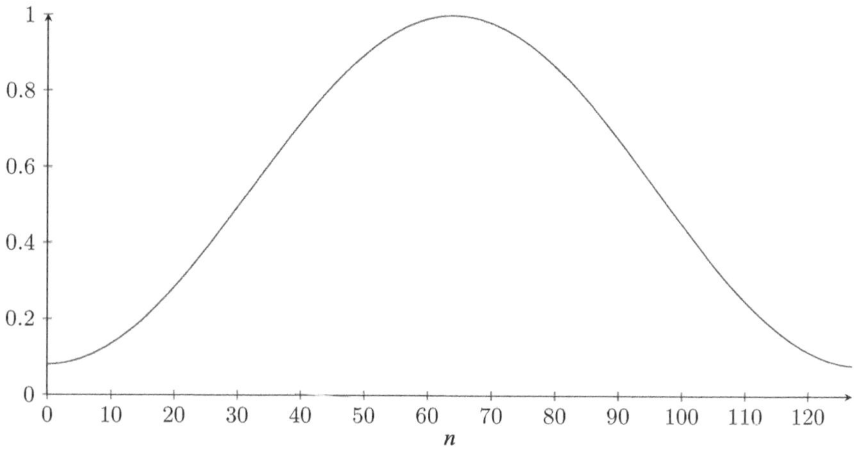

Figure 1.14 The Hamming window

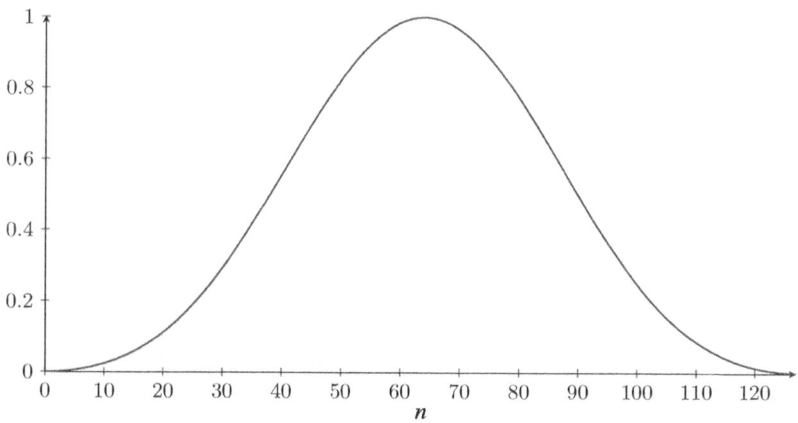

Figure 1.15 The Blackman window

The Blackman window: $w(n) = 0.42 - 0.5\cos\left(2\pi\frac{n}{N}\right) + 0.08\cos\left(4\pi\frac{n}{N}\right)$ with $0 \leq n \leq N$.

The impact of windows is very important in the estimation of a spectrum. In fact, their use modifies a lot the estimation of the spectrum. We highlight these meanings by the illustration of a real case of the stator current of an induction motor. To do that, the sampling frequency is 10kHz and the number of samples is 2^{18}. The acquisition time is 26.2144s. We firstly represent the spectrum using:

- a rectangular window;
- a Hamming window;

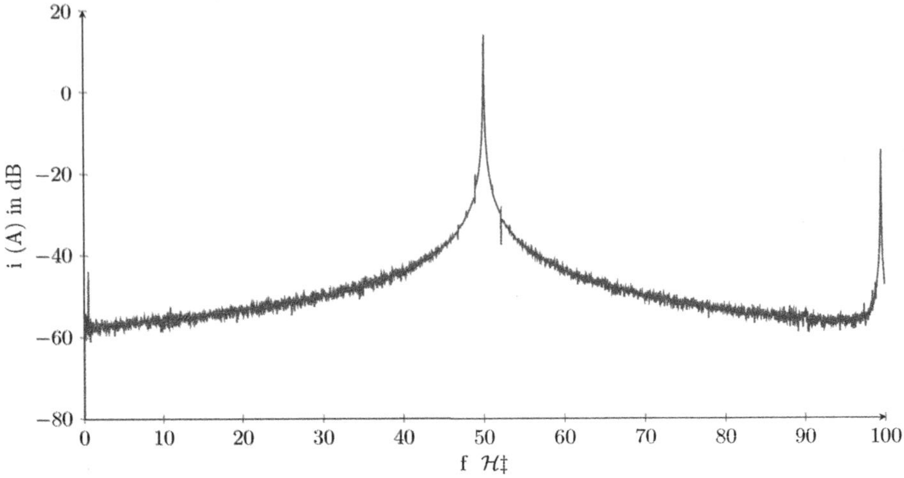

Figure 1.16 Spectrum without window

- a Hann window;
- a Blackman window.

We have to notice that the resolution of the acquisition board is on a 14 bits. So, discretization will produce noise. Moreover, we represent the amplitude of the stator current in A using a logarithmic scale, from +20dB to −80dB in order to show the impact of the use of a window function on the reduction of the noise in the spectrum.

As a conclusion, we must use a window function to better distinguish lines to each other in the spectrum. To do that, we can see that a window allows to better show the lines but affects the amplitude of each one.

(2) Fast Fourier transform (recursive Fourier transform)

Let's now address the recursivity aspect of the Fourier transform. The classical procedure consists in evaluating the spectral content at the instant $p\,T_e$. As a consequence, the spectrum is given by:

$$F_p[k] = \frac{1}{N}\sum_{n=0}^{N-1} x(p-n)W_N^{nk} \tag{1.20}$$

Now, we are going to use an artifact, which consists in adding and subtracting a term that will have the advantage of making an equation with recursivity appearing progressively.

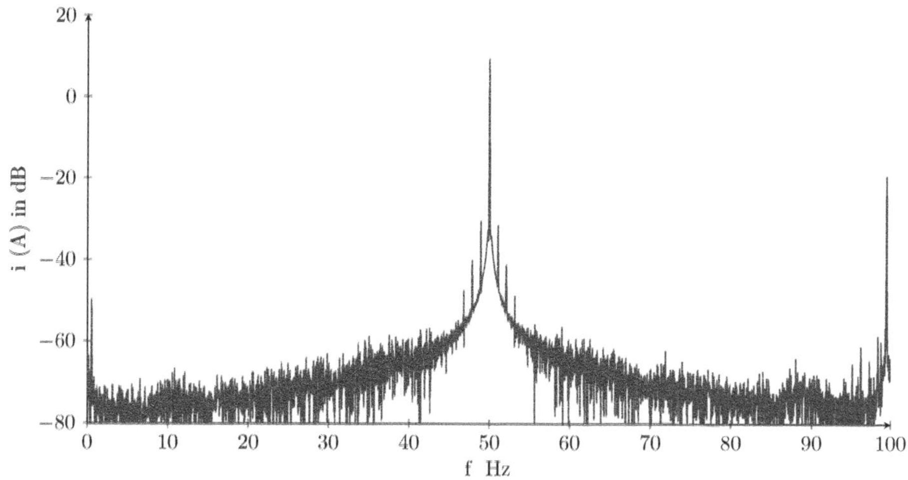

Figure 1.17 Spectrum using the Hamming window

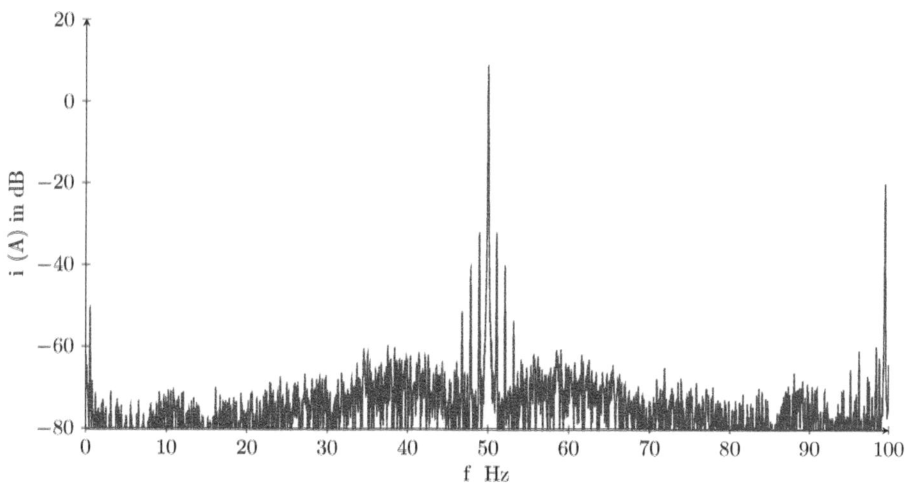

Figure 1.18 Spectrum using the Hann window

$$F_p[k] = \frac{1}{N}x(p-0) + \frac{1}{N}\sum_{n=1}^{N-1} x(p-n)W_N^{nk} \pm \frac{1}{N}x(p-N)W_N^{Nk} \quad (1.21)$$

This can be rewriten as:

$$F_p[k] = \frac{1}{N}x(p) + \frac{1}{N}\sum_{n=1}^{N} x(p-n)W_N^{nk} - \frac{1}{N}x(p-N) \quad (1.22)$$

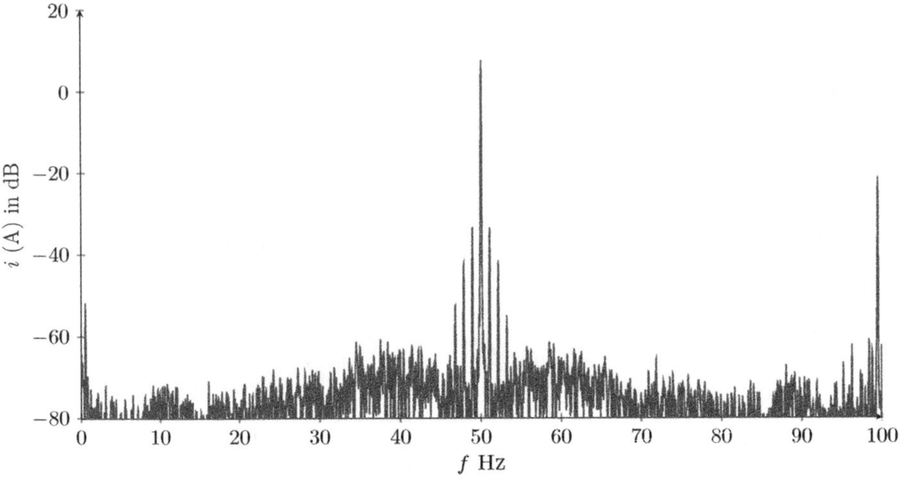

Figure 1.19 Spectrum using the Blackman window

$$F_p[k] = \frac{1}{N}x(p) + W_N^k.\frac{1}{N}\sum_{n=0}^{N-1} x(p-1-n)W_N^{nk} - \frac{1}{N}x(p-N) \tag{1.23}$$

$$F_p[k] = \frac{1}{N}x(p) - \frac{1}{N}x(p-N) + W_N^k\frac{1}{N}\sum_{n=0}^{N-1} x(p-1-n)W_N^{nk} \tag{1.24}$$

As at the instant $(p-1)\,T_e$, the spectrum is given by:

$$F_{p-1}[k] = \frac{1}{N}\sum_{n=0}^{N-1} x(p-1-n)W_N^{nk} \tag{1.25}$$

The recursive equation at the instant $p\,T_e$ is:

$$F_p[k] = \left(\frac{x(p) - x(p-N)}{N}\right) + W_N^k F_{p-1}[k] \tag{1.26}$$

A great advantage of the sliding discrete Fourier transform (SDFT) is the possibility to make the spectral estimation only on an interesting bandwidth like around the supply frequency for the monitoring of the induction motor.

$$x(p) = i_\alpha(p) + \jmath\, i_\beta(p) \tag{1.27}$$

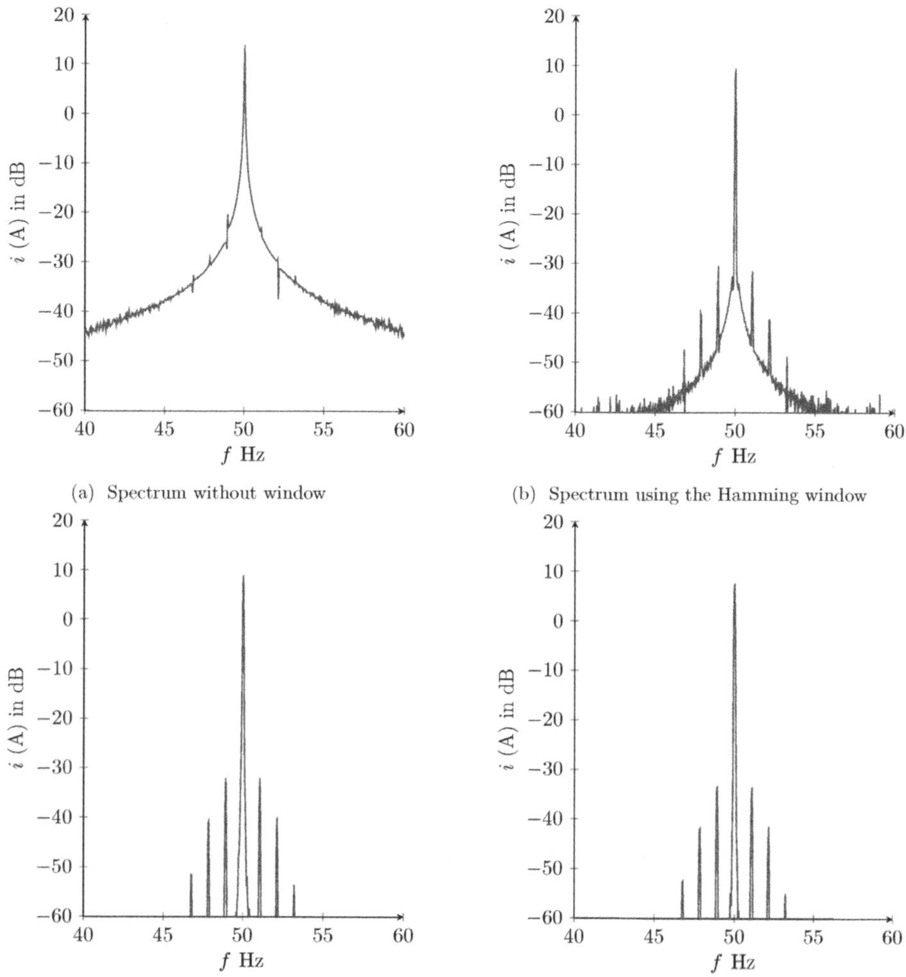

(a) Spectrum without window

(b) Spectrum using the Hamming window

(c) Spectrum using the Hann window

(d) Spectrum using the Blackman window

Figure 1.20 Spectra on the range 40–60Hz

The Discrete Fourier transform (DFT) can be written in the form as:

$$F_p[k] = \frac{1}{N} \sum_{n=0}^{N-1} \left[i_\alpha(p-n) + \jmath\, i_\beta(p-n) \right] W_N^{nk} \tag{1.28}$$

This expression is equal to:

$$F_p[k] = \frac{1}{N} \sum_{n=0}^{N-1} i_\alpha(p-n) W_N^{nk} + \jmath \frac{1}{N} \sum_{n=0}^{N-1} i_\beta(p-n) W_N^{nk} \tag{1.29}$$

Consequently, if we denote the DFT of the current i_α by F_{i_α} and by the same way, the current i_β by F_{i_β}, the relation can be rewritten as:

$$F_p[k] = F_{i_\alpha p}[k] + \jmath F_{i_\beta p}[k] \tag{1.30}$$

As the signals i_α and i_β are real, each Fourier transform has the propriety of symmetry. Thus, we can deduce the spectral density of each of them only by using the simple relations:

$$\begin{cases} F_{i_\alpha p}[k] = \frac{1}{2}\left(F_p[k] + F_p[k]^*\right) \\ F_{i_\beta p}[k] = \frac{\jmath}{2}\left(F_p[k]^* - F_p[k]\right) \end{cases} \tag{1.31}$$

where the asterisk denotes the complex conjugate part.

In conclusion, there is a possibility to estimate in real-time the spectral content of a signal thanks to the recursivity form. The advantage of this approach is that it does not require a lot of numerical calculations but has the disadvantage of not using a weighting window. The spectral estimation is done continuously but doesn't forget that the conditions must be such that the signal is stable; otherwise the estimation will be altered.

1.2.2.5 Wavelet Analysis

As far as we know, the concept of wavelet took its origin in the beginning of the previous century. Despite this, applications in signal analyses appear close to 1980. Let us be reminded that the Wavelet transform of a function $f(t)$ is a decomposition of it in a set of functions $\psi_{a,\tau}(t)$ named wavelets functions. A Wavelet transform is defined as follows:

$$W_\omega(a,\tau) = \int_{-\infty}^{+\infty} f(t)\,\psi^*_{a,\tau}(t)\,dt \tag{1.32}$$

where a is the scale factor and τ is the translation one and with $\psi_{a,\tau}(t)$ is the basic mother wavelet:

$$\psi_{a,\tau}(t) = \frac{1}{\sqrt{a}}\psi\left(\frac{t-\tau}{a}\right) \tag{1.33}$$

The analysis using the orthogonal wavelet allows to obtain a more compact representation of the signal. It permits the reconstitution of the original signal from its transformed one. Nevertheless, the analysis of a signal with a non-orthogonal wavelet is recommended for the continuous amplitude variation of the wavelet coefficients (Torrence and Compo [1998]).

Among all non-orthogonal wavelet families, we have: Gaussian, Morlet, and Mexican Hat. The most common functions used are the Morlet Hat and the

Mexican Hat. The expression of Wavelet Mother of Morlet function is described as follows:

$$\psi(t) = e^{jKt}e^{-\frac{t^2}{2}} \tag{1.34}$$

where the number of oscillation is depending on the coefficient K.

Although the Morlet function is non-orthogonal and complex, it allows to obtain information on the amplitude and the phase of signals. Thus, it is certainly better adapted for oscillatory systems. The Mexican Hat Mother Wavelet is a real function described in Equation (1.35):

$$\psi(t) = \left(1 - t^2\right)e^{-\frac{t^2}{2}} \tag{1.35}$$

This real function is particularly well adapted and so recommended for the observation of discontinuities into a signal.

However, the choice of the scale factor is not easy and a lot of work proposes criteria for the determination of this scale factor. It is possible to estimate the values minima and maxima of the scale factor to be used (see Torrence and Compo [1998]), or to determine the optimum value of scale factor (see Saleh and Rahman [2005]).

In our illustration, knowing the number of samples (N) of the signal to analyze, using the Wavelet Mexican Hat, and knowing that N is a power of two, the scale factor to be used is given by:

$$a = \log_2 N \tag{1.36}$$

where the minimal and maximal value are given by:

$$a_{\min} = \frac{1}{2}a \qquad \text{and} \qquad a_{\max} = \frac{3}{2}a \tag{1.37}$$

1.2.2.6 Instantaneous Amplitude and Frequency

(1) Instantaneous amplitude

To calculate the instantaneous amplitude of a real signal $x(t)$, it is possible to use the analytical signal $z(t)$ associated with $z(t)$:

$$z(t) = x(t) + j\,y(t) = a(t) * e^{j\varphi(t)} \tag{1.38}$$

Where $y(t)$ is the Hilbert transform of $x(t)$, defined by:

$$H[x(t)] = y(t) = \frac{1}{\pi}P\int_{-\infty}^{+\infty} \frac{x'(t)}{t - t'}\,dt' \tag{1.39}$$

where P is the principal value of Cauchy.

In 1.41, $a(t)$ corresponds to the instantaneous amplitude of $z(t)$:

$$a(t) = \sqrt{x^2(t) + y^2(t)} \tag{1.40}$$

(2) Instantaneous frequency

Instantaneous frequency is an important characteristic of the signal that is present in many areas. It is a concept related to time–frequency analysis, where it can be obtained from a time–frequency distribution (TFD). The frequency of a signal is usually a quantity that is defined independently of time. However, in the case of nonstationary signals, the introduction of the notion of instantaneous frequency can be useful for the analysis. Physically, it is important to note that the analyzed signal must imperatively be a signal with a single spectral component. In the case of multicomponent signals, the concept of instantaneous frequency is no longer valid (Loughlin and Tacer [1997]). Two basic definitions arc proposed in the literature for calculating instantaneous frequency. The first one is based on the time–frequency representation of the studied signal, and the second one is defined as the derivative of the phase of the analytical signal associated with the studied signal. The instantaneous frequency $f_i(t)$ of an analytical signal $z(t) = a(t) * e^{j\varphi(t)}$ containing a single component is defined as being the derivative of its phase:

$$f_i(t) = \frac{1}{2\pi}\frac{d\varphi(t)}{dt} \tag{1.41}$$

The analytical signal $z(t)$ is a complex signal containing the same information as the real signal $x(t)$. Unlike the real signal, the complex signal contains only the positive frequencies. It is built from the real signal using the Hilbert transform symbolized by the operator H:

$$z(t) = x(t) + jH[x(t)] \tag{1.42}$$

$$H(t) = \int_{-\infty}^{+\infty} \frac{x(t-\tau)}{\pi\tau}\,d\tau \tag{1.43}$$

1.2.2.7 Bilinear Time–frequency Distributions or Quadratic Time–frequency Distributions: Cohen's Class

Time analysis does not allow to represent precisely the harmonic components of the signal, while the frequency analysis does not allow them to be located temporally. Simultaneous representation in the time and frequency domain should achieve a compromise between the two previous approaches (Lebaroud [2007], Chatfield and Xing [2019], and Flandrin [1998]). However, the Heisenberg uncertainty principle excludes any precise temporal location of a frequency. This ambiguity has led to define a wide variety of time–frequency transformations.

1.2.2.7.a Uncertainty Principle of Heisenberg

For any signal $x(t)$ of finite energy, let us define the centroids in the time and frequency domains (eq. 1.44 and eq. 1.45):

$$<t>=\frac{1}{E}\int t|x(t)|^2 dt \tag{1.44}$$

and

$$<f>=\frac{1}{E}\int f|X(f)|^2 df \tag{1.45}$$

The time and frequency resolutions can be deduced from these equations (eq. 1.46 and eq. 1.47):

$$\Delta t = \sqrt{\frac{1}{E}\int (t-<t>)^2 \left|x(t)^2\right| dt} \tag{1.46}$$

and

$$\Delta f = \sqrt{\frac{1}{E}\int (f-<f>)^2 \left|x(f)^2\right| df} \tag{1.47}$$

Thus, we can write the Heisenberg's uncertainty principle as eq. 1.48:

$$\Delta t.\Delta f \geqslant \frac{1}{4\pi} \tag{1.48}$$

1.2.2.7.b General Representation

The time–frequency representations of Cohen's class (Cohen [1995]) are of the form (eq. 1.49):

$$C_x(t,f) = \int_{-\infty}^{+\infty}\int_{-\infty}^{+\infty} \varphi_{t-r}(t-\nu,\tau)x(\gamma+\frac{\tau}{2})x^*(\gamma-\frac{\tau}{2})e^{-j2\pi f\tau}d\gamma d\tau \tag{1.49}$$

where $\varphi_{t-r}(t-\nu,\tau)$ is the *kernel*.

Let us define the ambiguity function (eq. 1.50):

$$A_x(\tau,\xi) = \int_{-\infty}^{+\infty} x(t+\frac{\tau}{2})x^*(t-\frac{\tau}{2})e^{-j2\pi\xi t}dt \tag{1.50}$$

We can find two different forms of the previous equations:

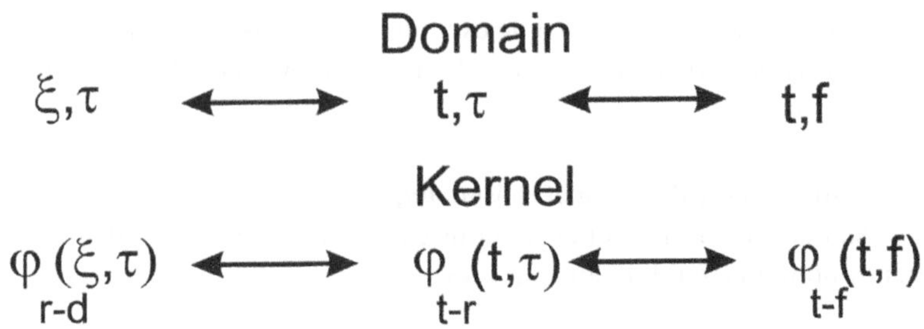

Figure 1.21 Rules of transformation between the time–frequency and Doppler-delay domains encoder

First form:

$$C_x(t,f) = \int_{-\infty}^{\infty}\int_{-\infty}^{\infty} \varphi_{r-d}(\tau,\xi)\, A_x(\tau,\xi)\, e^{j2\pi(t\,\xi - f\,\tau)}\, d\xi d\tau \tag{1.51}$$

where $A_x(\tau,\xi) = \int_{-\infty}^{+\infty} x(t+\frac{\tau}{2})x^*(t-\frac{\tau}{2})e^{-j2\pi\xi t}dt$ and $\varphi_{r-d}(\tau,\xi) = \int_{-\infty}^{\infty} \varphi_{t,\tau}(t,\tau)\, e^{-j2\pi\,\xi\,t}dt$

The two variables τ and ξ are called delay and Doppler frequency, respectively.

Second form:

$$C_x(t,f) = \int_{-\infty}^{\infty}\int_{-\infty}^{\infty} \varphi_{t-f}(t-\upsilon, f-\nu)\, W_x(\upsilon,\nu)\, d\upsilon\, d\nu \tag{1.52}$$

where $W_x(t,f) = \int_{-\infty}^{+\infty} x(t+\frac{\tau}{2})x^*(t-\frac{\tau}{2})e^{-j2\pi f t}d\tau$ and $\varphi_{t-f}(t,f) = \int_{-\infty}^{\infty} \varphi_{t,\tau}(t,\tau)\, e^{-j2\pi\,f\,\tau}d\tau$. The function $W_x(t,f)$ is the distribution of Wigner–Ville (Wigner [1932] and Ville [1948]).

The rules of transformation between the time–frequency and Doppler-delay domains are illustrated in Figure 1.21, which explains the type of *Kernel* to be associated with the type of variables. All these *Kernels* are detailed in the Tables 1.16 to 1.19.

1.2.2.7.c Properties

There is a wide variety of time–frequency representations based on the Cohen's class principle. They are distinguished by their properties. The research work carried out in this field (Hlawatsch and Boudreaux–Bartels [1992] and TA and Mecklenbräuker [1980]) has allowed associate properties and the different *kernels* of the time–frequency representations (TFR).

Let us consider the main properties in Table 1.12.

Table 1.12 First group of properties of the representations of the Cohen's class (Lebaroud [2007], Hlawatsch and Boudreaux–Bartels [1992], and Auger and Doncarli [1992]).

N	Property	Mathematical formulation	Condition
0	Invariance with respect to time and frequency translations	$\forall t_1, f_1,\ y(t) = x(t - t_1)e^{j2\pi f_1 t}$ $\Rightarrow C_y(t,f) = C_x(t - t_1, f - f_1)$	$\varphi_{r-d}(\tau,\xi)$ whatever is
1	Invariance with respect to changes in scale	$\forall a > 0,\ y(t) = a^{-1/2}x(t/a)$ $\Rightarrow C_y(t,f) = C_x(t/a, af)$	$\forall \tau,\xi,\forall a > 0,$ $\varphi_{r-d}(\tau,\xi) = \varphi_{r-d}(\tau/a, a\xi)$
2	Invariance with respect to time inversion	$y(t) = x(-t)$ $\Rightarrow C_y(t,f) = C_x(-t,-f)$	$\forall \tau,\xi$ $\varphi_{r-d}(\tau,\xi) = \varphi_{r-d}(-\tau,-\xi)$
3	Invariance with respect to complex conjugation	$y(t) = x^*(t)$ $\Rightarrow C_y(t,f) = C_x(t,-f)$	$\forall \tau,\xi$ $\varphi_{r-d}(\tau,\xi) = \varphi_{r-d}(-\tau,\xi)$
4	Real character of TFR	$\forall t, f,$ $C_x(t,f) = C_x^*(t,f)$	$\forall \tau,\xi$ $\varphi_{r-d}^*(\tau,\xi) = \varphi_{r-d}(-\tau,-\xi)$
5	Causality	$C_x(t,f)$ only depends on x(u) if $u < t$	$\forall t,\tau,$ $t < \lvert\tau\rvert/2 \Rightarrow \varphi_{t-r}(t,\tau) = 0$

where $\hat{x}(f) = \int_{-\infty}^{+\infty} x(t)e^{-j2\pi ft}dt$ is the Fourier transform of $x(t)$.

1.2.2.7.d Different Representations

Tables 1.16 to 1.19 give the different Cohen's class representations with the properties defined in the paragraph 1.2.2.7.c.

The Wigner–Ville representation is the only distribution that provides perfect localization of linearly frequency-modulated signals. But it has many interference terms, which makes it difficult to interpret.

The pseudo Wigner–Ville representation uses a window $h(\tau)$ that makes an easier computation.

The smoothed pseudo Wigner–Ville representation uses an additional window g(t), in order to perform a time domain smoothing independent of the frequency smoothing ($\varphi_{t,f}(t,f) = g(t)\hat{h}(f)$). This allows to have a good compromise between the localization of the different components and the attenuation of interference terms (Lebaroud [2007]) (Table 1.19).

Rihaczek Distribution provides a perfect localization of sinusoidal signals.
The distribution of Margenau-Hill is easier to represent graphically.
The Page's distribution is the only Cohen's class that simultaneously verifies the properties of causality (property 5), spectrum conservation (property

Table 1.13 Second group of properties of the representations of the Cohen's class (Lebaroud [2007], Hlawatsch and Boudreaux-Bartels [1992], and Auger and Doncarli [1992])

N	Property	Mathematical formulation	Condition
6	Energy conservation	$\int_{-\infty}^{+\infty}\int_{-\infty}^{+\infty} C_x(t,f)dtdf = \int_{-\infty}^{+\infty} x(t)\|^2 dt$	$\varphi_{r-d}(0,0) = 1$
7	Instantaneous power conservation	$\forall t,\ \int_{-\infty}^{\infty} C_x(t,f)\,df = \|x(t)\|^2$	$\forall \xi,\ \varphi_{r-d}(0,\xi) = 1$
8	Spectrum conservation	$\forall f,\ \int_{-\infty}^{\infty} C_x(t,f)\,dt = \|\hat{x}(f)\|^2$	$\forall \tau,\ \varphi_{r-d}(\tau,0) = 1$
9	Instantaneous frequency conservation	$\forall t,\ \int_{-\infty}^{\infty} fC_x(t,f)\,df = f_x(t)\,\|x(t)\|^2$ $f_x(t) = \frac{1}{2\pi}\frac{d}{dt}\arg(x(t))$ $[1mm]$	$\forall \xi, \varphi_{r-d}(0,\xi) = 1$ and $\frac{\partial}{\partial \tau}\varphi_{r-d}(0,\xi) = 0$
10	Group delay conservation	$\forall f,\ \int_{-\infty}^{\infty} t\,C_x(t,f)\,dt = t_x(f)\,\|\hat{x}(f)\|^2$ $t_x(f) = -\frac{1}{2\pi}\frac{d}{df}\arg(\hat{x}(f))$	$\forall \tau, \varphi_{r-d}(\tau,0) = 1$ and $\frac{\partial}{\partial \xi}\varphi_{r-d}(\tau,0) = 0$
11	Positivity	$\forall x(t), \forall t,\ f,\ C_x(t,f) \geqslant 0$	$\varphi_{r-d}(\tau,\xi) =$ $\int_{-\infty}^{\infty} c(\alpha)\,A_{h_\alpha}(\tau,\xi)\,d\alpha$ with $\forall \alpha, c(\alpha) \geqslant 0$

8), Moyal's formula (property 14) and the conservation of the signal domain (property 16) (Lebaroud [2007]) (Table 1.17).

The spectrogram is still widely used today, especially for analyzing signals that are slowly non-stationary or have a large number of components. For short and strongly non-stationary signals, a better compromise can then be obtained by the S representation (Lebaroud [2007]) (Table 1.18).

The use of product *kernels* reduces interference terms (Lebaroud [2007]) (Table 1.19).

We will illustrate the different Cohen's classes with the help of a chirp function (eq 1.53). It is a signal where the frequency increases or decreases with time.

$$y(t) = \Re e(a(t)e^{j\Phi(t)}) \tag{1.53}$$

Table 1.14 Third group of properties of the representations of the Cohen's class (Lebaroud [2007], Hlawatsch and Boudreaux-Bartels [1992], and Auger and Doncarli [1992]).

N	Property	Mathematical formulation	Condition
12	Compatibility with linear Filters	$y(t) = \int_{-\infty}^{\infty} h(u)\,x(t-u)\,du$ $\Rightarrow$ $C_y(t,f) = \int_{-\infty}^{\infty} C_h(u,f)\,C_x(t-u,f)\,du$	$\forall \tau, \tau', \xi$ $\varphi_{r-d}(\tau,\xi)\,\varphi_{r-d}(\tau',\xi) = \varphi_{r-d}(\tau+\tau',\xi)$
13	Compatibility with modulations	$y(t) = m(t)\,x(t)$ $\Rightarrow$ $C_y(t,f) = \int_{-\infty}^{\infty} C_m(t,\nu)\,C_x(t,f-\nu)\,d\nu$	$\forall \tau, \xi, \xi'$, $\varphi_{r-d}(\tau,\xi)\,\varphi_{r-d}(\tau,\xi') = \varphi_{r-d}(\tau,\xi+\xi')$
14	Preservation of scalar product	$\int_{-\infty}^{\infty}\int_{-\infty}^{\infty} C_x(t,f)\,C_y^*(t,f)\,dtdf =$ $\left\lvert \int_{-\infty}^{\infty} x(t)\,y^*(t)\,dt \right\rvert^2$	$\forall \tau, \xi, \lvert\varphi_{r-d}(\tau,\xi)\rvert^2 = 1$
15	Reversibility	$x(t)$ can be retrieved, within one phase, from its TFR	$\forall \tau, \xi$, $\varphi_{r-d}(\tau,\xi) \neq 0$

Table 1.15 Fourth group of properties of the representations of the Cohen's class (Lebaroud [2007], Hlawatsch and Boudreaux-Bartels [1992], and Auger and Doncarli [1992]).

N	Property	Mathematical formulation	Condition
16	Conservation of the signal domain	If $\forall\, t, \lvert t\rvert > T \Rightarrow x(t) = 0$ Then $\forall t,\ \lvert t\rvert > T \Rightarrow C_x(t,f) = 0$	$\forall t, \tau$, $\lvert\tau\rvert < 2\lvert t\rvert \Rightarrow \varphi_{r-d}(t,\tau) = 0$
17	Conservation of the null values of the signal	$x(t_0) = 0 \Rightarrow \forall\, f, C_x(t_0,f) = 0$	$\varphi_{r-d}(\tau,\xi) = h_1(\tau)\,e^{-j2\pi\xi\tau/2} + h_2(\tau)\,e^{j2\pi\xi\tau/2}$
18	Conservation of the the null values of the spectrum	$\hat{x}(f_0) = 0 \Rightarrow \forall\, t, C_x(t,f_0) = 0$	$\varphi_{r-d}(\tau,\xi) = G_1(\xi)\,e^{-j2\pi\xi\tau/2} + G_2(\xi)\,e^{j2\pi\xi\tau/2}$

Table 1.16 Cohen's class representations: Wigner–Ville Distribution and associated forms (Lebaroud [2007], Hlawatsch and Boudreaux–Bartels [1992], and Auger and Doncarli [1992]).

Name	Formula	Properties
Wigner Ville	$\int_{-\infty}^{\infty} x(t+\tau/2)\,x^*(t-\tau/2)\,e^{-j2\pi f\tau}d\tau$ ***Kernel:*** $\varphi_{r-d}(\tau,\xi)=1$ $\varphi_{t-r}(t,\tau)=\delta(t)$	0,1,2,3,4,6,7,8,9, 10,12,13,14,15,16
Pseudo Wigner–Ville	$\int_{-\infty}^{\infty} h(\tau)\,x(t+\tau/2)\,x^*(t-\tau/2)\,e^{-j2\pi f\tau}d\tau$ ***Kernel:*** $\varphi_{r-d}(\tau,\xi)=h(\tau)$ $\varphi_{t-r}(t,\tau)=\delta(t)\,h(\tau)$	0,2,3,4,6,7,9,16 if $h(t)$ is is real and even, $h(0)=1$, $h'(0)=0$
Smoothed Pseudo Wigner–Ville	$\int_{-\infty}^{\infty}\int_{-\infty}^{\infty} g(t-\upsilon)\,h(\tau)\,x(\upsilon+\tau/2)\,x^*(\upsilon-\tau/2)\,e^{-j2\pi f\tau}d\tau d\upsilon$ ***Kernel:*** $\varphi_{r-d}(\tau,\xi)=h(\tau)\,\hat{g}(\xi)$ $\varphi_{t-r}(t,\tau)=g(t)\,h(\tau)$	0,2,3,4,6 if $h(t)$ is real and even, $h(0)=1$ and if $g(t)$ is real and even, $\hat{g}(0)=1$

Table 1.17 Cohen class representations: Rihaczek Distribution and associated forms (Lebaroud [2007] and Auger and Doncarli [1992]).

Name	Formula	Properties
Rihaczek	$x(t)\,\hat{x}^*(f)\,e^{-j2\pi ft}$ ***Kernel:*** $\varphi_{r-d}(\tau,\xi)=e^{-j\pi\xi\tau}$ $\varphi_{t-r}(t,\tau)=\delta(t-\tau/2)$	0,1,2,6,7,8,12,13, 14,15,16,17,18
Margenau-Hill	$\int_{-\infty}^{\infty}\frac{1}{2}\left[x(t+\tau)\,x^*(t)+x(t)\,x^*(t-\tau)\right]e^{-j2\pi f\tau}=$ $\mathrm{Re}\{x(t)\,\hat{x}^*(f)\,e^{-j2\pi ft}\}$ ***Kernel:*** $\varphi_{r-d}(\tau,\xi)=\cos(\pi\xi\tau)$ $\varphi_{t-r}(t,\tau)=\frac{1}{2}\left[\delta(t-\tau/2)\right]+\delta(t+\tau/2)$	0,1,2,3,4,6,7,8, 9,10,16,17,18

(Continued)

Table 1.17 *(Continued)*

Name	Formula	Properties
Pseudo Margenau-Hill	$\int_{-\infty}^{\infty} \frac{h(\tau)}{2} [x(t+\tau)x^*(t) + x(t)x^*(t-\tau)] e^{-j2\pi f\tau} d\tau =$ $\mathrm{Re}\left\{x(t)\, TFCT_x^{h_*}(t,f)\, e^{-j2\pi ft}\right\}$ (h real and even) ***Kernel:*** $\varphi_{r-d}(\tau,\xi) = h(\tau)\cos(\pi\xi\tau)$ $\varphi_{t-r}(t,\tau) = \frac{h(\tau)}{2}[\delta(t-\tau/2)] + \delta(t+\tau/2)$	0,2,3,4,6,7,9,1617, if h(t) is is real even, $h(0) = 1\,, h'(0) = 0$
Page	$2\,\mathrm{Re}\left\{x(t)\left(\int_{-\infty}^{t} x(u)e^{-j2\pi fu}du\right)^* e^{-j2\pi ft}\right\}$ ***Kernel:*** $\varphi_{r-d}(\tau,\xi) = e^{j\pi\xi\lvert\tau\rvert}$ $\varphi_{t-r}(t,\tau) = \delta(t - \lvert\tau\rvert/2)$	0,3,4,5,6,7,16,17

Table 1.18 Cohen class representations: Spectrogram et S representation (Lebaroud [2007] and Auger and Doncarli [1992])

Name	Formula	Properties
Spectrogram	$\left\lvert\int_{-\infty}^{\infty} x(u)h^*(u-t)e^{-j2\pi fu}du\right\rvert^2$ ***Kernel:*** $\varphi_{r-d}(\tau,\xi) = A_h^*(\tau,\xi)$ $\varphi_{t-r}(t,\tau) = h^*(-t+\tau/2)\,h(-t-\tau/2)$	0,2,3,4,6,11 if $\int_{-\infty}^{\infty} \lvert h(t)\rvert^2 dt = 1$
S-representation	$\int_{-\infty}^{\infty} \hat{S}(\xi)\, TFCT_x^h(t, f+\xi/2)\, TFCT_x^{h_*}(t, f-\xi/2)\, e^{j2\pi\xi t} d\xi$ ***Kernel:*** $\varphi_{r-d}(\tau,\xi) = \int_{-\infty}^{\infty} \hat{S}(\nu) A_h^*(\tau, \xi-\nu)\, d\nu$ $\varphi_{t-r}(t,\tau) = S(t)\,h^*(-t+\tau/2)\,h(-t-\tau/2)$	0,2,3,4,6

where the amplitude is slowly varying compared to the variations of Φ. In our example, we will use a quadratic increase. Several distributions are presented in Table 1.20 for the same signal.

Table 1.19 Cohen class representations: Product-type or cone-shaped core representation (Lebaroud [2007] and Auger and Doncarli [1992]).

Name	Formula	Properties
Choi–Williams	$\int_{-\infty}^{\infty}\int_{-\infty}^{\infty}\frac{\exp\left(-\frac{\nu^2}{4\tau^2/\sigma}\right)}{\sqrt{4\pi\tau^2/\sigma}}x(t-\nu+\tau/2)\,x^*(t-\nu-\tau/2)\,e^{-j2\pi f\tau}d\nu d\tau$ ***Kernel:*** $\varphi_{r-d}(\tau,\xi)=\exp\left(-4\pi^2\xi^2\tau^2/\sigma\right)$ $\varphi_{t-r}(t,\tau)=\begin{cases}\delta(t) & if \quad \tau=0\\ \frac{\exp\left(-\frac{t^2}{4\tau^2/\sigma}\right)}{\sqrt{4\pi\tau^2/\sigma}} & if \quad \tau\neq 0\end{cases}$	0,1,2,3,4,6, 7,8,9,10,15
Born–Jordan	$\int_{-\infty}^{\infty}\int_{-\lvert\tau\rvert/2}^{\lvert\tau\rvert/2}\frac{1}{\lvert\tau\rvert}h(\tau)\,x(t-\nu+\tau/2)\,x^*(t-\nu-\tau/2)\,e^{-j2\pi f\tau}d\nu d\tau$ ***Kernel:*** $\varphi_{r-d}(\tau,\xi)=\sin c(\pi\xi\tau)$ $\varphi_{t-r}(t,\tau)=\begin{cases}\delta(t) & if \quad \tau=0\\ \frac{1}{\lvert\tau\rvert} & if \quad \lvert t\rvert<\lvert\tau\rvert/2\\ else \quad 0 & \end{cases}$	0,1,2,3,4,6, 7,8,9,10,16
Zhao–Atlas –Marks	$\int_{-\infty}^{\infty}\int_{-\lvert\tau\rvert/2}^{\lvert\tau\rvert/2}h(\tau)\,x(t-\nu+\tau/2)\,x^*(t-\nu-\tau/2)\,e^{-j2\pi f\tau}d\nu d\tau$ ***Kernel:*** $\varphi_{r-d}(\tau,\xi)=h(\tau)\sin c(\pi\xi\tau)$ $\varphi_{t-r}(t,\tau)=\begin{cases}0 & if \quad \tau=0\\ h(t) & if \quad \lvert t\rvert<\lvert\tau\rvert/2\\ else \quad 0 & \end{cases}$	0,2,3,4,16

Table 1.20 TFR examples for a chirp signal

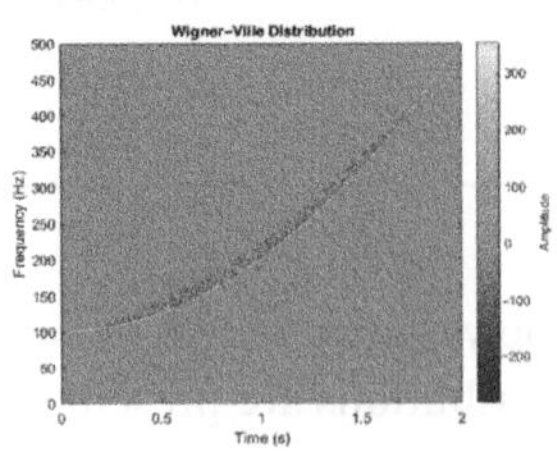 Wigner–Ville Distribution	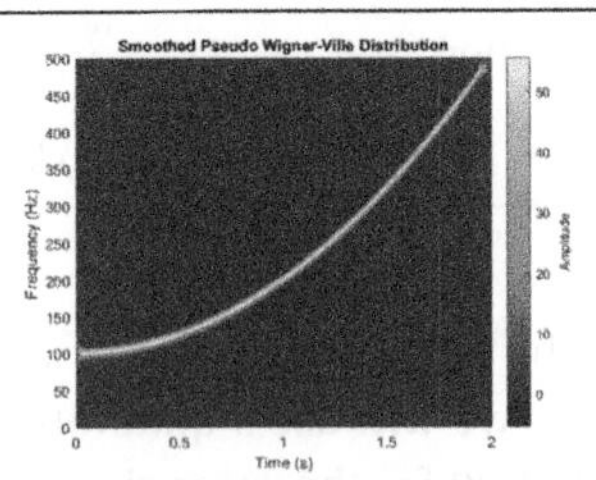 Smoothed Pseudo Vigner Ville Distribution

(Continued)

Table 1.20 *(Continued)*

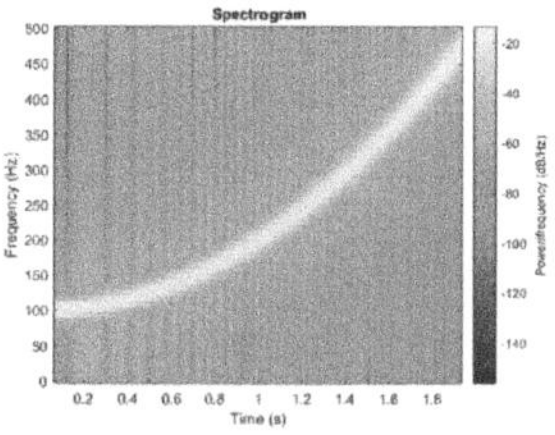 Spectrogram	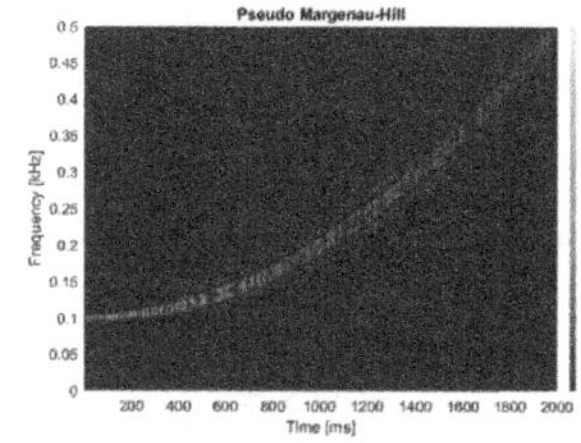 Pseudo Margenau-Hill
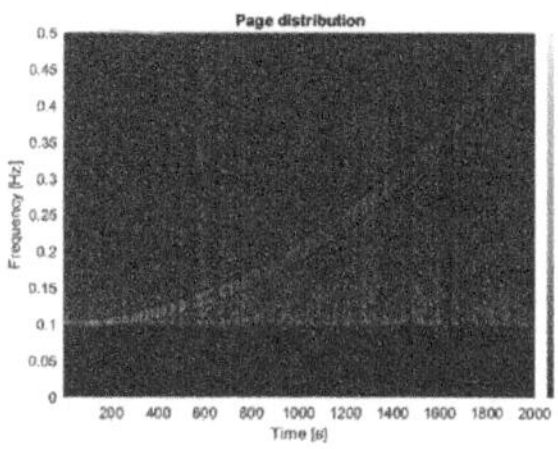 Page distribution	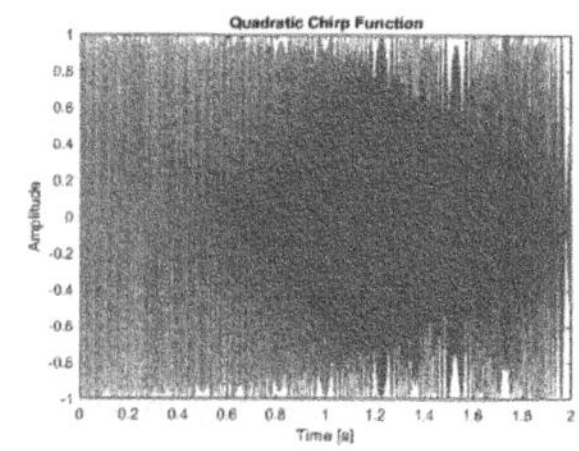 Quadratic Chirp Function

1.2.2.8 Statistic Features

Eight main statistic features are often used, in particular for vibration studies.

Let us consider a time series x_t of N samples. They are defined as follows:

- The energy

$$E = \sum_{\ell=1}^{N} x_\ell^2 \tag{1.54}$$

- The maximum

$$S_{PA} = \sup_{1 \leq \ell \leq N} |x_\ell| \tag{1.55}$$

- The mean value

$$\bar{x} = \frac{1}{N} \sum_{\ell=1}^{N} x_\ell \tag{1.56}$$

- The crest factor

$$F_c = {}^{S_{PA}}/_{RMS} \tag{1.57}$$

- The mean power

$$P = \frac{1}{N} \sum_{\ell=1}^{N} x_\ell^2 \tag{1.58}$$

- The standard deviation

$$RMS = \sqrt{\frac{1}{N} \sum_{\ell=1}^{N} (x_\ell - \bar{x})^2} \tag{1.59}$$

- The kurtosis

$$S_{kurt} = \frac{\frac{1}{N} \sum_{\ell=1}^{N} (x_\ell - \bar{x})^4}{RMS^4} \tag{1.60}$$

- The factor K

$$F_K = S_{PA}.RMS = F_c.RMS^2 \tag{1.61}$$

RMS value gives a measure of an overall vibration level. It is sensitive to the energy of the signal. The maximum value gives the same kind of information, but it is less adapted than RMS indicator.

Crest factor and kurtosis are suitable for shock measurement. They measure the impulsive nature of a signal. Kurtosis is often better. However, it can have a non-monotonous variation with the severity, in case of several defects.

One must beware of an overly simplistic interpretation of kurtosis (Balanda and Macgillivray [1988]). J. J. A. Moors [1986] explains that two causes involve a high values of the Kurtosis:

- The probability mass is concentrated around the mean and a few scattered values are far from the mean;
- The probability mass is concentrated in the tails of the distribution.

Factor K allows to evaluate the degree of severity of a rolling single defect (Boukar and Hamzaoui [2019]).

1.2.2.9 Cyclostationarity

(1) General description

The concept of cyclostationarity was introduced in the Russian literature by Gudzenko [1959] and Lebedev [1959]. It deals with signals which exhibit some hidden periodicity of their energy flow (see Figure 1.22). In other terms, cyclostationary processes, which result from the interaction of periodic and random phenomena, are characterized by periodic statistics (Antoni [2009]). The utilization of cyclostationarity is especially suitable for diagnosing the mechanical failures of rotating machines (Bonnardot [2004]), gears (Kidar [2015]), etc.

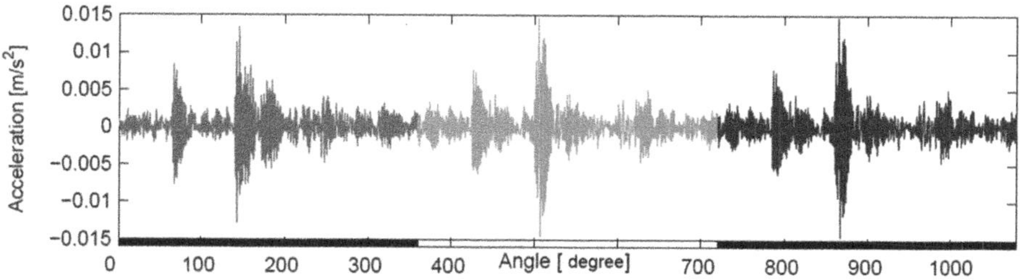

Figure 1.22 Stationary signal example (Bonnardot [2004])

We will now give some definitions of the cyclostationarity for rotating machines.

(2) nth-order strict cyclostationarity

Let us consider a rotating machine signal $X(\theta)$ with a stochastic behavior. For any θ, $X(\theta)$ is a random variable.

The nth-order distribution function is defined by:

$$F_{X(\theta+\theta_1)\ \ldots\ X(\theta+\theta_{N-1})\ X(\theta)}(\delta_1 \ldots \delta_{N-1}\delta_N) = P(X(\theta+\theta_1) \le \delta_1 \ldots X(\theta+\theta_{N-1}) \le \delta_{N-1} X(\theta) \le \delta_N) \tag{1.62}$$

The process $X(\theta)$ has a strictly n^{th}-order cyclostationary of period θ if and only if its nth-order distribution function is periodic of period Θ.

$$\forall\left[\ \delta_1, \ \ldots, \ \delta_N\ \right] \in R^N, \forall\left[\ \theta_1, \ \ldots, \ \theta_N\ \right] \in R^N, \forall\theta \in R,$$
$$F_{X(\theta+\theta_1), \ \ldots \ X(\theta+\theta_{N-1}, \ \theta)}(\delta_1, \ \ldots \ \delta_{N-1}, \ \delta_N)$$
$$F_{X(\Theta+\theta+\theta_1), \ \ldots \ X(\Theta+\theta+\theta_{N-1}), \ X(\Theta+\theta)}(\delta_1, \ \ldots \ \delta_{N-1}, \ \delta_N) \tag{1.63}$$

(3) nth-order wide sense cyclostationarity

The process $X(\theta)$ has a wide sense 2nd-order cyclostationary of period θ if and only if it satisfies the following properties:

- Periodicity of the 1st-order moment

$$\forall \theta, \quad E(x(\theta + \Theta)) = E(x(\theta)) \tag{1.64}$$

where E is the expected value.
- Periodicity of the 2nd-order moment

$$\begin{aligned} &\forall \theta, \forall \delta, \; R_X(\theta, \delta) = R_x(\theta + \Theta, \delta) \\ &\text{with} \\ &R_X(\theta, \delta) = E(X(\theta)X(\theta + \delta)) \end{aligned} \tag{1.65}$$

As we obtain a cyclic function, and under certain conditions of convergence, we can compute the coefficients of the Fourier series.

$$E(X(\theta)) = \sum_{k=-\infty}^{+\infty} m_{X,\Theta,k} e^{j2\pi\left(\frac{k}{\Theta}\right)\theta} \text{ with } m_{X,\Theta,k} = \frac{1}{\Theta} \int_{-\frac{\Theta}{2}}^{+\frac{\Theta}{2}} E(X(\theta)) e^{-j2\pi\left(\frac{k}{\Theta}\right)\theta} d\theta \tag{1.66}$$

and

$$R_X(\theta, \delta) = \sum_{k=-\infty}^{+\infty} R_{X,\Theta,k}(\delta) e^{j2\pi\left(\frac{k}{\Theta}\right)\theta} \text{ with } R_{X,\Theta,k}(\delta) = \frac{1}{\Theta} \int_{-\frac{\Theta}{2}}^{+\frac{\Theta}{2}} R_X(\theta, \delta) e^{-j2\pi\left(\frac{k}{\Theta}\right)\theta} d\theta \tag{1.67}$$

However, these different parameters and their harmonic decomposition allow the construction of efficient fault signatures in the vibratory domain. These techniques can require angular resampling of signals in order to transform time representation to angular one.

The ergodicity of some cyclostationary signals makes it possible to evaluate these static parameters by splitting the time series into successive blocks of length Θ (cyclic period) and making a synchronous average (Bonnardot [2004]). A signal is ergodic if its statistical characteristics, computed from the values at the same instant of a large number of different realizations of the system, are identical with those computed from the successive values in time of any of these realizations.

1.2.3 Model Approach

Some relevant features for diagnosis come from a model of the system (Clerc and Marques [2011]). They are two different approaches:

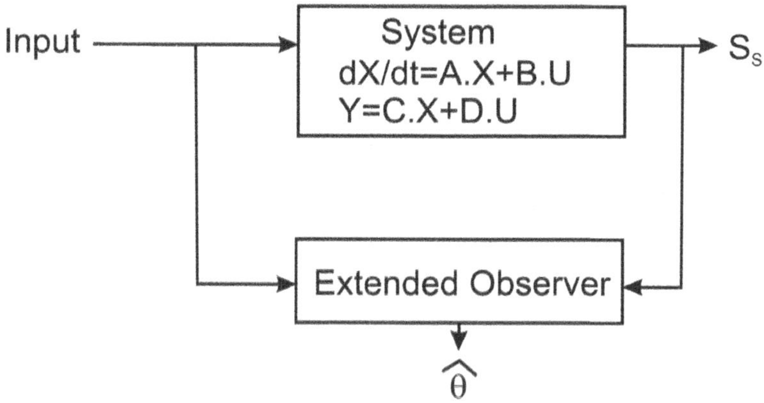

Figure 1.23 Estimation of the parameters

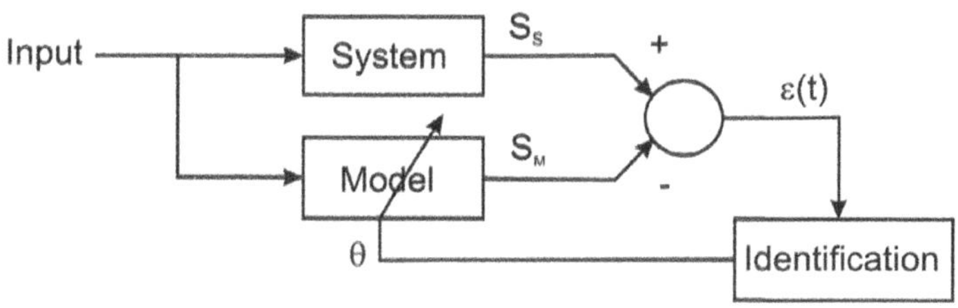

Figure 1.24 Estimation of the parameters

- The first one is based on the estimation of the model parameters.
- The second one is based on an analysis of the residuals between the model and the actual process.

In case of estimated parameters, these later can be identified with off-line algorithms or estimated by extended observers. Table 1.21 gives an overview of the main characteristics of these methods.

We will now describe the main concept of observer. It allows the reconstruction of non-measurable state quantities (rotor flow, motor slip ...) or difficult to measure (torque, speed ...). It associates an estimator and a state correction from the model output error. The estimator is based on the direct solving of the equations associated with the model. It operates in open loop. The observer is an estimator where the estimated quantity is corrected by a term that depends on an error built from measurable quantities.

The observer makes it possible to rebuild the state of an observable system from the measurement of inputs and outputs. It is used in state feedback control when all or part of the state vector cannot be measured. In its extended version,

Table 1.21 Some characteristics of the model approach

	Advantages	Disadvantages
Parameters identification	-Taking into account more complex models that can represent defective behaviors. -Can be used in variable speed drives with converter. -Numerous algorithms that can be adapted to the application.	-Mostly used off-line
Extended observers	-Can be used online. -Can be integrated in controls laws where the state must be rebuilt such as state feedback, etc ... -Can be used in variable speed drives with converter	-Requires simple models. -Requires slowly varying parameters

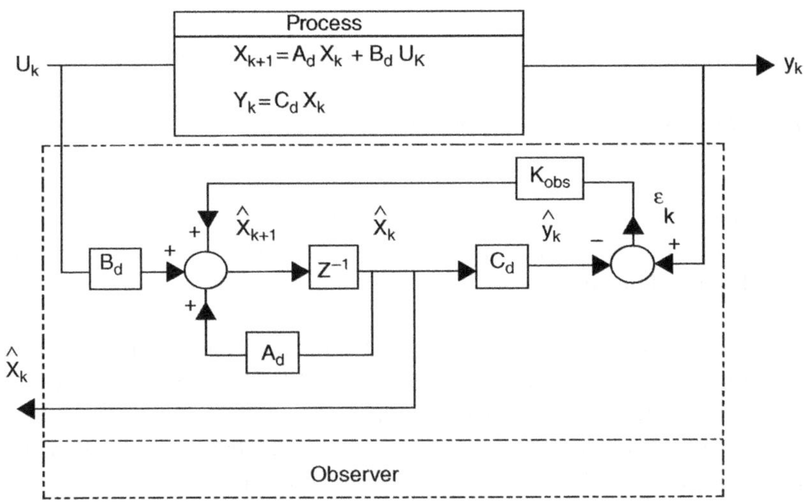

Figure 1.25 Observer

it allows to estimate variable or unknown parameters of a system. It can be represented by the following block diagram (Figure 1.25):

The different types of observers differ on the gain synthesis.

To be able to build an observer, the system must be observable. A discrete linear Single Input Single Output (SISO) system is observable if and only if the initial state of the state vector $X(k.T_{sampling_0})$ can be reconstructed from the observation of its input and output over a finite number of sampling periods.

A linear SISO described by the following state equations

$$\begin{cases} X_{k+1} & = & A.X_k + B.U_k \quad \text{(state equation)} \\ Y_k & = & C.X_k + D.U_k \quad \text{(output equation)} \end{cases} \tag{1.68}$$

with the state matrix of dimension n, is observable if and only if the matrix

$$\begin{bmatrix} C \\ CA \\ \dots \\ CA^{n-1} \end{bmatrix} \tag{1.69}$$

is of maximum order (Its determinant is not zero).

In case of nonlinear systems, readers will find an approach to the observability in Gauthier and Kupka [1994] and Hwang and Seinfeld [1972]. We assume the conditions of observability verified in this book, in particular for the extended observers.

1.2.3.1 Kalman Observer

The Kalman filter (Kalman [1960]) gives a realization of the random variable $\hat{X}_k$ representing the state of the system at instant k, knowing the measurement vector $Y = \{y_1, y_2, ..., y_{k-1}\}$, which minimizes the a priori variance of the estimation error $E(\tilde{X}_k^T \tilde{X}_k)$ with $\tilde{X}_k = X_k - \hat{X}_k$.

The discretized process is modeled by the following equations:

$$\begin{aligned} X_{k+1} & = A_k.X_k + B_k.U_k + W_k \\ Y_k & = C_k.X_k + V_k \end{aligned} \tag{1.70}$$

W_k and V_k are uncorrelated centered white noises characterized by their covariance matrix.

The noises are centered, so $E(W_i) = 0$ and $E(V_i) = 0$. The covariance matrices are noted:

- $\Sigma_{W_{k_i} W_{k_j}} = \left[E(W_{k_i} W_{k_j}) \right] = Q_k \delta_{ij}$ where δ_{ij} is is the operator defined by $\delta_{ij} = 1$ if $i = j$ and $\delta_{ij} = 0$ if $i \neq j$ and Q is a positive-definite symmetrical matrix.

- $\Sigma_{V_{k_i}V_{k_j}} = \left[E(V_{k_i}V_{k_j})\right] = R_k\delta_{ij}$ where Q is a positive-definite symmetrical matrix.

W_k and V_k are uncorrelated, thus $E(W_{k_i}V_{k_j}) = 0$.

Let us note $\hat{X}_{k|n}$ the instantaneous estimation, at the sample $k.T_e$, minimizing the variance of the estimation error $E((X_k - \hat{X}_k)^T(X_k - \hat{X}_k))$ knowing the measures $y_1, ..., y_n$

- if $k < n, \hat{X}_{k|n}$ is obtained by smoothing;
- if $k = n, \hat{X}_{k|n}$ is obtained by filtering;
- if $k > n, \hat{X}_{k|n}$ is obtained by a prediction.

The Kalman algorithm minimizing the a priori conditional variance $(n = k - 1)$ is decomposed into two steps which are:

1. Prediction or propagation step

$$\begin{aligned} \hat{X}_{k+1|k} &= A_k.X_{_{k+1|k}} + B_k.U_k \\ P_{k+1|k} &= A_k.P_{k+1|k}.A_k^T + Q_k \end{aligned} \tag{1.71}$$

   where $P_{k|k}$ is the a priori covariance matrix
2. Correction state

$$\begin{aligned} \hat{X}_{k|k} &= A_k.\hat{X}_{k|k-1} + K_k(Y_K - C_k\hat{X}_{k|k-1}) \\ P_{k|k} &= ([I] - K_k.C_k)P_{k|k-1} \\ K_k &= P_{k|k-1}.Cd_k{}^T.(Cd_k.P_{k|k-1}Cd_k{}^T + R_k)^{-1} \end{aligned} \tag{1.72}$$

   where $P_{k|k}$ is the a postiori covariance matrix

This filter has several advantages:

- If the output matrix C_k is constant and the noises are stationary (their stochastic properties are not time-dependent), the computation of the Kalman gain matrix can be done off-line.
- It is recurrent, which reduces the volume of stored data.

1.2.3.2 Extended Observer

(1) Kalman oberver for nonlinear system

Let us consider a nonlinear continuous system

$$\begin{aligned} \frac{dX}{dt} &= f_c(X(t), U(t)) + W(t) \\ Y(t) &= h(X(t)) + V(t) \end{aligned} \tag{1.73}$$

This system is discretized at a sampling step T_e:

$$\hat{X}_{k+1|k} = \hat{X}_{k|k} + \int_{kT_e}^{(k+1)T_e} f(X(t|t_k), u(t))dt \tag{1.74}$$

The following state equations are obtained:

$$\begin{aligned} X_{k+1} &= f(X_k, U_k) + W_k \\ Y_k &= h(X_k) + V_k \end{aligned} \tag{1.75}$$

The filter linearizes the system at each step around the operating point defined at the previous step. Let us define:

$$\begin{aligned} F_k &= \left(\frac{\partial f}{\partial X}\right)_{X_{k-1|k-1}, U_k} \\ H_k &= \left(\frac{\partial h}{\partial X}\right)_{X_{k-1|k-1}} \end{aligned} \tag{1.76}$$

To calculate $P_{k+1|k}$ the system has been linearized around $\hat{X}_{k|k}$. Thus, the algorithm becomes:

1. Prediction or propagation step

$$\begin{aligned} \hat{X}_{k+1|k} &= f(\hat{X}_{k|k}, U_k) \\ P_{k+1|k} &= F_k P_{k|k} {F_k}^T + Q_k \end{aligned} \tag{1.77}$$

2. Correction state

$$\begin{aligned} K_{k+1} &= P_{k+1|k}.{H_k}^T\left[H_k.P_{k+1|k}.{H_k}^T + R_k\right]^{-1} \\ P_{k+1|k+1} &= ([I] - K_{k+1}H_k)P_{k+1|k} \\ \hat{X}_{k+1|k+1} &= \hat{X}_{k|k+1} + K_{k+1}.(y_k - h(\hat{X}_{k+1|k})) \end{aligned} \tag{1.78}$$

(2) Extended observer

The extended filter (EKF) can be used to identify variable parameters. Two important applications are the monitoring of rotor time constants or the estimation of rotor speed. To achieve this goal, an augmented system is constructed from the state of the system X_k and the unknown parameters θ_k. A new state vector is defined:

$$\breve{X}_k = \begin{bmatrix} X_k \\ \theta_k \end{bmatrix} \tag{1.79}$$

By modeling the variation of the parameters by a noise W_{θ_k} and assuming a slow variation of θ_k, the equations of the augmented system are:

$$\begin{bmatrix} X_{k+1} \\ \theta_{k+1} \end{bmatrix} = \begin{bmatrix} A(\theta_k) & 0 \\ 0 & I \end{bmatrix} \begin{bmatrix} X_k \\ \theta_k \end{bmatrix} + \begin{bmatrix} B(\theta_k) \\ 0 \end{bmatrix} U_k + \begin{bmatrix} W_{X_k} \\ W_{\theta_k} \end{bmatrix}$$

$$Y_{k+1} = \begin{bmatrix} C(\theta_k) & 0 \end{bmatrix} \begin{bmatrix} X_k \\ \theta_k \end{bmatrix} + V_k \tag{1.80}$$

Then, a Kalman observer can be designed for this nonlinear augmented system with:

$$F(X, u, kT_e) = \begin{bmatrix} A(\theta_k) & \frac{\partial(A(\theta_k)X_k + B(\theta_k)U_k)}{\partial\theta_k} \\ 0 & I \end{bmatrix}$$

$$H(X, kT_e) = \begin{bmatrix} C(\theta_k) & \frac{\partial(C(\theta_k)X_k)}{\partial\theta_k} \end{bmatrix} \tag{1.81}$$

For instance, let us consider the following system (Rayyam et al. [2015]):

$$X_{k+1} = f(X_k, U_k) + W_k$$

$$Y_k = h(X_k) + V_k$$

with

$$X = \begin{bmatrix} i_{sd} & i_{sq} & \varphi_{rd} & \varphi_{rq} & R_r \end{bmatrix}$$

$$U = \begin{bmatrix} u_{sd} \\ u_{sq} \end{bmatrix} \tag{1.82}$$

$$Y = \begin{bmatrix} i_{sd} \\ i_{sq} \end{bmatrix}$$

The state discrete equations are:

$$f(X_k, U_k) = \begin{bmatrix} i_{sd}(1 - T_e\left(\frac{R_s}{\sigma L_s} + \frac{1-\sigma}{\sigma\tau_r}\right)) + i_{sq}T_e\omega_s + \varphi_{rd}\frac{T_e L_m}{\sigma L_s L_r \tau_r} + \varphi_{rq}T_e\frac{L_m\omega_{sl}}{\sigma L_s L_r} + u_{sd}\frac{T_e}{\sigma L_s} \\ i_{sd}T_e\omega_s + i_{sq}(1 - T_e\left(\frac{R_s}{\sigma L_s} + \frac{1-\sigma}{\sigma\tau_r}\right)) - \varphi_{rd}T_e\frac{L_m\omega_{sl}}{\sigma L_s L_r} + \varphi_{rq}\frac{T_e L_m}{\sigma L_s L_r \tau_r} + u_{sq}\frac{T_e}{\sigma L_s} \\ i_{sd}T_e\frac{L_m}{\tau_r} + \varphi_{rd}(1 - \frac{T_e}{\tau_r}) + \varphi_{rq}T_e\omega_m \\ i_{sq}T_e\frac{L_m}{\tau_r} - \varphi_{rd}T_e\omega_m + \varphi_{rq}(1 - \frac{T_e}{\tau_r}) \\ R_r \end{bmatrix}$$

$$h(X_k) = \begin{bmatrix} 1 & 0 & 0 & 0 & 0 \\ 0 & 1 & 0 & 0 & 0 \end{bmatrix}$$

with

$$\tau_r = \frac{L_r}{R_r}$$

$$\omega_{sl} = \omega_s - \omega_m \tag{1.83}$$

Thus we obtain the Jacobian matrix F_k:

$$F_k = \begin{bmatrix} (1 - T_e\left(\frac{R_s}{\sigma L_s} + \frac{1-\sigma}{\sigma\tau_r}\right)) & T_e\omega_s & \frac{T_e L_m}{\sigma L_s L_r \tau_r} & T_e\frac{L_m\omega_{sl}}{\sigma L_s L_r} & T_e\frac{L_m\varphi_{rd} - L_m^2 i_{sd}}{L_r^2 L_s - L_m^2 L_r} \\ T_e\omega_s & (1 - T_e\left(\frac{R_s}{\sigma L_s} + \frac{1-\sigma}{\sigma\tau_r}\right)) & -T_e\frac{L_m\omega_{sl}}{\sigma L_s L_r} & \frac{T_e L_m}{\sigma L_s L_r \tau_r} & T_e\frac{L_m\varphi_{rq} - L_m^2 i_{sq}}{L_r^2 L_s - L_m^2 L_r} \\ T_e\frac{L_m}{\tau_r} & 0 & (1 - \frac{T_e}{\tau_r}) & T_e\omega_m & T_e\frac{L_m i_{sd} - \varphi_{rd}}{L_r} \\ 0 & T_e\frac{L_m}{\tau_r} & -T_e\omega_m & (1 - \frac{T_e}{\tau_r}) & T_e\frac{L_m i_{sq} - \varphi_{rq}}{L_r} \\ 0 & 0 & 0 & 0 & 1 \end{bmatrix} \tag{1.84}$$

Then, we obtain the rotor time constant (See Figure 1.26).

1.2.3.3 Unscented Kalman Filter

Unscented Kalman filter (UKF) (Julier and Uhlmann [1997]) deals with online state estimation of a discrete-time nonlinear system. It uses suitable chosen X-state samples (called sigma points), which are passed directly into the state

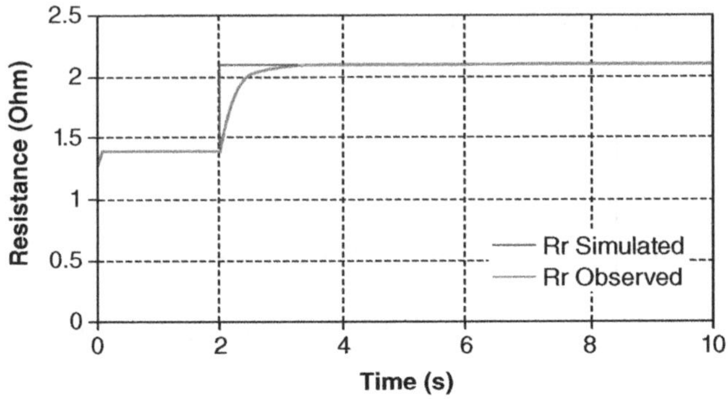

Figure 1.26 Extended rotor resistance observer

function f and whose statistics are extracted from it, rather than linearising it, as in the classical Kalman filter.

Readers could find the algorithm in Wan and Van Der Merwe [2000]. See Figure 1.27.

1.2.4 Parity Space

This method was introduced by Chow and Willsky [1984]. It deals with the generation of residuals based on nonlinear analytical redundancies. In the absence of a failure, the residuals should converge to zero. In the presence of a fault, the residue deviates from zero. The threshold is chosen in order to avoid unsuspected alarms. We will describe it for linear systems (Chow and Willsky [1984]). Let us consider the state equations:

$$\begin{aligned} x(k+1) &= A.x_k + \sum_{j=1}^{q} b_j u_j(k) \\ y_j(k) &= c_j x(k) \text{ for } j = 1..M \end{aligned} \tag{1.85}$$

where

- x a state vector of dimension N,
- A is a constant state matrix of dimension NxN,
- b_j is a constant input vector of dimension Nx1,
- b_j is a constant output vector of dimension 1xN,
- u_j is the known scalar input to the jth actuator
- y_j is the scalar output of the jth sensor.

Initialize with:

$$\hat{\mathbf{x}}_0 = E[\mathbf{x}_0]$$

$$\mathbf{P}_0 = E[(\mathbf{x}_0 - \hat{\mathbf{x}}_0)(\mathbf{x}_0 - \hat{\mathbf{x}}_0)^T]$$

$$\hat{\mathbf{x}}_0^a = E[\mathbf{x}^a] = [\hat{\mathbf{x}}_0^T\ \mathbf{0}\ \mathbf{0}]^T$$

$$\mathbf{P}_0^a = E[(\mathbf{x}_0^a - \hat{\mathbf{x}}_0^a)(\mathbf{x}_0^a - \hat{\mathbf{x}}_0^a)^T] = \begin{bmatrix} \mathbf{P}_0 & \mathbf{0} & \mathbf{0} \\ \mathbf{0} & \mathbf{P_v} & \mathbf{0} \\ \mathbf{0} & \mathbf{0} & \mathbf{P_n} \end{bmatrix}$$

For $k \in \{1, \ldots, \infty\}$,

Calculate sigma points:

$$\boldsymbol{\mathcal{X}}_{k-1}^a = \left[\hat{\mathbf{x}}_{k-1}^a\ \ \hat{\mathbf{x}}_{k-1}^a \pm \sqrt{(L+\lambda)\mathbf{P}_{k-1}^a}\right]$$

Time update:

$$\boldsymbol{\mathcal{X}}_{k|k-1}^x = \mathbf{F}[\boldsymbol{\mathcal{X}}_{k-1}^x, \boldsymbol{\mathcal{X}}_{k-1}^v]$$

$$\hat{\mathbf{x}}_k^- = \sum_{i=0}^{2L} W_i^{(m)} \mathcal{X}_{i,k|k-1}^x$$

$$\mathbf{P}_k^- = \sum_{i=0}^{2L} W_i^{(c)} [\mathcal{X}_{i,k|k-1}^x - \hat{\mathbf{x}}_k^-][\mathcal{X}_{i,k|k-1}^x - \hat{\mathbf{x}}_k^-]^T$$

$$\boldsymbol{\mathcal{Y}}_{k|k-1} = \mathbf{H}[\boldsymbol{\mathcal{X}}_{k|k-1}^x, \boldsymbol{\mathcal{X}}_{k-1}^n]$$

$$\hat{\mathbf{y}}_k^- = \sum_{i=0}^{2L} W_i^{(m)} \mathcal{Y}_{i,k|k-1}$$

Measurement update equations:

$$\mathbf{P}_{\tilde{\mathbf{y}}_k \tilde{\mathbf{y}}_k} = \sum_{i=0}^{2L} W_i^{(c)} [\mathcal{Y}_{i,k|k-1} - \hat{\mathbf{y}}_k^-][\mathcal{Y}_{i,k|k-1} - \hat{\mathbf{y}}_k^-]^T$$

$$\mathbf{P}_{\mathbf{x}_k \mathbf{y}_k} = \sum_{i=0}^{2L} W_i^{(c)} [\mathcal{X}_{i,k|k-1} - \hat{\mathbf{x}}_k^-][\mathcal{Y}_{i,k|k-1} - \hat{\mathbf{y}}_k^-]^T$$

$$\mathcal{K} = \mathbf{P}_{\mathbf{x}_k \mathbf{y}_k} \mathbf{P}_{\tilde{\mathbf{y}}_k \tilde{\mathbf{y}}_k}^{-1}$$

$$\hat{\mathbf{x}}_k = \hat{\mathbf{x}}_k^- + \mathcal{K}(\mathbf{y}_k - \hat{\mathbf{y}}_k^-)$$

$$\mathbf{P}_k = \mathbf{P}_k^- - \mathcal{K}\mathbf{P}_{\tilde{\mathbf{y}}_k \tilde{\mathbf{y}}_k}\mathcal{K}^T$$

where, $\mathbf{x}^a = [\mathbf{x}^T\ \mathbf{v}^T\ \mathbf{n}^T]^T$, $\boldsymbol{\mathcal{X}}^a = [(\boldsymbol{\mathcal{X}}^x)^T\ (\boldsymbol{\mathcal{X}}^v)^T\ (\boldsymbol{\mathcal{X}}^n)^T]^T$, λ=composite scaling parameter, L=dimension of augmented state, $\mathbf{P_v}$=process noise cov., $\mathbf{P_v}$=measurement noise cov., W_i=weights as calculated in Eqn. 15.

Figure 1.27 Unscented Kalman Filter Algorithm (Wan and Van Der Merwe [2000]) with permission of IEEE.

Some of the outputs can be linear dependent of the others. They can also depend on some inputs and outputs. These links can provide redundancy relations and, thus, residuals. The general method to generate general redundancy relations is described below.

Let us consider:

$$C_j(k) = \begin{bmatrix} c_j \\ c_j A \\ \dots \\ c_j A^k \end{bmatrix} \text{ for } j = 1..M \tag{1.86}$$

It exist a n_j such as $1 \leq n_j \leq N$ and

$$rank[C_j(k)] = \begin{cases} k+1 \text{ for } k < n_j \\ n_j \text{ for } k \geq n_j \end{cases} \tag{1.87}$$

Les us define a nonzero raw vector $\omega = [\omega^1, \dots, \omega^M]$ of dimension $n = \sum_{i=1}^{M} (n_i + 1)$ such as

$$[\omega^1, \dots, \omega^M] \begin{bmatrix} C_1(n_1) \\ C_2(n_2) \\ \\ C_M(n_M) \end{bmatrix} x(k) = 0 \text{ with } x(k) \in R^N \tag{1.88}$$

Assuming that the system 1.85 is observable, there are only n-N linearly independent ω satisfying 1.88. They define a (n-N) x N matrix noted Ω. A n-N vector $P(k)$ called generalized parity vector is finally built:

$$P(k) = \Omega\left(\begin{bmatrix} Y_1(k, n_1) \\ \dots \\ Y_M(k, n_M) \end{bmatrix} - B\begin{pmatrix} B_1(n_1) \\ \dots \\ B_M(n_M) \end{pmatrix} U(k, n_0)\right)$$

with

$$Y_j(k, n_j) = \begin{bmatrix} y_j(k) \\ \dots \\ y_j(k + n_j) \end{bmatrix}$$

$$B_j(n_j) = \begin{bmatrix} 0 & 0 & \dots & \dots & \dots & \dots & 0 \\ c_jB & 0 & \dots & \dots & \dots & \dots & 0 \\ \dots & c_jB & 0 & \dots & \dots & \dots & 0 \\ \dots & \dots & \dots & \dots & \dots & \dots & \dots \\ c_jA^{n_j-1}B & c_jA^{n_j-2}B & \dots & \dots & \dots & c_jB & 0 \end{bmatrix} \tag{1.89}$$

$$B = \begin{bmatrix} b_1 & \dots & b_q \end{bmatrix} \text{ with q the number of actuators}$$

$$u(k) = \begin{bmatrix} u_1(k) & \dots & u_q(k) \end{bmatrix}^T$$

$$n_0 = \max(n_1, \dots, n_M)$$

$$U(k, n_0) = \begin{bmatrix} u(k)^T & \dots & u(k + n_0)^T \end{bmatrix}^T$$

$P(k)$ is nonzero if and only if a failure occurs.

But parity space is not limited to linear systems. Zhong et al. [2010] have developed this strategy for linear discrete dime-varying systems.

Nguang et al. [2006] described an extension of parity space to nonlinear systems with an LMI approach.

Amrane et al. [2017] applied it on nonlinear systems described in Takagi–Sugeno form by using a transformation by sector nonlinearity.

These methods can be implemented in numerous applications. For instance, they can be used:

- to the fault detection and isolation of an Induction Machine (Amrane et al. [2017]).
- to detect current sensor faults (Berriri et al. [2011]).

- the time-domain synchronous generator diagnosis (Lalami and Wamkeue [2012]).

1.3 Feature Reduction, Principal Component Analysis

The performance of diagnostic and prognostic methods depends very strongly on the choice of pattern vector (fault signature). Its components must be sufficiently informative without being too numerous, at the risk of overloading the useful information and degrading the performance of the classifier. Thus, the selection of the pattern vector components requires:

- selecting informative parameters for diagnosis: Shannon entropy indicator;
- analyzing correlated parameters in order to keep only the relevant ones: Pearson coefficient;
- selecting monotonic parameters for prognostic: Spearman indicator;
- reducing the dimension of the research space for supervised classifier;
- reducing the dimension of the research space for unsupervised classifier: Laplacian score;
- determining the number of classes;
- evaluating the quality of the classification.

First of all, we will present the principal component analysis (PCA), which allows to reduce the representation space. It is used both to visualize the different classes and to perform an unsupervised classification. However, it does not reduce the number of parameters to be calculated.

1.3.1 Principal Component Analysis: A Space Reduction and an Unsupervised Classification

Principal component analysis (PCA) is a multivariate exploratory analysis method introduced by Karl Pearson in 1901 (Pearson [1901]), formalized by Hotelling in 1933 (Hotelling [1933]), and taken up in the work of Rao [1964]. A more detailed history of PCA has been given in Mishra et al. [2017].

The general objective of this method is to summarize the initial variables in the raw data using a reduced number of synthetic variables called principal components. These new variables are a linear combination of the initial variables, allowing the representation of the data to be condensed while maintaining their overall structure (Eke [2018]).

The PCA can then be used:

- to visualize the structure of data;

- to realize an unsupervised classification;
- as a preconditioning step for other classifiers;
- reducing the dimension of the research space for supervised classifier.

The principle of PCA is to reduce the size of the initial data by replacing the initial p variables $[X_1, X_2, ..., X_p]^t$ by q ones $[X_1^{new}, X_2^{new}, ..., X_q^{new}]^t$ with $q < p$ with X^t the transposed vector of X.

These later q factors are linear combination of the initial variables.

$$\begin{aligned} X_1^{new} &= \alpha_1^1 X_1 + \alpha_1^2 X_2 + ... + \alpha_1^p X_p \\ &... \\ X_q^{new} &= \alpha_q^1 X_1 + \alpha_q^2 X_2 + ... + \alpha_q^p X_p \end{aligned} \tag{1.90}$$

They are called factors or principal components.

Equation 1.90 can be rewritten as follows:

$$X^{new} = P_J.X \tag{1.91}$$

with P_J a projection matrix.

The choice of the principal components is made by maximizing the dispersion of individuals according to these factors.

The dispersion of a quantitative variable is generally measured by its variance (or standard deviation). When a cloud of points is available in several dimensions, we speak about inertia, (the sum of the variances of the variables under consideration). Thus, the factors selected must have a maximum variance.

In other words, PCA aims to find the directions of maximum variances in a high p dimensional data and project it into a q lower dimensional subspace without too much loss of information (Figure 1.28).

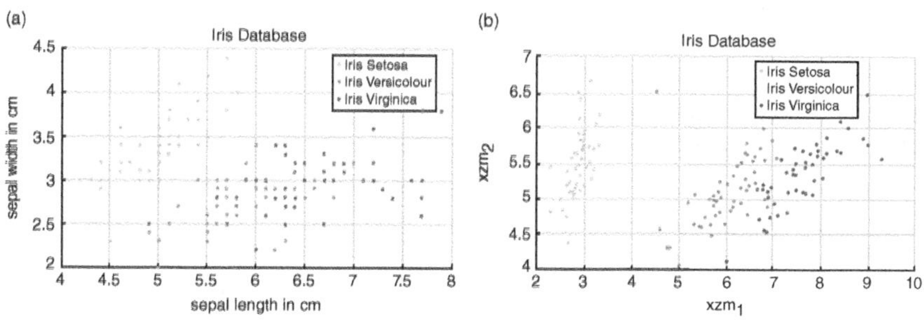

Figure 1.28 Iris data base (a) initial values (b) after PCA.

PCA involves the following steps:

- step 1 - Collect m different data in a matrix X. Each column represents a parameter, each line represents a pattern (data);
- step 2 - Standardize the data: $X_m = \frac{X-\bar{X}}{\sigma_X}$ with $\bar{X}$ the mean of each column of X, and σ_X the variance;
- step 3 - Obtain the covariance matrix of X_m: $\Sigma_X = \frac{1}{p} X_m {X_m}^T$. Diagonal terms are the variances of each parameter. Non-diagonal terms are the covariances between two parameters. A parameter with a high variance carries information . A high covariance indicates highly redundant parameters;
- step 4 - Calculate Eigen vectors and Eigen values. Diagonalize the matrix in order to maximize the variances and minimize the covariances. Eigen vectors are given with $det(\Sigma_X - \lambda.I) = 0$. If we note λ_i the Eigen values and V_i the Eigen vectors, then $\Sigma_X.V_i = \lambda_i.V_i$;
- step 5 - Sort the Eigen vectors. Take the eigenvalues λ_1,λ_2,…,λ_p and sort them from largest one to smallest one. Then, sort the eigenvectors V_i in the same way in order to build the matrix P_J;
- step 6 - Compute the new vector: $X^{new} = P_J.X$;
- step 7 - Drop uninformative features from the new set to keep q parameters. For that, compute the inertia of each axe: $\frac{\lambda_i}{p}$ then the cumulated inertia $\sum_{i=1}^{r} \frac{\lambda_i}{p}$ where r is the rank of the main component. Finaly compute the proportion of variance explained, which is the sum of the eigenvalues of the features you kept until rank r divided by the sum of the eigenvalues of all features $\frac{\sum_{i=1}^{r} \lambda_i}{\sum_{i=1}^{p} \lambda_i}$. Keep the q most representative ones.

1.3.2 Intercorrelation

Intercorrelation represents the statistical link between two variables. It doesn't deal with causal interaction. In our case, intercorrelation is used to keep the minimum number of parameters in a set of correlated ones and remove redundant features.

1.3.2.1 Pearson Coefficient "r"

The Pearson coefficient (Pearson and Galton [1895] describes the correlation between two variables by assuming their linear variation. This coefficient always varies between -1 and 1 (Tufféry [2012] and Ben Marzoug [2020]).

Given paired data $\{(X_1, Y_1), \dots, (X_n, Y_n)\}$ with n the sample size, this coefficient can be calculated according to the following equations eq. 1.92 and eq. 1.93.

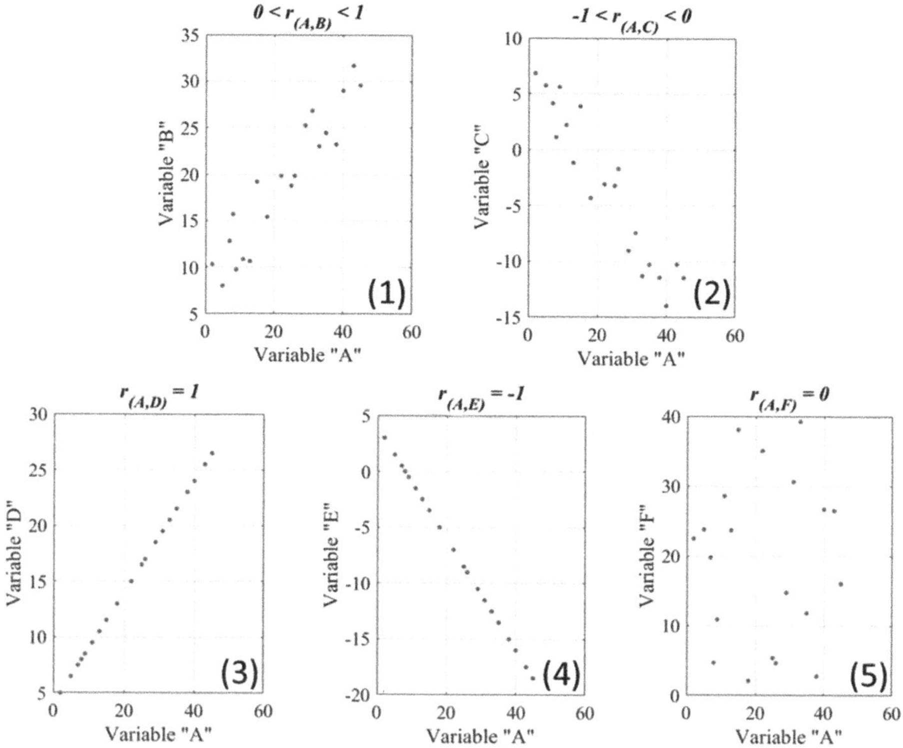

Figure 1.29 Pearson coefficient *Source:* Tufféry [2012] and Ben Marzoug [2020]

$$r = \frac{\sigma_{XY}}{\sigma_X \sigma_Y} = \frac{\frac{1}{n}\sum_{i=1}^{n}(X_i - \bar{X})(Y_i - \bar{Y})}{\sqrt{\frac{\sum_{i=1}^{n}(X_i - \bar{X})^2}{n}}\sqrt{\frac{\sum_{i=1}^{n}(Y_i - \bar{Y})^2}{n}}} \tag{1.92}$$

with $\bar{X}$ and $\bar{Y}$, the sample mean of the points X_i and Y_i.

$$r = \frac{1}{n}\sum_{i=1}^{n}\left(\frac{(X_i - \bar{X})}{\sigma_X}.\frac{(Y_i - \bar{Y})}{\sigma_Y}\right) = \frac{1}{n}\sum_{i=1}^{n}(XCR_i.YCR_i) \tag{1.93}$$

1.3.2.2 Spearman Coefficient "ρ "

The Spearman's index (Spearman [1904]) identifies and quantifies the existence of a monotonic link, linear or not, between the variables under consideration (Gray [2013] and Ben Marzoug [2020]). It deals with the ranks of the variables under consideration. The following example may explain the procedure for calculating this coefficient. If we consider a variable $X = [7, 10, 26, 20, 23, 2, 5]$, X contains seven observations. The vector $rank_X$ can be obtained as follows: we start by sorting variable X in ascending order. Then, we replace the values taken by variable X by their order number. This gives $rank_X = [3, 4, 7, 5, 6, 1, 2]$. Now, to calculate the Spearman's coefficient for variables X and Y, we can use the (eq. 1.94):

$$\rho = 1 - \frac{6 \sum_{i=1}^{n} d_i^2}{n^3 - n}, \tag{1.94}$$

with n the number of observations and di the difference between $rank_{X_i}$ and $rank_{Y_i}$ Alternatively, the following expression can be used (eq. 1.95):

$$\rho = \frac{cov(rank_X, rank_Y)}{\sigma_{rank_X} . \sigma_{rank_Y}} \tag{1.95}$$

with σ_{rank_X} and σ_{rank_Y} standard deviations of the rank variables.

Figure 1.30 shows the difference between the Pearson's coefficient and the Spearman's index.

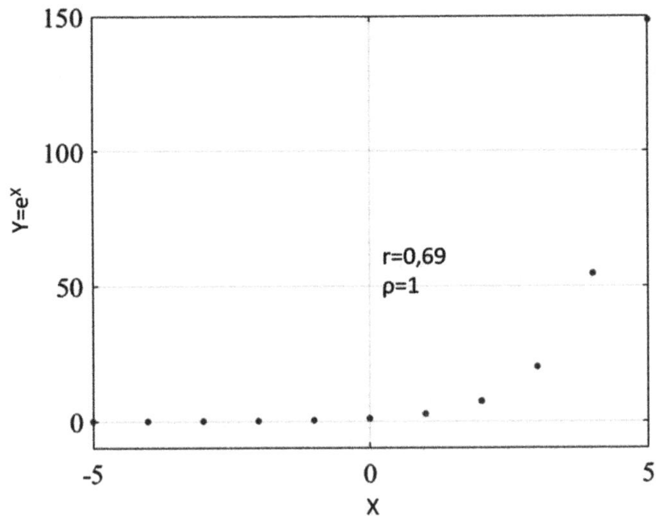

Figure 1.30 Difference between Pearson and Spearman coefficients *Source:* Gray [2013] and Ben Marzoug [2020]

1.3.3 Information Content: Shannon Entropy

Shannon's entropy (Shannon [1948]) measures the amount of information in a signal. The larger the entropy of a signal is, the more nonredundant information it contains (Gray [2013] and Ben Marzoug [2020]). It is used to remove irrelevant features and keep informative ones.

Given a discrete random variable X, with m possible outcomes $x_1, ..., x_m$, which occur with probability $P_1, ..., P_m$, the entropy of X is defined as (eq. 1.96):

$$H(X) = -\sum_{i=1}^{m} P_i.\log_2(P_i) \tag{1.96}$$

For instance, let us consider X a signal of n individuals, containing m different symbols with $m < n$.

Example: $X = (a, b, a, c, a, b, d, c)$.

In this case, the number of individuals is $n = 8$ and in the vector X, there are four different symbols ("a", "b", "c", and "d") from where $m = 4$.

To measure the quantity of information, we must first calculate the probability of appearance of each of the symbols i with $i \in [1, m]$, we call it P_i. For our example $P = (3/8; 2/8; 1/8)$. As for the calculation of the entropy, we apply equation (eq. 1.96). For our example, we obtain $H(X) = 1.9056$.

1.3.4 Pattern Sizing Reduction for a Supervised Classification

A too large number of parameters involves too many computations and, in some cases, can decrease the performance of the classifier.

To reduce the number of parameters and, thus, the size of the display space, parameter selection methods can be used (Ondel [2006]).

These methods assume that the class structure of the learning set is known,that is it used in supervised mode.

The objective is to select the d' most representative parameters among the initial d, keeping as well as possible the information on the class structure. Thus, we build a subspace E' from the initial space E.

1.3.4.1 Selection Criteria

The criterion used for the determination of the subspace E', must represent as well as possible the classes which are supposed to be known. Intuitively, this approach is based on:

- grouping of the points of the same class, that is a minimizing the intra-class variance (notion of compactness);
- separating the different classes, that is a maximizing the inter-class variance (notion of separability);

The intra-class and inter-class variance matrices deal with these two conditions. The intra-class variance matrix involves the samples of each class as well as the centroid (eq. 1.97):

$$\Sigma_w = \frac{1}{N}\sum_{c=1}^{M}\sum_{\upsilon=1}^{N_c}\left(\underline{X}_{c\upsilon} - \underline{m}_c\right).\left(\underline{X}_{c\upsilon} - \underline{m}_c\right)^t \tag{1.97}$$

while the inter-class variance matrix characterizes the dispersion of classes among themselves (eq. 1.98).

$$\Sigma_B = \frac{1}{N}\sum_{c=1}^{M}\left(\underline{m}_c - \underline{m}\right)\left(\underline{m}_c - \underline{m}\right)^t \tag{1.98}$$

with:

- $\underline{m}$: overall center of gravity;
- M: number of clusters;
- $\underline{m}_c$: center of gravity of cluster Ω_c;
- $\underline{X}_{c\upsilon}$: υ^{th} vector of the cluster Ω_c;
- N_c: number of vectors of the cluster Ω_c;
- N: total number of vectors of the learning base.

We can then define the variance-covariance matrix Σ by (eq. 1.99):

$$\Sigma = \Sigma_w + \Sigma_B \tag{1.99}$$

Two criteria can be defined (eq. 1.100 and eq. 1.101):

$$J_1 = trace(\Sigma_w{}^{-1}.\Sigma_B) \tag{1.100}$$

$$J_2 = \frac{\det(\Sigma)}{\det(\Sigma_w)} \tag{1.101}$$

where $trace(X)$ is the trace of the matrix X (sum of the diagonal elements).

The Fisher ratio can also be applied as a criterion. In the case of a two-class problem, the criterion is expressed, for a parameter (eq. 1.102):

$$J_3(\alpha) = \frac{m_1(\alpha) - m_2(\alpha)}{N_1\sigma_1{}^2(\alpha) - N_2\sigma_2{}^2(\alpha)} \tag{1.102}$$

where

- $m_c(\alpha)$ with $c = 1, 2;$, center of gravity of the class Ω_c, considering only the parameter α

$$\underset{c=1,2}{m_c(\alpha)} = \frac{1}{N_c}\sum_{\upsilon=1}^{N_c}\underline{X}_{c\upsilon}(\alpha) \tag{1.103}$$

- $\sigma_c^2(\alpha)$: variance of the component α of the vectors of the class Ω_c;

$$\sigma_c^2(\alpha) = \frac{1}{N_c}\sum_{v=1}^{N_c}\left[\underline{X}_{cv}(\alpha) - m_c(\alpha)\right]^2 \tag{1.104}$$

In the general case of a problem with M classes, $J_3(\alpha)$ is expressed as (eq. 1.105):

$$J_3(\alpha) = \sum_{c=1}^{M}\sum_{r=1}^{M-1}\frac{m_c(\alpha) - m_r(\alpha)}{N_c\sigma_c^2(\alpha) - N_r\sigma_r^2(\alpha)} \tag{1.105}$$

The greater the criterion is, the more separate the classes are.

1.3.4.2 Sequential Backward Feature Selection and Sequential Forward Feature Selection

The aim of the Feature Selection is to select d parameters from an initial set of m parameters with $d < m$. If the number of features is too great, it is not possible to test all the subset of parameters. Thus, the number of possible combinations is (eq. 1.106)

$$\text{Number of combinations} = \sum_{k=1}^{m}\frac{m!}{k!(m-k)!} \tag{1.106}$$

For instance, for 10 features, we have 1023 subsets and for 11 features, 2047 subsets.

Thus, it is necessary to use sub-optimal approaches. Two of the main feature selection algorithms come from the work of Efroymson [1960]:

- Sequential Backward Feature Selection (SBS);
- Sequential Forward Feature Selection (SFS).

The Sequential Backward Feature Selection (SBS) starts with the full set of the m features. A selection criterion is computed with all the m-1 sets of features extracted from the original one. The combination with the optimal criterion is kept. This removal stage is repeated until a break appear on the criterion. The best combination before this break gives the final set of features. Thus, the features are sequentially removed from the given m features list to reach the list of d features (see Figure 1.31).

In the case of SFS, the algorithm starts with singletons and at each step, all combinations of each remaining feature from the original set with the previous optimal subset are tested (see Figure 1.32).

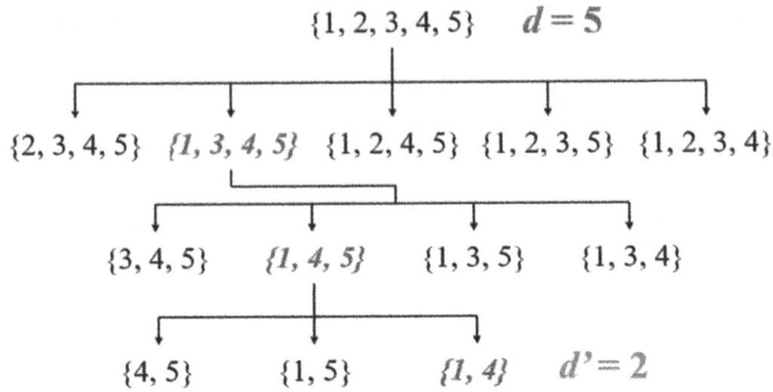

Figure 1.31 Sequential backward feature selection

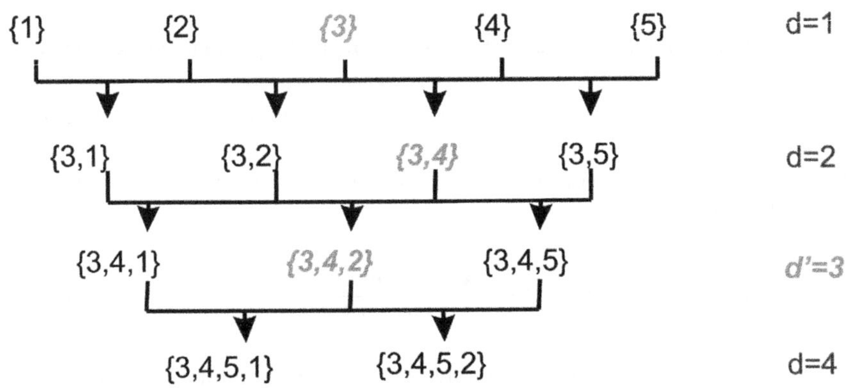

Figure 1.32 Sequential forward feature selection

1.3.5 Pattern Sizing Reduction for an Unsupervised Classification: Laplacian Score

The Laplacian Score (He et al. [2005]) determines the importance of a feature by evaluating its ability to preserve the locality (Eke [2018]).

Let us consider I_r the Laplacian score of the rth variable of the initial pattern vector and f_{ri} the ith value of the rth variable with $i = 1, ..., n$.

The Laplacian score is built with the following method:

- Construction of the nearest neighbor graph with n nodes. The ith node corresponds to the point x^i. A boundary is built between node i and j if x^i and x^j are close points, that is x^i is among the k closest neighbors to x^j or x^j is among the k closest neighbors to x^i (when label information is available, a boundary can be put between two nodes sharing the same label).

- Construction of the weighting matrix S of the connected nodes. If the nodes i and j are connected, associate $S_{ij} = e^{-\frac{\|x_i - x_j\|^2}{t}}$ where t is a suitable constant and associate $S_{ij} = 0$ in the other case. The matrix S models the local structure of the data representation space. S_{ij} evaluates the similarity between the ith and the jth node.
- For the rth variable, define:

$$
\begin{aligned}
&f_r = [f_{r1}, f_{r2}, \dots, f_{rn}]^t \\
&\mathbf{1} = [1, 1, \dots, 1]^t \\
&D = diag(S\mathbf{1}) \\
&\tilde{f}_r = f_r - \frac{f_r^t D\mathbf{1}}{\mathbf{1}^t D\mathbf{1}}\mathbf{1} \\
&L = D - S.
\end{aligned}
\tag{1.107}
$$

Then, compute

$$
L_r = \frac{\tilde{f}_r^t L \tilde{f}_r}{\tilde{f}_r^t D \tilde{f}} \tag{1.108}
$$

1.3.6 Choice of the Number of Classes for an Unsupervised Classification

In this case, the label of the different pattern in the learning base and the number of clusters are unknown.

1.3.6.1 Choice of the Number of Classes with a PCA

Two methods are mainly used:

- the elbow criterion;
- the Kaiser's law (Kaiser [1960]).

Both techniques are based on the calculation of inertia (inertia measures the dispersion of the point cloud) explained by the principal component axes. The total inertia of a point cloud is the sum of the squares of the distances of the points from the center of gravity. The average inertia is equal to the total inertia divided by the number of variables.

When the data are projected on the axes of the PCA, the inertia explained by each of the main components can be calculated (Gonzalez [2020]). The variance of a principal component i (variance of the coordinates of the projected points on the axis under consideration) is equal to the inertia λ_i carried by the principal axis i associated with it.

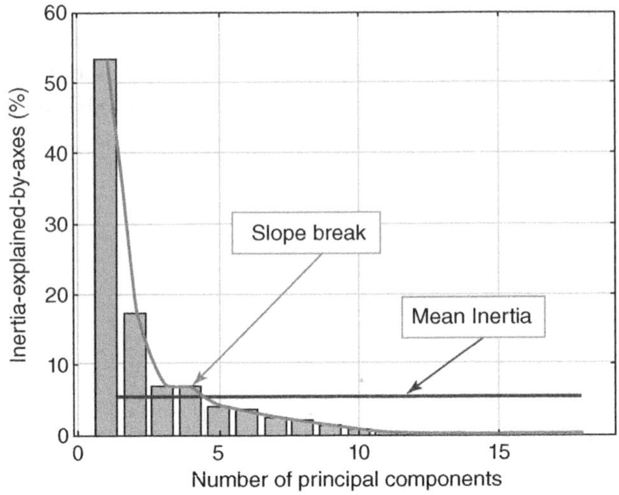

Figure 1.33 Inertia explained by axes

Kaiser's law recommends to retain all axes which have an inertia greater than or equal to the mean inertia (Beavers et al. [2013]).

The elbow criterion is a visual method that proposes to plot the percentage of inertia explained by each component and keep the axes located before the slope break.

The graph of Figure 1.33 shows the inertia explained by each axis of the PCA in relation to the total inertia.

In this example, we have kept four axes, taking into account 80% of the total inertia.

Herrera and Gac [2002] define the quality of the projection of the variables (eq. 1.109):

$$QV_n = \sum_{i=1}^{n} corr^2(V, Ax_i) \tag{1.109}$$

where QV_n is the quality index of the projection of the variable V in n axes of the PCA. cor^2 is the square of the Pearson's correlation. A_i is the data vector of the ith axis of the PCA.

1.3.6.2 General Case

One of the most commonly used techniques is to test an increasing number of classes and assess the number of false positives or false negatives on a given test base. One can built for instance a confusion matrix (Figure 1.34).

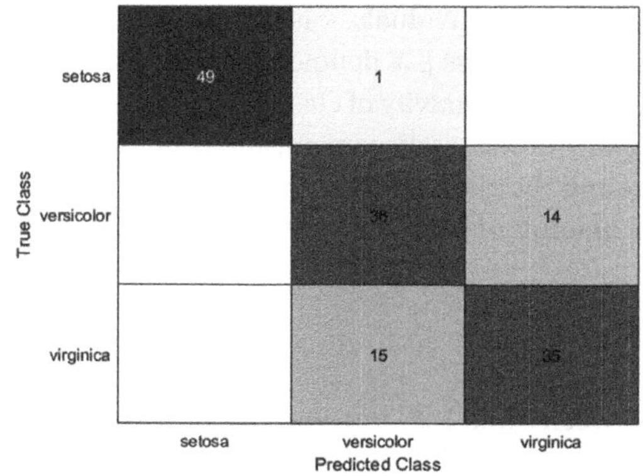

Figure 1.34 Confusion matrix after a discriminant analysis of Fisher's iris data

The confusion matrix allows visualization of the performance of an supervised classifier. Each row of the matrix represents the instances in a predicted class, while each column represents the instances in an actual class (or vice versa).

1.3.7 Other Quality Criteria of a Classification

Many indices have been presented in Desgraupes [2017] in order to evaluate the quality of the classification. We will present the main ones:

- R^2 index;
- Calinski–Harabasz index;
- Silhouette index;
- Dunn.

1.3.7.1 R^2 index

R^2 index is based on the calculation of the inertia that is the variance explained by the classes.

According to the Huygens relation and as we have already seen (eq. 1.99), the total inertia can be decomposed into intra-class and inter-class-inertia, Figure 1.35:

$$\sum_{i=1}^{n} d^2(X,G) = \sum_{j=1}^{k}\sum_{l=1}^{m_j} d^2(X_l,G_j) + \sum_{j=1}^{k} m_j d^2(G_j,G) \tag{1.110}$$

$$I_T = I_w + I_B$$

where n is the number of individuals, k is the number of classes, and m_j is the number of individuals in class j. X denotes individuals, G for the global centre of gravity, G_j is the centre of gravity of class j. I_T deals with the total dispersion, I_w represents the within classes dispersion, and I_B denotes the dispersion of the group centers around the global center.

R^2 is given the amount of inertia explained with this partition (1.111):

$$R^2(k) = \frac{I_B}{I_T} \tag{1.111}$$

R^2 is always between 0 and 1. R^2 is 0 when we have only one class and 1 when the number of classes equals the number of individuals. To choose the optimal number of classes, R^2 must be as close as possible to 1 without having too many classes and by plotting the curve of R^2 according to the number of classes, we must stop after the last important jump (as in the method descibed for PCA).

1.3.7.2 Calinski–Harabasz Index

The Calinski–Harabasz criterion (Caliski and Harabasz [1974]) measures the separation between all classes (Desgraupes [2017]). Let us note n is the number of individuals, k is the number of classes. The Calinski–Harabasz, also called pseudo-F is given by (eq. 1.112):

$$PseudoF(k) = \frac{\frac{R^2}{k-1}}{\frac{1-R^2}{n-k}} = \frac{\frac{I_B}{k-1}}{\frac{I_W}{n-k}} \tag{1.112}$$

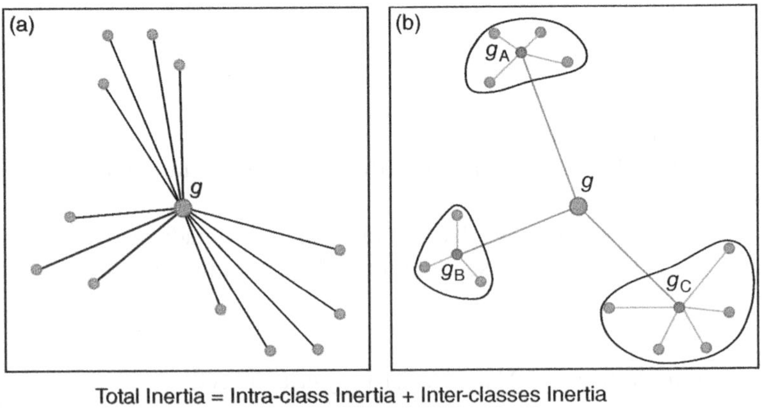

Figure 1.35 Illustration of Huygens relation

1.3.7.3 Davies–Bouldin Index

This criterion (Davies and Bouldin [1979]) compares intra-class dispersions with distances between classes. It measures the similarity between classes. Thus, the "best" choice of the number of clusters is the one that minimizes this similarity (Desgraupes [2017]).

Davies–Bouldin index is given by eq. 1.113:

$$DB(k) = \frac{1}{k}\sum_{i=1}^{k} \max_{a,b\neq a}(\frac{S_a + S_b}{d_{ab}}) \tag{1.113}$$

1.3.7.4 Silhouette Index

Silhouette index has been by defined by Rousseeuw [1987].

Let us design the data matrix

$$X = \begin{bmatrix} x_{11} & \dots & x_{1p} \\ \dots & \dots & \dots \\ x_{N1} & \dots & x_{Np} \end{bmatrix} \tag{1.114}$$

The set of points belonging to the k_{th} class is defined in eq. 1.115:

$$C_k = \{i \in [1, N]/class(i) = k\} \tag{1.115}$$

According to Wikipédia [2019] and Desgraupes [2017], the silhouette coefficient or score is first defined on a point i, whose group is $k = class(i)$ (see eq. 1.116).

$$a(i) = \frac{1}{|C_k| - 1}\sum_{j\in C_k, j\neq i} d(x_i, x_j) \tag{1.116}$$

where $|C_k|$ is the number of elements of C_k.

The average distance from the point to its neighboring group is given by (eq. 1.117):

$$b(i) = \min_{k'\neq k}\frac{1}{|C_{k'}|}\sum_{i'\in C_{k'}} d(x_i, x_{i'}) \tag{1.117}$$

The silhouette index of point i is provided by (eq. 1.118):

$$s_{silhouette}(i) = \frac{b(i) - a(i)}{\max(a(i), b(i))} \tag{1.118}$$

It can be averaged group by group to compare their homogeneity: those with the highest silhouette coefficients are the most homogeneous. Over the whole classification, it will be expressed as follows (eq. 1.119):

$$S_{silhouette}(i) = \frac{1}{K}\sum_{k=1}^{K}\frac{1}{|C_k|}\sum_{i\in C_k} s_{silhouette}(i) \tag{1.119}$$

1.3.7.5 Dunn Index

This index (Dunn [1974]) looks for the minimum distance between two classes in the partition while taking into account the distribution of elements within the classes (eq. 1.120). The greater the distance is, the better the partition. (Desgraupes [2017] and Eke [2018]))

$$ID = \min_{1\le i\le n}\left(\min_{1\le j\le n, j\ne i}\frac{d(w_i w_j)}{\min_{1\le k\le n} d'(w_k)}\right) \tag{1.120}$$

where

- w_i and w_j are two distinct classes;
- $d(w_i w_j) = \min_{1\le i\le n, 1\le j\le n}\left(d(x,y)\ with\ x\in w_i\ and\ \ y\in w_j\right)$;
- $d(w_k) = \max_{x,y\in C_k}(d(x,y)\ )$, maximum distance between two elements of the class.

1.4 Classification Methods

1.4.1 Generalities

In order to diagnose the different states of the electrical drive, clustering techniques can be used. They are used to organize raw data in sets (classes) with similar characteristics. The different classes are organized in order to maximize the intra-class compactness and the inter-class distance (Raschka and Mirjalili [2017], Russell et al. [2010], and Cornuéjols et al. [2018]).

Clustering generally consists of two steps:

- A learning phase where the boundaries of the different classes are defined from a database called the learning database or the training database;
- A classification phase where unknown data are assigned to the different classes.

Rejection rules can be implemented in order not to classify ambiguous data (either too far away from already classified data or too close to two different classes).

This rejected data can also be used to determine the appearance of new classes (Figure 1.36).

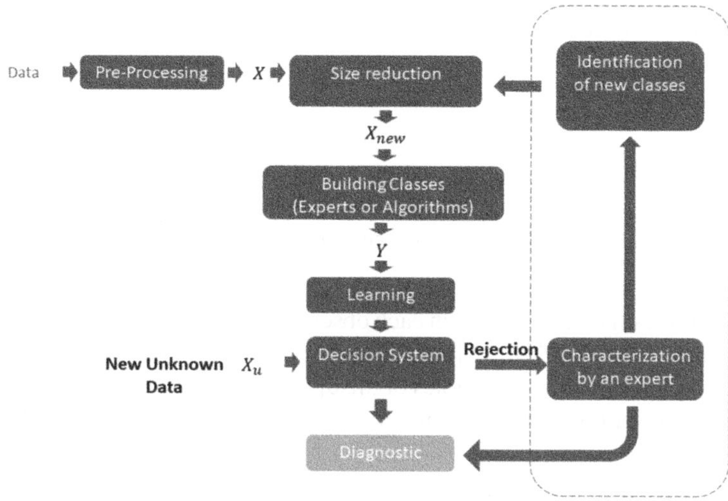

Figure 1.36 General process of classification

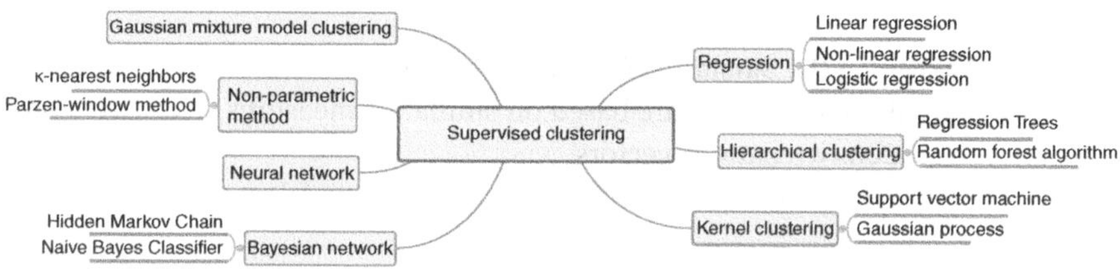

Figure 1.37 Examples of supervised clustering

1.4.1.1 Supervised and Unsupervised Clustering

Supervised clustering uses a learning data based with labeled patterns. It is used to classify unknown raw data in known classes or to build a regression (Bishop [2007], Barber [2011] and Dreyfus et al. [2002]).

Figure 1.37 gives some examples of supervised techniques.

We will present in detail three methods of supervised clustering in Section 1.4.3:

- k nearest neighbors (KNN);
- Support vector machine (SVM);
- Recurrent neural network (RNN).

Unsupervised clustering uses learning data based on unlabeled patterns, but with some diversity. It is used to organize data in sets with similar

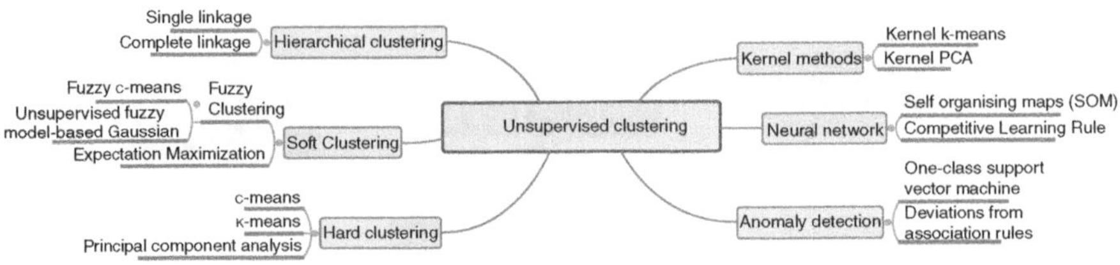

Figure 1.38 Examples of unsupervised clustering

characteristics. The class to which each observation belongs is not known (Yang [2018] and Aldrich and Auret [2013]).

Figure 1.38 gives some examples of unsupervised techniques.

We will present in detail three methods of unsupervised clustering in Section 1.4.3:

- K-means and centroid clustering;
- Hierarchical classification;
- Self-organizing map.

1.4.1.2 Measuring the Similarity: Different Distances

Pattern recognition classifiers are based on similarity measures. These rely on distances between two pattern vectors.

Let us define the points (eq. 1.121) of a set E of objects (elements, patterns):

$$x_i = \begin{bmatrix} x_{i1} \\ x_{i2} \\ \dots \\ x_{id} \end{bmatrix} \tag{1.121}$$

with Σ^{-1} the inverse of the covariance matrix of the descriptive variables.

The distance is a function from EXE to R^+, which fulfills the following properties:

$$\begin{cases} d(x_i, x_j) = d(x_j, x_i) \\ d(x_i, x_j) \geq 0 \\ d(x_i, x_i) = 0 \\ d(x_i, x_j) \leq d(x_i, x_k) + d(x_k, x_j) \end{cases} \tag{1.122}$$

The main distances which are used for classifiers are given in Table 1.22:

Table 1.22 Main distances

Euclidian distance	$d(x_i, x_j) = \left(\sum_{k=1}^{d} (x_{ik} - x_{jk})^2 \right)^{\frac{1}{2}}$
Manhatann distance or city-block	$d(x_i, x_j) = \sum_{k=1}^{d} (\lvert x_{ik} - x_{jk} \rvert)$
Tchebychev distance d_∞	$d(x_i, x_j) = \max_{1 \le k \le d} \lvert x_{ik} - x_{jk} \rvert$
Mahalanobis distance d_∞	$d(x_i, x_j) = (x_1 - x_2)^t \Sigma^{-1} (x_1 - x_2)$

Figure 1.39 shows some isovalues for Euclidean and Mahalanobis distances:

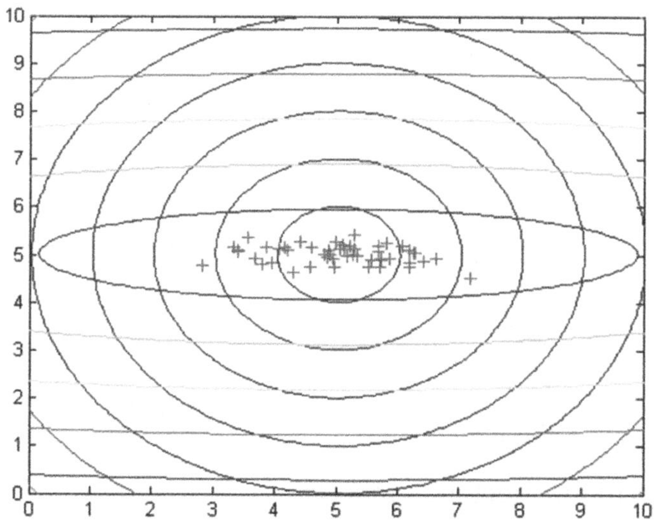

Figure 1.39 Isovalues of Euclidean (circles) and Mahalanobis (ellipse) distances from the center of gravity of the class Ω

As we can see, the Mahanalobis distance takes into account of the variance on each axis of the Ω class, which is not realized by the Euclidean distance.

1.4.2 Supervised Clustering

Wu et al. [2008] have written a synthesis paper dealing with algorithms in data mining in the research community in 2008. They present ten of these algorithms.

1.4.2.1 k Nearest Neighbors

The choice of k is sensitive because larger k would give a reliable estimate, but this choice remains arbitrary. In fact, if k is too small, the result can be sensitive to noise, and if k is too large, the result may include other classes. Consequently, assigning a new observation to a known class can lead to a diagnostic error. It is therefore advisable to choose a rejection class. To do this, the decision rule can include two rejection possibilities which can be based on distance rejection or ambiguous rejection Figure 1.40.

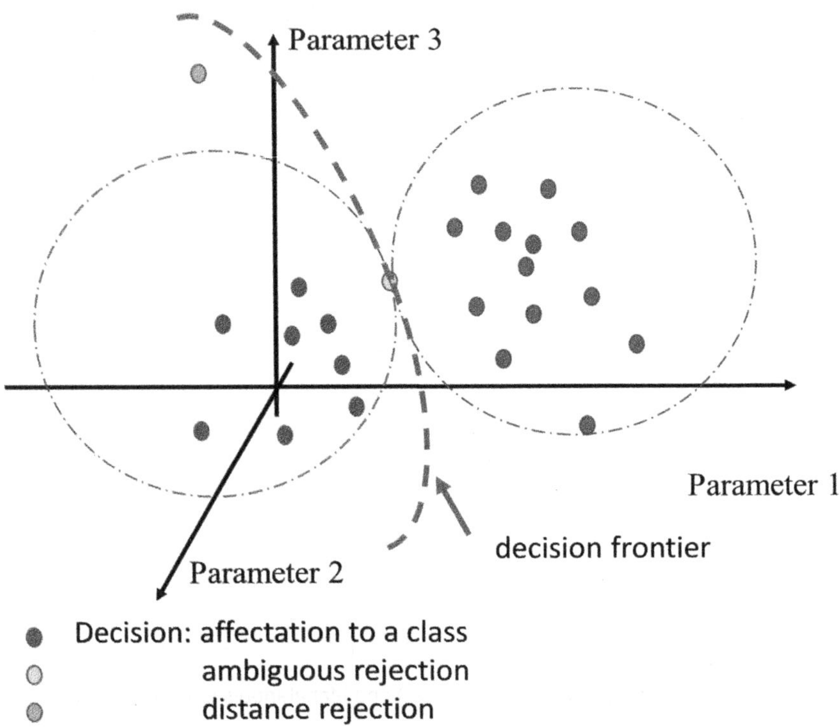

Figure 1.40 kNN rules applied to a new observation

In the case an observation is close to two classes and at the edge of the decision boundary, the decision to reject due to ambiguity is the safest. For this purpose, this new observation can be assigned to a specific and fictitious class of rejection due to ambiguities. However, for a new observation, which is located in an unexplored area, a new specific class can be attributed to it. It is based on distance rejection. Of course, the distance threshold is arbitrarily defined by the user. However, it can be defined thanks to the distance between the observation and the center of gravity of the assignment class.

There is, however, a weight disadvantage with this approach. This method requires the calculation of distances to all observations in the learning set. This can be cumbersome in terms of computation time, especially if the size of the observations is large. To do this, a mesh-based approach can be efficient. For example, this would involve checking the location of a new observation in a class area. Thus, the calculation of distances will no longer be necessary with vectors. The algorithm is then faster.

When two classes overlap because they are not sufficiently distant from each other, one proposal would be to assign the new observation to the class that is predominantly represented among its closest neighbors Figure 1.41. A nonexclusive decision rule may be interesting in that an observation located at the border of a class or at the border of contiguous or disjoint classes may signify

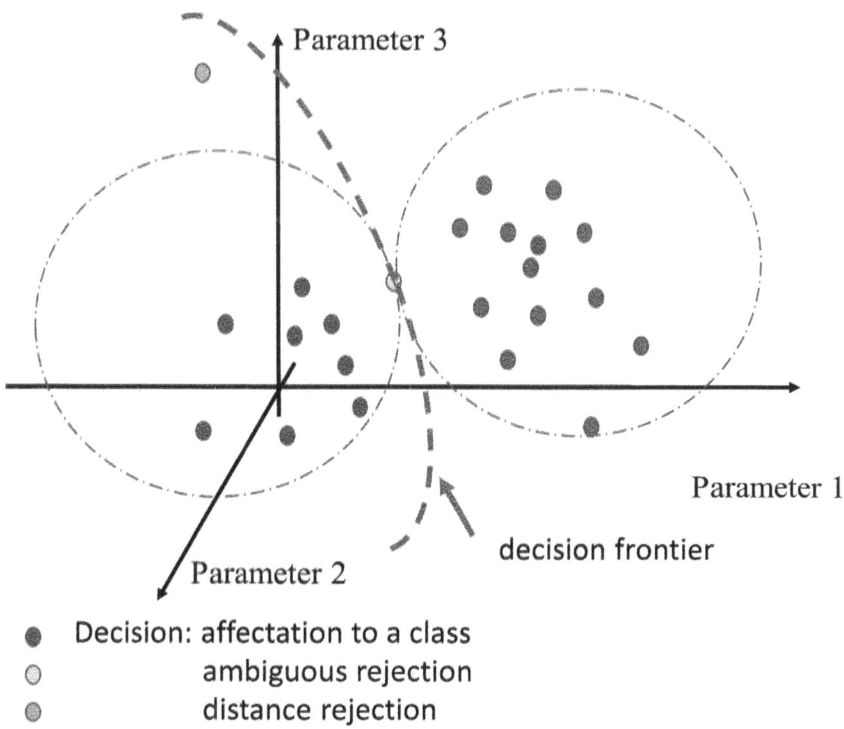

Figure 1.41 Class overlap

a transitional state between classes. The transition from exclusive to nonexclusive assignment must be characterized by a gradation between absolute nonmembership and full membership.

Since the similarity of an object with a reference form is often relative, in general, a membership function characterizing the degree of compatibility between the observation and a class is introduced. There is no general method to construct the membership function associated with a given set. The form of this function is closely related to the context. However, it would be interesting, for a new observation, to take into account not only the distance with the different center of gravity of each class, but also to give an advantage to the classes which contain the greatest number of closest neighbors.

In any case, a new observation located between two classes can correspond to an intermediate state of this mode. By using the exclusive or nonexclusive decision rule with the reject options, the observation can be rejected. It is therefore necessary to take into account the evolution of the different operating modes of the system.

A new observation can be rejected due to ambiguity or due to distance, even though it belongs to a mode of operation defined by its classes, due to poor feedback. The learning set does not exhaustively cover all modes of operation. Also, in order to establish an efficient diagnosis, it will be necessary to take into account the evolution of the operating modes of the system to be diagnosed.

An exclusive classification does not give strong information about the observation except that it belongs to a class. This observation can be at the edge of the class as well as inside the class. In the same way, what about an observation at the border of two contiguous classes or disjoint classes? A way should be implemented to follow the evolution of an observation inside a class and to associate to it a degree of confidence of belonging to this class. The transition from exclusive to nonexclusive assignment should be made by establishing a degree of belonging that could be modulated through the notion of proximity.

Instead of a binary classification, the use of membership functions, as used in fuzzy logic, will allow a gradual notion to be introduced. One possibility is to use an C^{∞} function such as a Gaussian function. It would be centered on the center of gravity of the class. A threshold defined in relation to the standard deviation allows to define: the whole membership to a class ($\mu = 1$), the degree of membership to a class ($0 < \mu < 1$), or its non membership ($\mu = 0$).

1.4.2.2 Support Vector Machine

Among classifications techniques, one can find in the literature the support vector machine (SVM), which is characterised by its robustness. It was introduced by Vapnik [1995]. The goal of the SVM is to find a hyperplane separating two classes of data thanks to support vectors. A new datum coming from observation, is classified by evaluating the distance from this datum to a hyperplane. So, a label is generated, which can have two types of value based on the distance. If the distance is positive, thus the label is +1, and if the distance is negative, the label is −1.

The equation of the hyperplane is:

$$\omega . x + b = 0 \tag{1.123}$$

with ω a normal vector and b a constant.

However, a margin distance d has to be defined as the sum of two distances. The first one is obtained as the distance from the hyperplane and the closest positive datum (observation), the second one as the closest negative datum (observation). The training process of the SVM consist to maximise the margin, which is equivalent to minimise $||\omega||$ knowing that all data should satisfy the condition:

$$y_i\,(\omega . x_i + b) \geq 0 \tag{1.124}$$

Figure 1.42 shows a hyperplane which separates data into classes.

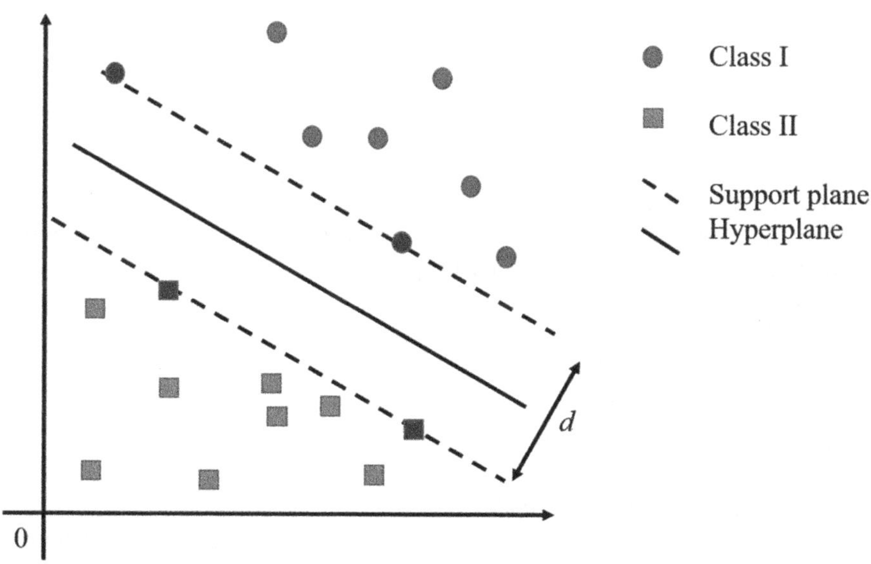

Figure 1.42 Binary SVM

The hyperplane can be obtained by solving a quadratic optimization problem. So, a dual Lagrangian formulation should be used as follows:

$$L_D = \sum_{i=1}^{M} \alpha_i + \sum_{i,j=0}^{M} \alpha_i \alpha_j y_i y_j x_i x_j \tag{1.125}$$

knowing that constraints of this problem are:

$$C_M \geq \alpha_i \geq 0$$

and

$$\sum_{i=1}^{M} \alpha_i y_i = 0$$

where the parameter C deals with the tolerance to errors and the coefficient α_i is the Lagrangian multipliers which is greater than zero if the datum x_i is in the support vector, otherwise α_i is equal to zero.

In order to take into account nonlinear boundaries, a *kernel* can be used. This allows a projection of the research space into a space of higher dimension where the border becomes an hyperplane.

However, in case of nonlinear *kernel*, the expression of the formulation becomes:

$$L_D = \sum_{i=1}^{M} \alpha_i - \sum_{i,j=0} \alpha_i \alpha_j y_i y_j K\left(x_i x_j\right) \tag{1.126}$$

The discriminant function for a new datum is based on:

$$f\left(x_{new}\right) = \sum_{i=1}^{M} \alpha_i y_i K\left(s_i, x_{new}\right) + b \tag{1.127}$$

where $s_i \in [1 \ldots M]$.

Among all *kernel* functions, one can find in the literature these different expressions Breuneval R. and Mansouri [2018]:

- $K\left(x_i x_j\right) = x_i . y_j$
- $K\left(x_i x_j\right) = \left(x_i . y_j + 1\right)^p$
- $K\left(x_i x_j\right) = \exp \frac{\|x_i - y_j\|^2}{2\sigma^2}$
- $K\left(x_i x_j\right) = \tan\left(a . x_i . y_j + b\right)$

1.4.2.3 Recurrent Neural Network

The structure of a neural network is simple. It is based on layers which are composed of perceptrons (mathematical functions). Each output of perceptron is connected to an other perceptron. Figure 1.43 shows a neural network

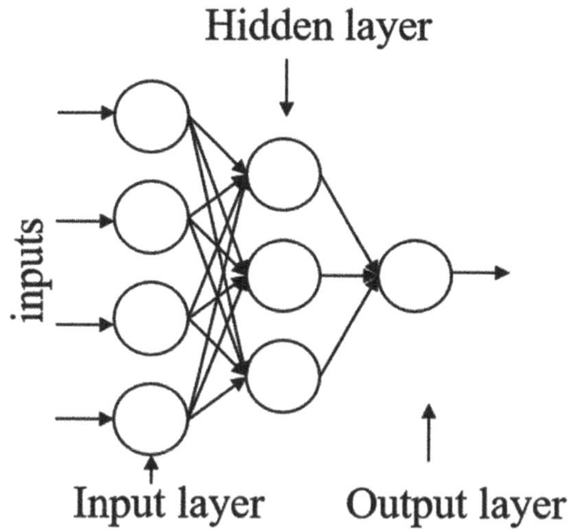

Figure 1.43 A basic neural network with one hidden layer

composed of three layers. The first one is the input layer composed of four perceptrons, the second layer is called hidden layer which is composed of three perceptrons and one output layer composed of one perceptron. The number of layers as the number of perceptrons in each layer depends on the user Figure 1.43.

Hayashi et al. [2002] has used a neural network composed of 240 neurons in the input layer, 30 neurons in the hidden layer and one neuron in the output layer. His neural network was dedicated to fault diagnosis of a machine based on the spectrum of vibrations from it.

The learning of the diagnosis network is performed with some normal and abnormal data coming from the process to monitor. Usually, the learning process is based on a gradient method as described by Rumelhart et al. [1986]. However, there are other optimization algorithms such as the Levenberg–Marquardt, which combines two strategies which are: acceleration of the convergence at the beginning of the optimization then slow convergence when we are close to the optimum, Korbicz [1997]. as presented in the Figure 1.44, an error is evaluated thanks to a comparison of the output of a process with the output of the neural network. Based on this error, the optimization algorithm modifies all coefficients of the neural network in order to minimize the error.

The process can be a multiple input multiple output system. We can use a neural network whose inputs would take into account the past and would then be written in the form:

$$u = [u_0 \; u_1 \; u_2 \; \dots \; u_p]$$

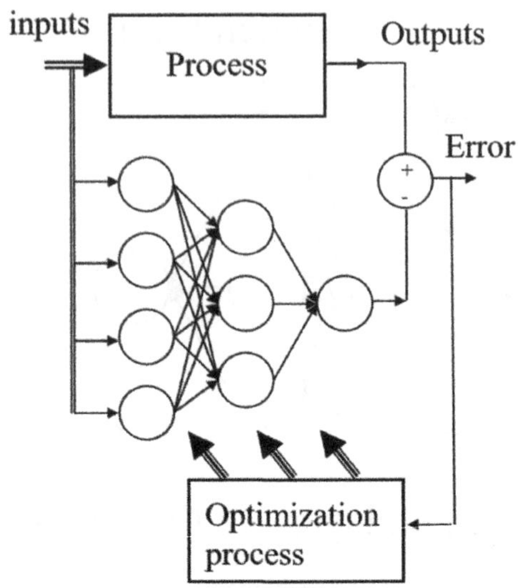

Figure 1.44 Optimization process of a basic neural network

Then, the dynamics of the system can be taken into account by also taking into account the past of the output in the form:

$$y = [y_1\ y_2\ y_3\ \ldots\ u_q]$$

Therefore, the neural network should duplicate the dynamic behavior of the process which may be nonlinear. By way of illustration, we represent a neural network duplicating a process using three samples as input u and three samples for the output y Figure 1.45.

Of course, once the learning process is complete, the coefficients (weights) within the neural network will be fixed. So, we do have a recurrent neural network.

Neurons, or perceptrons, are nonlinear functions called "sigmoid." The equation governing the neuron is simple and is written in the following form:

$$y = f(\Sigma w_i.x_i) \tag{1.128}$$

where w is the connecting weight between two perceptrons of two consecutive layers, and x is the output of the previous perceptron. The sigmoid function can be written as:

$$f(\Sigma w_i.x_i) = \frac{1}{1 + \exp^{-\Sigma w_i.x_i}} \tag{1.129}$$

An application of neural networks can also be the detection of defects in processes. To do this, the inputs will be residuals, for example, residuals formed

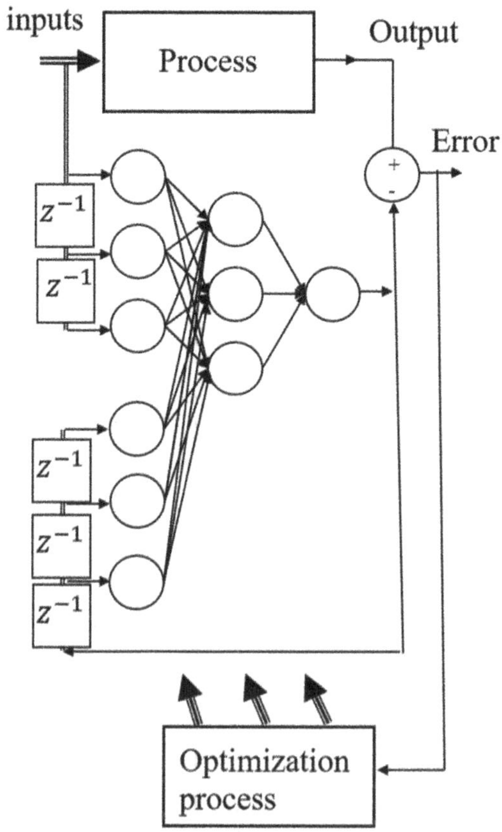

Figure 1.45 A basic recurrent neural network

by the difference in behavior between the output of a process and its model. Consequently, the x variables at the input of the neural network will be the residuals. It is necessary to implement a learning phase in order to train the neural network to detect defects. Therefore, a knowledge and database is mandatory. We represent in Figure 1.46 one structure among many others.

1.4.3 Unsupervised Clustering

There are several methods of unsupervised classification, also known as coalescence methods. If we limit ourselves to exclusive assignment, these methods are divided into two parts (Jain et al. [1999]):

- Hierarchical classification methods (ascending or descending);
- Classification by partition.

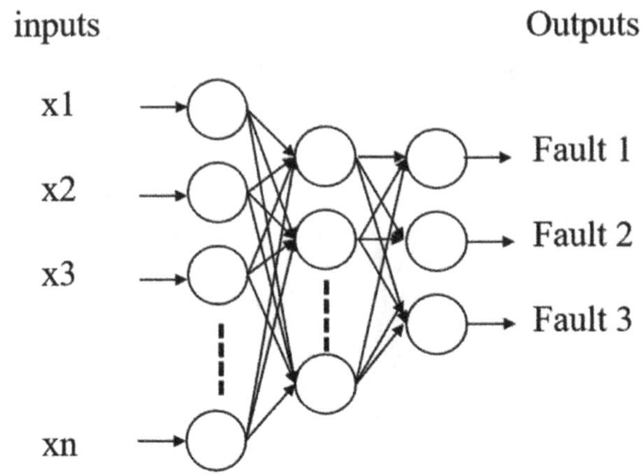

Figure 1.46 A basic neural network for faults detection

1.4.3.1 Hierarchical Classification

Hierarchical classification produces a hierarchical set of classes according to a linkage criterion. The classes, which are nested, are linked recursively either in agglomeration mode as Ascending Hierarchical Classification (AHC) or in division mode as Descending Hierarchical Classification (DHC). In the hierarchical ascending classification, the classes are determined iteratively in four steps:

- Step 1: the classification algorithm begins by considering each observation as a class. The initialization phase is defined by the variable:

$$\Omega_i^{\ell} = \{x_i\} \qquad \forall i \quad \text{and} \quad \ell = 0 \tag{1.130}$$

  where the variable Ω_i^{ℓ} denotes the membership class of observation x_i at the ℓth iteration, with ℓ=0 for the initialization phase.
- Step 2: the similarity between two classes is defined by a link criterion. This criterion evaluates two classes using a similarity measure which evaluates the similarity between two classes. The "Sim" similarity between two classes is defined as follows:

$$Sim\left(\Omega_l^{\ell}, \Omega_h^{\ell}\right) = \beta\left(\Omega_l^{\ell}, \Omega_h^{\ell}\right) \tag{1.131}$$

  where the link criterion is represented by the function β which can be a link criterion by minimum jump, maximum jump, average distance or by the intra-class inertia represented by the Ward criterion. Equation 1.131

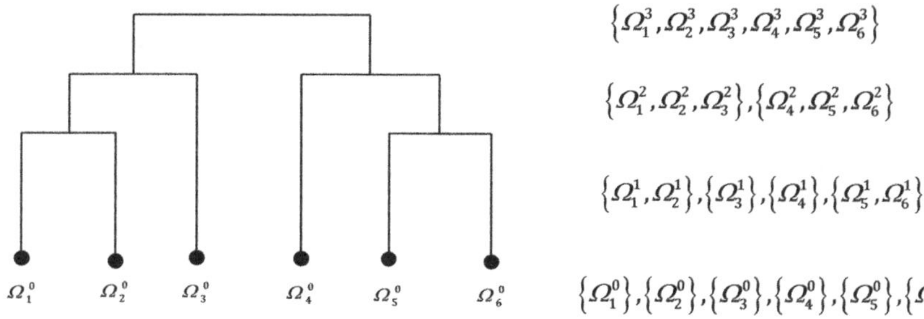

Figure 1.47 Example of a dendrogram

indicates that all observations of class (ℓ) should be compared with all observations of class (h). The link criterion β returns a similarity between the two classes.

- Step 3: select the pairs of classes which have a very great similarity $Sim\left(\Omega_l^\ell, \Omega_h^\ell\right)$. Let Ω_{l*}^ℓ and Ω_{h*}^ℓ two classes selected as follows:

$$\left[\Omega_{l*}^\ell, \Omega_{h*}^\ell\right] = \underset{l,h \in [1-M];\, l \neq k}{\arg\max}\ Sim\left(\Omega_l^\ell, \Omega_h^\ell\right) \tag{1.132}$$

- Step 4: merge the two classes with an index $l*$ and $h*$

$$\Omega_{l*}^{\ell+1} = \left(\Omega_{l*}^\ell \cup \Omega_{h*}^\ell\right) \tag{1.133}$$

$$\Omega_{h*}^{\ell+1} = \{\} \tag{1.134}$$

Steps 2–4 are repeated until all the classes are merged into one class (see Figure 1.47). In the standard case, the aggregation stops when a stop criterion is reached. For example, the stopping criterion is obtained by a maximum distance between the different classes as well as by a low dispersion of observations in each class. Class fusion sequences can be represented graphically by a tree diagram, called a dendrogram. The dendrogram is a multilevel hierarchy with the merging of two classes at each level (see Figure 1.47).

(1) Linkage criterion

The merging of classes is based on the similarity between them. This similarity is calculated from a linkage criterion. The criterion of connection by minimal jump $\beta_{\text{jump-min}}$ is the first criterion which can be taken into account. It seeks

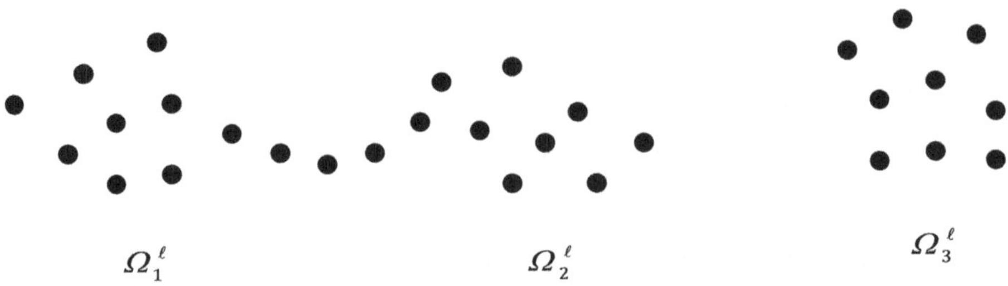

Figure 1.48 Chain effect

the greatest similarity between two observations x_i and x_j which are located in the classes Ω_l^ℓ and Ω_h^ℓ. It is expressed as follows:

$$\beta_{jump-\min}\left(\Omega_l^\ell,\Omega_h^\ell\right)=\max\left\{d^{-1}\left(x_i,x_j\right)\right\}\ \ \forall\ x_i\in\Omega_l^\ell\ ,\ x_j\in\Omega_h^\ell\ \text{and}\ l\neq h \tag{1.135}$$

The variable x_i denotes the $i-th$ observation and x_j denotes the $j-th$ observation. The similarity distances proposed in the literature can be used for the calculation of the link criterion. However, the linkage criterion by minimal jump is ill-suited in the case where the classes have a stretched shape which has the disadvantage of producing chain effects as presented in Figure 1.48.

- Maximum hop link criterion
  The linkage criterion by maximum jump $\beta_{\text{jump}-\max}$ calculates a dissimilarity between the classes based on the maximum distance between two observations which are located in the classes Ω_l^ℓ and Ω_h^ℓ. It is expressed as follows:

$$\beta_{jump-\max}\left(\Omega_l^\ell,\Omega_h^\ell\right)=\min\left\{d^{-1}\left(x_i,x_j\right)\right\}\ \ \forall\ x_i\in\Omega_l^\ell\ ,\ x_j\in\Omega_h^\ell\ \text{and}\ l\neq h \tag{1.136}$$

- Connection criterion by average distance
  The average distance linkage criterion $\beta_{\text{jump-avg}}$ calculates the similarity between two classes by finding the average distance between all pairs of observations in classes Ω_l^ℓ and Ω_h^ℓ. It is expressed as follows:

$$\beta_{dist-avg}\left(\Omega_l^\ell,\Omega_h^\ell\right)=\frac{1}{\left|\Omega_l^\ell\right|}\frac{1}{\left|\Omega_h^\ell\right|}\sum_{x_i\in\Omega_l^\ell}\sum_{x_j\in\Omega_h^\ell}d\left(x_i,x_j\right)\quad l\neq h \tag{1.137}$$

  $\left|\Omega_l^\ell\right|$ and $\left|\Omega_h^\ell\right|$ denote the number of observations in the classes Ω_l^ℓ and Ω_h^ℓ, respectively. This criterion can be seen as a compromise between $\beta_{\text{jump}-\min}$ and $\beta_{\text{jump}-\max}$.

- Ward's criterion
  The Ward's criterion β_{ward} corresponds to the intra-class inertia which results from the fusion of Ω_l^ℓ and Ω_h^ℓ. It is expressed as follows:

$$\beta_{ward}\left(\Omega_l^\ell, \Omega_h^\ell\right) = \frac{\left|\Omega_l^\ell\right| \cdot \left|\Omega_h^\ell\right|}{\left|\Omega_l^\ell\right| + \left|\Omega_h^\ell\right|} d_{Euc}^2\left(g_l, g_h\right) \quad l \neq h \tag{1.138}$$

  where g_l and g_h are the centers of gravity of the classes Ω_l^ℓ and Ω_h^ℓ, respectively; $d_{Euc}^2\left(g_l, g_h\right)$ is the square of the Euclidean distance between the centers g_l and g_h.

1.4.3.2 K-means and Centroid Clustering

Classification by partition subdivides a set of observations into several subsets. Each subset represents a class containing at least one observation. Each class is represented by a function (F) which assigns observations to these classes based on their similarities. Let $\Omega_1^\ell, \Omega_2^\ell, \ldots, \Omega_M^\ell$ be a set of M-classes. The function (F) which affects the set of N-observations $x_i \;_{i\in(1,\ldots,N)}$ in M-classes is given as follows:

$$F\left(x_i\right) = \underset{1 \leq k \leq M}{\arg\min} \left\{d\left(x_i, g_k\right)\right\} \tag{1.139}$$

where the metric $d\left(x_i, g_k\right)$ represents a measure of similarity between the i-th observation xi and the center of gravity of the class $\Omega_k^\ell \;_{k\in(1,\ldots,M)}$. The variable ℓ designates the ℓ-th iteration.

The K-means algorithm of Forgy [1965] is the most used algorithm in partition classification. It has four stages:

- Step 1: Choose a number of M-observations among the set of N-observations to be partitioned and consider them to be the centers of gravity of the M-classes.
- Step 2: Assign each observation xi to its closest center of gravity using eq. 1.139.
- Step 3: When all the observations are affected, recalculate the centers of gravity of the M-classes by:

$$g_k = \frac{1}{\left|\Omega_k^\ell\right|} \sum_{x_i \in \Omega_k^\ell} x_i \;, \quad 1 \leq k \leq M \tag{1.140}$$

  where $\left|\Omega_k^\ell\right|$ denotes the number of observations in the class Ω_k^ℓ.
- Step 4: Repeat steps 2 and 3 until the centers of gravity of the M-classes no longer vary.

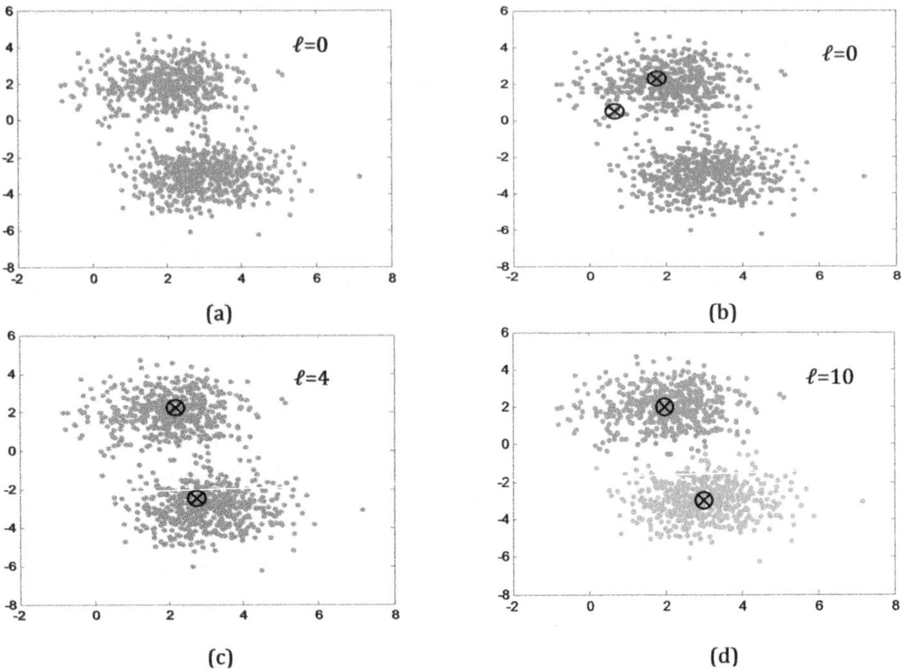

Figure 1.49 K-means algorithm for the partitioning of two classes

An example showing the execution of the K-means algorithm is given in Figure 1.49. At each iteration, the observations are assigned to the closest center of gravity (shown in color, the observations take on the same color as the center of gravity of the class to which they are assigned). The center of gravity of each class is then recalculated as a function of the average of the observations attributed to it. The final partitioning gives us a good separation between the two classes by giving a distinct color for each of them. As we can see in Figure 1.49, observations are represented by points and the centers of gravity of the two classes are represented by crosses. (a) represents the set of observations. (b) represents randomly chosen centers of gravity. (c) and (d) illustrate the distribution of the set of observations after four and ten iterations of the K-means algorithm.

1.4.3.3 Self-organizing Map

Designed by McCulloch and Pitts in 1943, the neuron model derived from the biological model experienced a major boom in the 1980s. This emergence is closely linked to its ability to process incomplete, uncertain or noisy data and also to its application in various fields ranging from decision support, planning and even to the fields of pattern recognition and classification (Dumolard [1994]). Kohonen self-organizing networks are competitive and dynamic networks, in the sense that they tend to elect a winning neuron and favor it, on the one hand, and create or destroy neurons, on the other hand. The topological map model proposed by Kohonen is a self-organization process that will project the initial data onto a discrete and regular space of low dimension (generally 1, 2, or 3D). The spaces used are regular trellises. Each node is occupied by a neuron. The notion of neighborhood between neurons then follows directly from the structure and defines a topology of the map. Thanks to the self-organization process, the topology which links the initial data is preserved at the level of the responses proposed by the network (IFTENE and Mahi [2003] and Kohonen [1989]).

Kohonen topological maps consist of two layers:

- the input layer is only used for the presentation of the shapes to be classified and the states of all its neurons are forced to the values of the input data;
- the adaptation layer is formed by a network of neurons. The choice of the geometry of the network used is made a priori. The neurons used at this level are linear neurons, each of them plugged into all elements of the input layer. In order to enable the process of self-organization, the weights that link the two layers are adaptively determined.

(1) Structure and operation

In Figure 1.50, we present a network organized in a one-dimensional layer. Each output neuron $NE_{j\ j\in(1,\ldots,K)}$ is connected to a number $x_{i\ i\in(1,\ldots,N)}$ of input neurons through N connections $(K * N)$ of respective weight $W_{j,i}$. When presenting an input, a neuron on the map is selected based on the minimum distance separating it from that entry.

(2) Description of the algorithm

Here we describe the 5 steps of the algorithm which are composed of:

- Step 0:
  - Initialization of the weights of the output neurons (low random values);
  - Choice of map shape;
  - Initialization of the learning parameters:
    * Learning rate $\alpha(t)$:

$$\alpha(t) = \frac{\alpha_0}{1 + K_\alpha t}$$

K_α, α_0 : Positive constants.

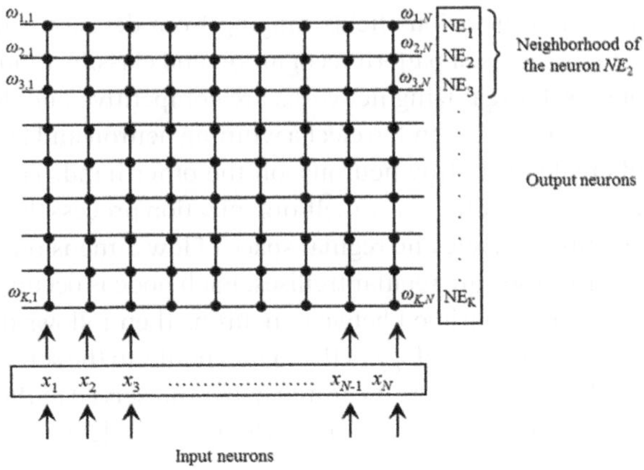

Figure 1.50 Architecture of a self-organizing map (1D network)

* $\nu(s, \mathrm{NE_f}, t)$: Neighborhood function.
  s: Neuron neighborhood of $\mathrm{NE}_{\mathrm{f}\ \mathrm{f}\in(1,\ldots,\ K)}$.
  The neighborhood function $\nu(s, \mathrm{NE_f}, t)$ is a Gaussian, evolving with time, of the form:

$$\nu(s, \mathrm{NE_f}, t) = e^{-\frac{d^2(s,\mathrm{NE_f})}{2\sigma^2(t)}}$$

  d $(s, \mathrm{NE_f})$: distance between the winning neuron $\mathrm{NE_f}$ and its neighborhood identified by the index s defined by:

$$\left\| \omega_{NE_f} - \omega_s \right\|$$

  $\sigma(t)$ is defined by:

$$\sigma(t) = \sigma_{initial}\left(\frac{\sigma_{final}}{\sigma_{initial}}\right)^{\frac{t}{t_{\max}}}$$

  where t_{max} is the total number of iterations.

- Step 1: For $t = 1$ to t_{max}
- Step 2: For each input vector $x_{i\ i\in(1,\ldots,N)}$
- Step 3: Find in x_i the most closest output neuron $\mathrm{NE}_{\mathrm{f}\ \mathrm{f}\in(1,\ldots,\ K)}$ according to the following criterion:

$$NE_f\ :\ \left\|\omega_{f,i} - \mathrm{x}_i\right\| \leq \left\|\omega_{j,i} - \mathrm{x}_i\right\|\ \ \forall j \in \{1, \ldots, \mathrm{k}\}$$

$\omega_{f,i}$ is the weight associated with the output neuron $\mathrm{NE_f}$.

- Step 4: All neurons in the neighboring of the winning neuron are adapted:

$$\omega_{j,i}(t+1) = \omega_{j,i}(t) + \alpha(t).\nu(s, \mathrm{NE_f}, t).(x_i - \omega_{j,i}(t))$$

  $\alpha(t)$ is a learning function which decreases slowly and linearly with time ($0 < \alpha\,(\mathrm{t}) < 1$) (see Figure 1.51).
- Step 5: Adjust the learning parameters.

1.5 Prognosis Methods

1.5.1 Prognosis Process

The prediction of the remaining useful life or the forecasting of impeding failures allows for maintenance scheduling. This has been the subject of many research works. The theoretical basis will be presented in this section. Many prognosis methods have been developed. Figure 1.52 gives an overview of the different approaches:

They can be classified in three categories:

- model based methods;
- data driven techniques;
- hybrid prognostics approach.

Model-based methods deal with physical representation of the aging process. They do not require historical date but they need a accurate representation of the degradation process. These models can then be exploited by Kalman filters, particle filters, etc. These methods are used, in particular, in mechanics or for electronic components.

Data-driven methods extrapolate the behavior of the system from historical data which come from similar components and working conditions. It can be used to determine the remaining useful life (RUL). They do not require any specific physical knowledge or model. These data, based on the history, are used to extract the parameters which are sensitive to the faults. These later must be able to vary monotonically with aging. The prognosis can be a combination of the extrapolation of these parameters and a diagnosis of the future state from these extrapolated parameters (Figure 1.53).

These methods can be based on belief rule (Zhou et al. [2014]), neural networks (Soualhi et al. [2018], Cui et al. [2017]), Bayesian networks and their derivatives (Bartram and Mahadevan [2013], Medjaher et al. [2012]), on similarity-instance based approach (Khelif et al. [2014]), core methods such as support vector machines (Tao and Zhao [2016], Satishkumar and Sugumaran [2016]).

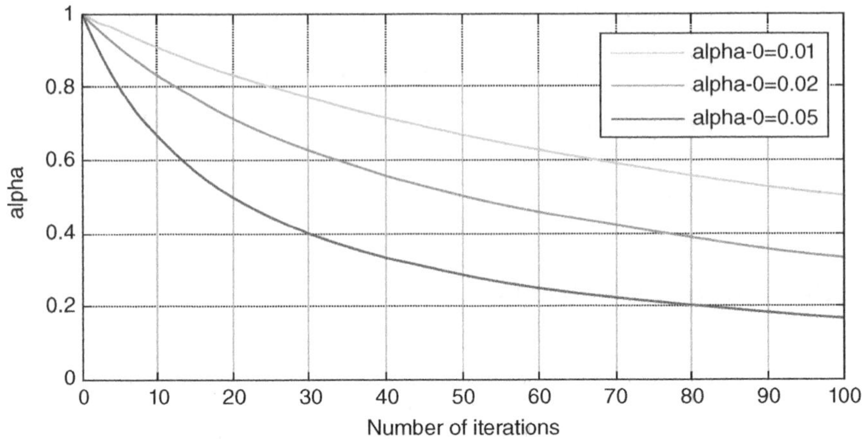

Figure 1.51 Evolution of the learning rate over time

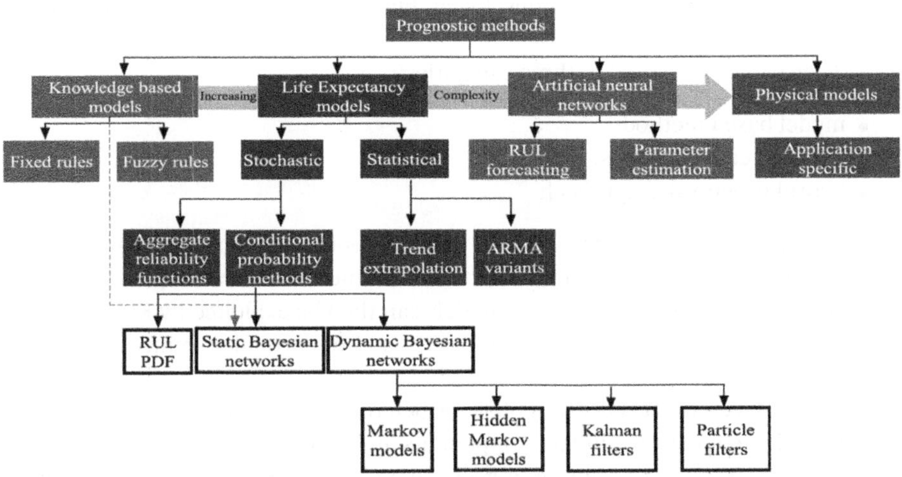

Figure 1.52 Prognosis methods (Soualhi et al. [2020]) with the permission of Elsevier

Hybrid prognostics approach combines model-based and data-driven methods (Figure 1.54).

In the serial hybrid method, the parameters of a physical model are estimated by a data-based method (Mangili [2013]). In the parallel hybrid method, the outputs of a physical model and a data-based method are combined to estimate the RUL (Sachin Kumar et al. [2008]).

We will detail in the next paragraphs some of these techniques:

- the extrapolation of time series (Subsection 1.5.2);

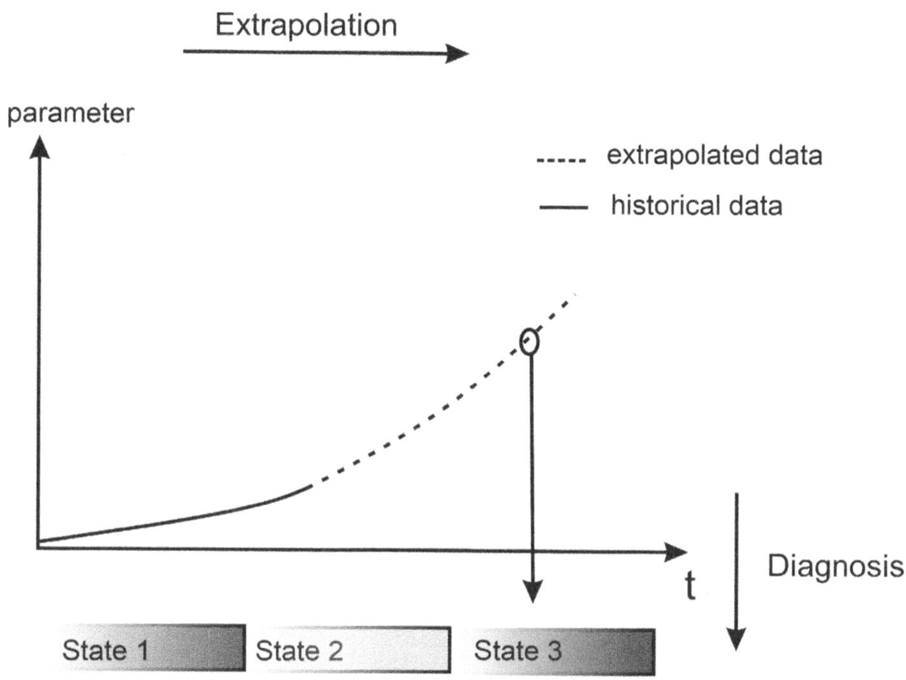

Figure 1.53 Prognosis process

- the Bayesian inference (Subsection 1.5.3);
- the Markov chains (Subsection 1.5.4);
- the rain flow (Subsection 1.5.6).

Diagnostic methods based on supervised classifiers (1.4.2) and Neural Networks (Subsection 1.4.2.3) have been previously described.

1.5.2 Time Series Extrapolation Methods

Time series allows to evaluate the evolution of a variable over the time. The first objective is to find a model that best describes or represents the data. It consists of extracting the signal structure and eliminating the noise. The second objective is to predict the behavior of the process. On the basis of the observations $(X_1, \dots X_T)$ of the time series at the first T time steps, the value X_{T+h} with $h > 0$ has to be predicted.

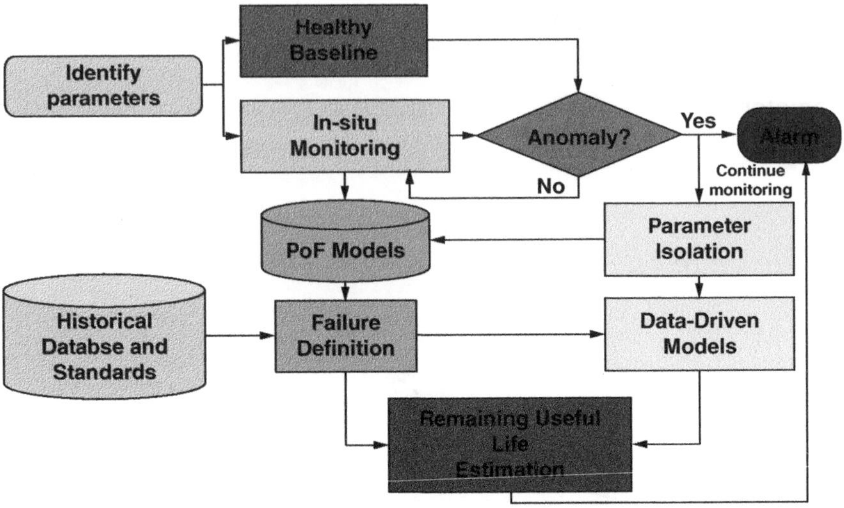

Figure 1.54 Hybrid Prognostics approach. (Pecht and Jaai [2010]) with the permission of Elsevier

(1) Definition

A time series is a series of data points indexed in time order (Wikipedia contributors [2020b]) and Chatfield and Xing [2019]). A time series is a random process such as:

$$X : \Omega X\, T \to \mathbf{R} \tag{1.141}$$

where Ω is a set of events and $T \subset \mathbf{R}$ a discrete set or an interval of $\mathbf{R}$.

Let us note X_t the event at time $t \in T$, a time series (or chronological series). It is a sequence of events in ascending order. The set of measurements of a random variable are the realizations of the random variable X. They are noted:

$$(X_i)_{1 \leq i \leq T} \tag{1.142}$$

Figure 1.55 gives an example of time series from the Time Series Data Library. All the functions are written with MATLAB [2020]. These functions are also available in R Core Team [2020].

A time series can be represented with an additive model and split into three terms:

$$X_t = m_t + S_t + \varepsilon_t \tag{1.143}$$

where:

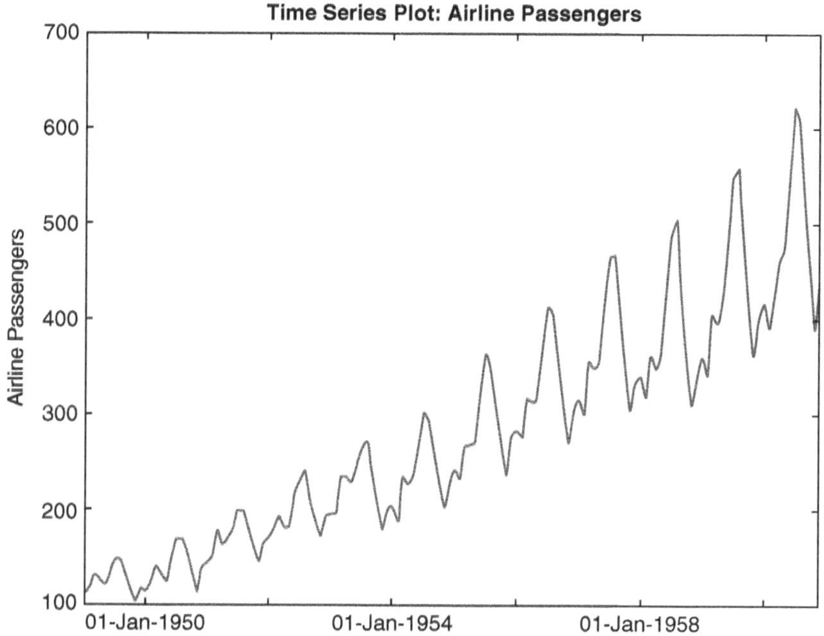

Figure 1.55 Example of time series with monthly counts of airline passengers (Hyndman [2005])

- m_t a deterministic function (called the trend) which describes the "long term" behavior such linear, exponential, logarithmic increase or decrease ... ;
- S_t is a periodic deterministic function (called seasonality) of period m with $\sum_{i=1}^{m} S_{t+i} = 0,\ \forall t \in T$;
- ε_t is a random variation in the series and it could be autocorrelated $(cov(\varepsilon_t, \varepsilon_{t+h}) \neq 0)$.

Next, Figure 1.56 gives the linear trend plus a seasonal component of the previous time series (Hyndman [2005]).

The time series is stationary if:

$$F_X(X_{t_1}, X_{t_2}, \dots, X_{t_n}) = F_X(X_{t_1+h}, X_{t_2+h}, \dots, X_{t_n+h}),\ \forall (t_1, \dots, t_n) \in T^n, \forall h \in N \tag{1.144}$$

where $F_X(X_{t_1}, X_{t_2}, \dots, X_{t_n})$ is the cumulative distribution function.

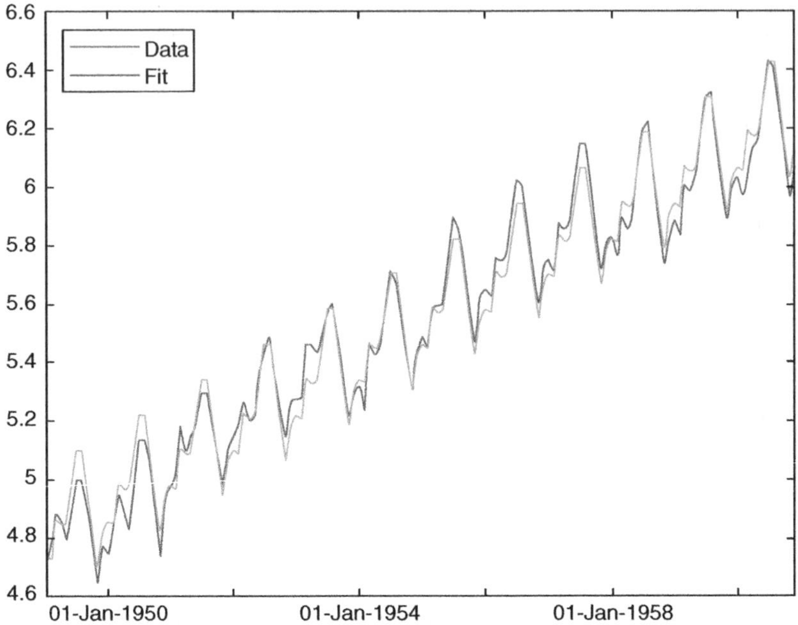

Figure 1.56 Example of the trend and seasonability of a time series (MATLAB [2020])

The time series is weakly stationary if:

$$\begin{aligned} &E(X_t) = \mu \\ &cov(X_t, X_{t+h}) = \nu_k \end{aligned} \tag{1.145}$$

where μ is constant and ν_k is independent of the time t.

To study a time series, it is necessary to:

1. Write a statistical model (eliminate the trend, periodicity, and test if it is a stationary model);
2. Estimate the parameters of the model;
3. Perform statistical tests to validate/reject the hypotheses;
4. Predict the series of events $(X_{1+T}, X_{2+T}, \dots, X_{n+T})$ from $(X_1, X_2, \dots, X_n)$.

(2) Estimation of the trend (Weber [2001])

(2)1 Time series in the absence of seasonality The objective is to discriminate the trend from the time series. **First approach**: the trend can be evaluated by assuming that it is a linear combination of known functions:

$$m_t = \sum_{i=1}^{n} \alpha_i m_t{}^{(i)} \tag{1.146}$$

where α_i are unknown coefficients and the $m_t{}^{(i)}$ are time functions with a known structure.

1. linear trend: $m_t = \alpha_0 + \alpha_1 t$;
2. polynomial trend: $m_t = \alpha_0 + \alpha_1 t + ... + \alpha_n t^n$;
3. exponential trend: $m_t = c_0 + c_1 \alpha^t$;
4. Gompertz's trend: $m_t = c_0.e^{c_1.e^{c_2.t}}$;
5. Gompertz's trend: $m_t = \frac{1}{c_0+c_1 e^{c_2.t}}$;
6. or a combination of these different functions.

The unknown coefficients can be determined by the method of least squares.

Second approach: the trend can can also be estimated by moving average filtering. The moving average is defined by the operator:

$$Y_t = \sum_{i=-m_1}^{m_2} \theta_i Y_{t+i} \text{ with } (m_1, m_2) \in \mathbf{N}x\mathbf{N} \text{ and } \theta_i \in \mathbf{R} \tag{1.147}$$

It is often given by:

$$Y_t = \frac{1}{2l+1} \sum_{i=-l}^{l} Y_{t+i} \tag{1.148}$$

The moving average is a non-parametric estimator of trend, that is trend has no a priori structure.

Figure 1.57 gives an example of the output from the moving average of a time series.

Third approach: to eliminate the k_{th} order polynomial trend of a time series, a differentiation of order k can be made. The differentiation operator is defined:

$$\Delta Y_t = Y_t - Y_{t-1} \tag{1.149}$$

and for the k_{th} order differentiation operator is:

$$\Delta^k Y_t = \Delta(\Delta^{k-1} Y_t) \tag{1.150}$$

(2)2 Time series with trend and seasonality The previous methods can be extended in the case of trend with seasonality. In the first approach, the time series can be represented by:

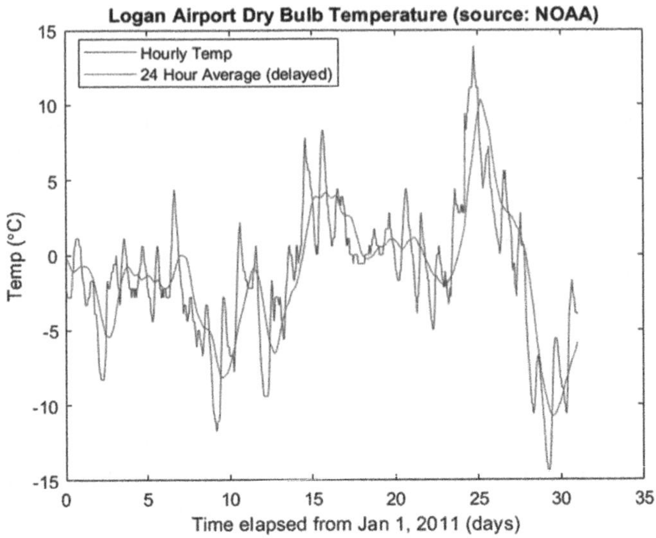

Figure 1.57 Example of a time series moving average filtering (MATLAB [2020])

$$X_t = \sum_{i=1}^{n} \alpha_i m_t^{(i)} + \sum_{j=1}^{p} \beta_j s_t^{(j)} + \varepsilon_t \text{ with } E(\varepsilon_t) = 0 \tag{1.151}$$

where $s_t^{(j)}$ are periodic time functions with a known structure.

(2)3 Auto-Regressive Models Many time series can be modeled by auto-regressive process such as:

- Moving average process;
- Auto-Regressive-Integrated-Moving Average process;
- Auto-Regressive-Integrated-Moving Average;
- ...

An auto-regressive model is defined by the equation 1.152.

$$X_t = \sum_{r=1}^{q} \Phi_r X_{t-r} + \varepsilon_t \tag{1.152}$$

where $\{\varepsilon_t\}$ are uncorrelated random variables following a Normal law ($\varepsilon_t \sim N(0, \sigma^2)$) and Φ_r are constant parameters.

Let us consider some of the auto-regressive processes:

MA : Mean average (eq 1.153).

$$X_t = \sum_{s=0}^{q} \theta_s \varepsilon_{t-s} \tag{1.153}$$

where $\{\varepsilon_t\}$ are uncorrelated random variables following a Normal law ($\varepsilon_t \sim N(0, \sigma^2)$) and θ_i are constant parameters with $\theta_0 = 1$.

ARMA - Auto-regressive moving average process (1.154).

$$X_t = \sum_{r=1}^{p} \phi_r X_{t-r} + \sum_{s=0}^{q} \theta_s \varepsilon_{t-s} \tag{1.154}$$

where $\{\varepsilon_t\}$ is a white noise. This process should be stationary for appropriate θ and ε.

ARIMA - Auto-regressive integrated moving average Average. If X_i the is not stationary, we can consider $Y_t = \nabla X_t = X_t - X_{t-1}$ or $Y_t = \nabla^2 X_t = X_t - 2X_{t-1} + X_{t-2}$ or an higher order differentiation.

If the differenced process is a stationary process, X_t is said to be an auto-regressive integrated moving average process.

Figure 1.58 gives an example of an ARIMA model of a scalar time series with linear trend.

(2)4 Other approaches Many other methods can be used to predict the behavior of time series such as Kalman filters (paragraph 1.2.3.1) with state model, neural network (paragraph 1.4.2.3). They also can be associated with clustering techniques (Maharaj et al. [2019]).

1.5.3 Bayesian Inference

Bayesian networks, initiated by Judea Pearl in the 1980s, are among the models of probabilistic analysis. They propose a mathematical formalism and solid theoretical bases for the modeling of complex systems. A Bayesian network is made up of two components:

- A graphic component which consists of a directed graph without circuit. The nodes represent the relevant variables of the domain and the arcs represent the dependency relationships between the variables.
- A numerical component which consists of a set of conditional probability distributions of each node in the context of its parents.

The graph is always directed and acyclic. Directed arcs represent a direct dependency relationship (most of the time this is causation). Thus an arc going from the variable X to the variable Y will express the fact that Y depends

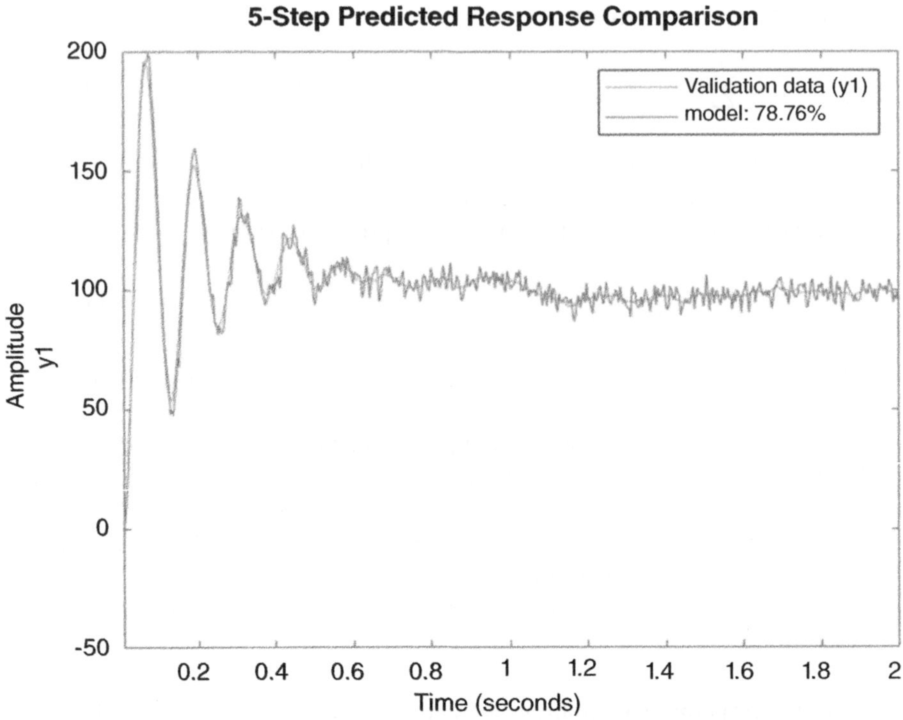

Figure 1.58 Example of ARIMA model (MATLAB [2020])

directly on X. The absence of an arc only informs about the nonexistence of a direct dependence. The parameters express the weights given to these relations and are the conditional probabilities of the variables knowing their parents (example: $P(Y|X)$ or the a priori probabilities if the variable has no parents.

It is possible to achieve efficient classifiers thanks to Bayesian networks (Friedman [1989]; Langley et al. [1992]; Pernkopf [2005]). A Bayesian classifier of a problem with p observations (variables) has the particularity of having $p + 1$ nodes. All Bayesian classifiers model the membership of a class by a discrete node called "class node," noted Ω, Ω is a multinomial discrete node with k modalities, where k represents the number of classes of the problem $(\Omega_1, \Omega_2, \ldots, \Omega_k)$. This class node does not have a parent node. The other nodes, p-numbered, represent the observations of the classification problem and are noted Oi. There are two main types of Bayesian classifiers: naive Bayesian network naive Bayesian network augmented by a tree.

The naive Bayesian network has the simplest structure. Also called the Bayes classifier, it is called naive because it makes the very strong assumption that each observation is, conditionally on the class, independent of the others. The observation independence assumption makes it easy to write the posterior probability of each class as shown in the following equation:

$$P(\Omega_i|O) = P(\Omega_i)\prod_{j=1}^{N} P(O_j|\Omega_i) \tag{1.155}$$

N is the number of observations.

The structure of a naive Bayesian network can be represented as in Figure 1.59:

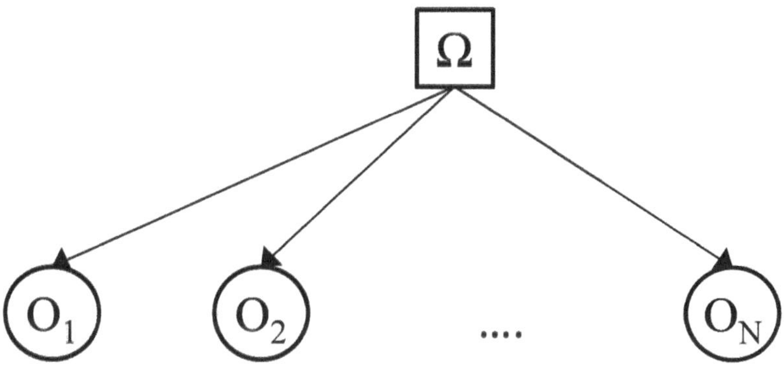

Figure 1.59 Naive Bayesian network

In order to improve the performance of the naive Bayesian network in the case of not completely independent observations, Friedman [1989] suggests adding arcs between the different descriptive observations of the naive classifier. The algorithm starts from a naive Bayesian network and adds an arc between the observations that share the most important mutual information. But, in order to respect the topology of the tree, the algorithm prohibits each node from having more than two parents (i.e. one parent in addition to the class node). Thus, the solution proposed by Friedman [1989] results in a complete structure, that is all nodes are taken into account in the model. The structure of an augmented naive Bayesian network can be represented as in Figure 1.60:

1.5.4 Markov Chain

In this model, a series of random states $X = \{X_t, X_{t-1}, ..., X_0\}$ in a finite state space E is a Markov chain if the following property is satisfied (Markov

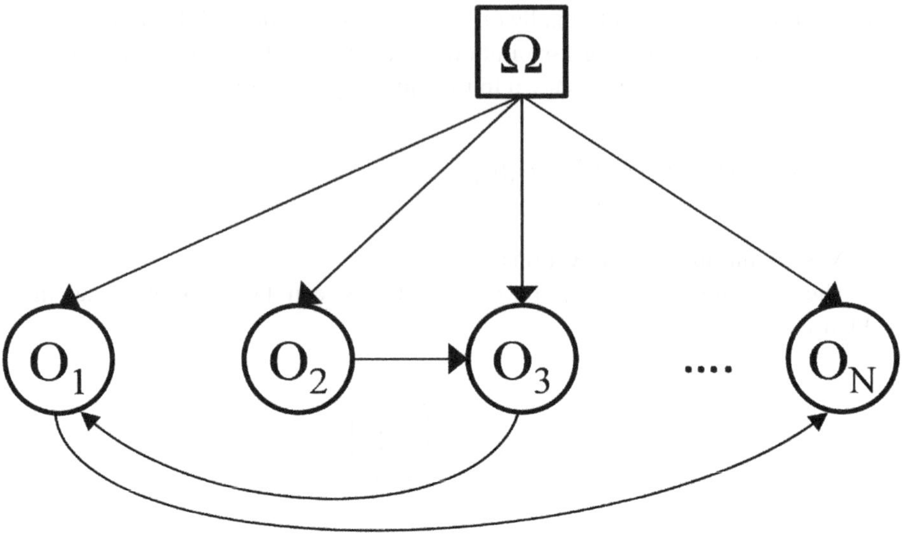

Figure 1.60 Augmented naive Bayesian network

property):

$$P(X_t|X_0, ..., X_{t-1}) = P(X_t = j|X_{t-1} = i) \qquad \text{for all i,j } \in E \tag{1.156}$$

This definition generalizes the notion of a deterministic dynamic system: the probability distribution of the present state X_t depends only on the immediately past state X_{t-1}. It follows that a Markov chain is a sequence of states X which is fully characterized by:

The initial state probability $\pi = (\pi_{\mathrm{i}})$

$$\pi_i = P\,(X_0 = i) \qquad \text{for all } i \in \mathrm{E} \tag{1.157}$$

The transition matrix $A = \left(A_{ij}\right)$

$$A_{ij} = P\,(X_t = \ j \mid X_{t-1} = \ i) \qquad \text{for all } i,\ j \in \mathrm{E} \tag{1.158}$$

Therefore, the probability distribution of the Markov chain in a sequence of states X, of initial state probability π and a transition matrix A, is given by:

$$\begin{cases} P(X_0, ..., X_{t-1}, X_t) = P(X_t|X_0, ..., X_{t-1})P(X_0, ..., X_{t-1}) \\ \qquad = A_{ij}^t \ P(X_0, ..., X_{t-1}) \\ \qquad = \pi_i^0 A_{ij}^0, \ ..., \ A_{ij}^{t-1}, A_{ij}^t \end{cases} \tag{1.159}$$

1.5.5 Hidden Markov Models

In this model, we do not directly observe the sequence X_t, but we have observations o_t with values in a finite observation space O, or in R^d. We suppose that conditionally to the states X_t, observations o_t are mutually independent, and each observation o_t depends only on the state X_t at the same time. This property is expressed as follows:

$$P(o_0, ..., o_T | X_0, ..., X_T) = \prod_{t=0}^{T} P(o_t | X_t) \tag{1.160}$$

It follows that a hidden Markov model for (X_t, o_t) is entirely characterized by:

- The initial state probability: $\pi_i = P(X_0 = i)$ for each $i \in E$
- The transition matrix: $A = (A_{ij})$ with $A_{ij} = P(X_{t+1} = j | X_t = i)$ for each $i,j \in E$
- The observation probability: $b = (b_0, ..., b_t)$ with $b_t = P(o_t | X_t)$

A local datum is therefore sufficient (the transition probabilities between two successive instants, and the emission probabilities at a given time) to characterize a hidden Markov model globally.

The probability distribution of (X_t, o_t), with an initial state probability π, a transition matrix A and observation probabilities b is given by:

$$P(X_0 = i, ..., X_t = j, o_0, ..., o_t) = \pi_i^0 A_{ij}^0 ... A_{ij}^t b_0 ... b_t \tag{1.161}$$

We denote by $\lambda = (\pi, A, b)$ the characteristic parameters of a hidden Markov model. We are interested in the following three problems:

1. Evaluate the model λ: This involves efficiently calculating the probability distribution of the series of observations $(o_0, ..., o_t)$ (or likelihood function) according to the parameters of the model.
2. Estimate the state of the observation series: Given a series of observations $(o_0, ..., o_T)$, the objective is the estimation of the present state X_T, or the estimation of an intermediate state X_t for $t = 0, ..., T$, or to estimate globally the sequence of states $(X_0, ..., X_T)$, for a given model λ. The answer to this problem is provided by the forward and backward equations of Baum which allow to calculate the conditional probability distribution of state X_T knowing the observations $(o_0, ..., o_T)$. The answer to the second problem is provided by a dynamic programming algorithm, the Viterbi algorithm, which maximizes the conditional probability distribution of the sequence of states $(X_0, X_1, ..., X_T)$.

3. Identify the model λ: Given a series of observations $(o_0, ..., o_T)$, it is a question of calculating the maximum likelihood estimator for the unknown parameters of the model. The answer to this problem is provided by the Baum–Welsh re-estimation formulas, which define an iterative algorithm to maximize the likelihood function.

To justify the introduction of Baum's forward and backward equations, we start by presenting a first method to calculate the probability distribution of the observations $(o_0, ..., o_T)$.

The probability distribution of the observations $(o_0, ..., o_T)$ is given by:

$$P(o_0, ..., o_T) = \sum_{i,j \in E} \pi_i^0 \, A_{ij}^0 ... A_{ij}^T \; b_0 ... b_T \tag{1.162}$$

Equation 1.161 is used to calculate the probability distribution of the observations $(o_0, ..., o_T)$. We obtain:

$$\begin{aligned} P(o_0, ..., o_T) &= \sum_{\substack{i,j \\ \in E}} P(X_0, ..., X_T \; o_0, ..., o_T) \\ &= \sum_{i,j \in E} \pi_i^0 \, A_{ij}^0 ... A_{ij}^T \; b_0 ... b_T \end{aligned} \tag{1.163}$$

This method provides a first expression for the conditional probability distribution of the state's sequence $(X_0, ..., X_T)$ knowing the observations $(o_0, ..., o_T)$:

$$P(X_0, ..., X_T | o_0, ..., o_T) = \frac{\pi_i^0 \, A_{ij}^0 ... A_{ij}^T \; b_0 ... b_T}{\sum_{i,j \in E} \pi_i^0 \, A_{ij}^0 ... A_{ij}^T \; b_0 ... b_T} \tag{1.164}$$

And for the likelihood function of the model (obtained by using the series of observations $(O = o_0, ..., oT)$:

$$P(O/\lambda) = L_T = \sum_{i,j \in E} \pi_i^0 \, A_{ij}^0 ... A_{ij}^T \; b_0 ... b_T \tag{1.165}$$

Remark:

The number of operations necessary to calculate the probability distribution of the observations $(o_0, ..., o_T)$ from the given formulas (1.141) is considerable. For each possible trajectory from a state (i) to a state (j) of the Markov chain, we must perform the product of 2(T+1) terms and there are $|E|^{T+1}$ different possible trajectories. The total number of elementary operations (additions and multiplications) to be performed is therefore of the order of: 2(T+1) $|E|^{T+1}$. This number grows exponentially with the number T of observations.

- **Forward equation**: We define the forward variable $\alpha_t(i)$ by:

$$\alpha_t(i) = P(o_0, ..., o_t, X_t = i|\lambda) = P(X_t = i|o_0, ..., o_t)\ L_t \quad \text{for each } i \in E \tag{1.166}$$

The sequence α_t satisfies the following recurrent equation:

$$\alpha_{t+1}(j) = b_j^{o_{t+1}} \sum_{i\in E} A_{ij}\, \alpha_t(i) \tag{1.167}$$

with the initial condition: $\alpha_0(i) = \pi_i\ b_0$ for each $i \in$ E

- **Backward equation** Conditionally to $(X_t = i)$, the sequence $X_{t+1}, X_{t+2}, ...,$ is a Markov chain, with an initial probability state π_i and transition matrix A. We then deduce that the probability distribution of the observations $(o_{t+1}, .., oT)$ given $(X_t = i)$ is expressed as follows:

$$P(o_{t+1}, ..., o_T, X_i|\lambda) = \sum_{i,j\in E} A_{ij}^{t+1} ... A_{ij}^{T}\ b_{t+1} ... b_T \tag{1.168}$$

And we define the backward variable as the likelihood function of the model (obtained by using the series of observations $(o_{t+1}, ..., o_T)$.

$$\beta_t(i) = \sum_{i,j\in E} A_{ij}^{t+1} ... A_{ij}^{T}\ b_{t+1} ... b_T \tag{1.169}$$

And especially:

$$\beta_{T-1}(i) = \sum_{j\in E} A_{ij}^{T} b_T \tag{1.170}$$

$$\beta_t(i) = \sum_{j\in E} A_{ij}^{t} b_{t+1} \beta_{t+1}(j) \tag{1.171}$$

with the initial condition: β_T(i)= 1 for each $i \in$ E.

- **Viterbi algorithm (dynamic programming)**

It results that the forward and backward variables, studied previously, make it possible to calculate the conditional probability distribution of the present state X_T, or of the state X_t at an intermediate instant, knowing the observations

$(o_0, ..., o_T)$, defined by:

$$P(X_T = i|o_0, ..., o_T) = \frac{1}{L_T}\alpha_T(i) \tag{1.172}$$

and

$$P(X_T = i|o_0, ..., o_T) = \frac{1}{L_T}q_t(i) \tag{1.173}$$

with

$$L_T = \sum_{i\in E}\alpha_T(i) = \sum_{i\in E}\alpha_t(i)\beta_t(i) = \sum_{i\in E}q_t(i) \tag{1.174}$$

We define the posterior maximum estimator, which minimizes the probability of the estimation error given the observations $(o_0, ..., o_T)$ of the present state by:

$$X_T^{LMAP} = \arg\max_{i\in E} P(X_T = i|o_0, ..., o_T) = \arg\max_{i\in E} \alpha_T(i) \tag{1.175}$$

,

and for the state at an intermediate instant by:

$$X_t^{LMAP} = \arg\max_{i\in E} P(X_t = i|o_0, ..., o_t) = \arg\max_{i\in E} q_t(i) \tag{1.176}$$

However, it can happen that the sequence $(X_0^{LMAP}, \dots, X_T^{LMAP})$ thus generated is inconsistent with the model, in the following way: it can happen that one obtains $X_t^{LMAP} = i$ and $X_{t+1}^{LMAP} = j$ for two successive times, while $A_{ij} = 0$ for this same pair (i, j). This means that the transition from state i to state j is impossible for the model. For this reason, we instead use another estimator defined by:

$$(X_0^{LMAP}, ..., X_T^{LMAP}) = \arg\max_{i_0,...,i_T\in E} P(X_0, ..., X_T|o_0, ..., o_T), \tag{1.177}$$

which minimizes the probability of the estimation error of the sequence of hidden states knowing the observations $(o_0, ..., o_T)$. The efficient calculation of this estimator is provided by a dynamic programming algorithm, called the Viterbi algorithm.

If the following sequence, $(i_0^*, ..., i_t^*)$ reaches the maximum of the function:

$$(i_0^*, ..., i_t^*) \to P(X_0 = i_0, ..., X_t = i_t|o_0, ..., o_t) \tag{1.178}$$

Then necessarily i_0^* reaches the maximum of the function:

$$i \to \max_{i_0,...,i_{t-1}} P(X_0 = i_0, ..., X_{t-1} = i_{t-1}, X_t = i|o_0, ..., o_t) \tag{1.179}$$

and we therefore define:

$$V_t^i \to \max_{i_0,...,i_{t-1}} P(X_0 = i_0, ..., X_{t-1} = i_{t-1}, X_t = i|o_0, ..., o_t)\, L_t \tag{1.180}$$

The sequence V_t checks the following recurrence:

$$V_{t+1}^j = b_j^{o_{t+1}} \max_{i \in E} [A_{ij} \; V_t^i] \tag{1.181}$$

with the initial condition: $V_0^i = \pi_i \, b_i^{o_0}$

We deduce that the optimal trajectory ends up in the state : $X_T^{LMAP} = \arg\max_{i \in E} V_T^i$.

• **Baum–Welch re-estimation formulas**

It follows from eq. 1.165 that the likelihood function of the model $\lambda = (\pi, A, b)$ admits the following expression: $L_T = \sum_{i,j \in E} \pi_i^0 \, A_{ij}^0 ... A_{ij}^T \; b_0 ... b_T$ and we propose to study an iterative algorithm to maximize the likelihood function L_T with respect to the parameters $\lambda = (\pi, A, b)$ of the model.

Let $\lambda' = (\pi', A', b')$ be another model, for which we have already evaluated the likelihood function:

$$L'_T = \sum_{i,j \in E} \pi'^0_i \, A'^0_{ij} ... A'^T_{ij} \; b'_0 ... b'_T \tag{1.182}$$

The likelihood ratio between the model λ and the model λ' can be written as:

$$\frac{L_T}{L'_T} = \frac{\sum_{i,j \in E} \pi_i^0 \, A_{ij}^0 ... A_{ij}^T \; b_0 ... b_T}{\sum_{i,j \in E} \pi'^0_i \, A'^0_{ij} ... A'^T_{ij} \; b'_0 ... b'_T}$$
$$= \frac{\sum_{i,j \in E} \pi'^0_i \, A'^0_{ij} ... A'^T_{ij} \; b'_0 ... b'_T \frac{\pi_i^0 \, A_{ij}^0 ... A_{ij}^T \; b_0 ... b_T}{\pi'^0_i \, A'^0_{ij} ... A'^T_{ij} \; b'_0 ... b'_T}}{\sum_{i,j \in E} \pi'^0_i \, A'^0_{ij} ... A'^T_{ij} \; b'_0 ... b'_T} . \tag{1.183}$$

the log likelihood checks

$$Q_T = \log \frac{L_T}{L'_T} = \log \frac{\sum_{i,j \in E} \pi_i^0 \, A_{ij}^0 ... A_{ij}^T \; b_0 ... b_T}{\sum_{i,j \in E} \pi'^0_i \, A'^0_{ij} ... A'^T_{ij} \; b'_0 ... b'_T}$$
$$\geq \frac{\sum_{i,j \in E} \pi'^0_i \, A'^0_{ij} ... A'^T_{ij} \; b'_0 ... b'_T \log[\frac{\pi_i^0 \, A_{ij}^0 ... A_{ij}^T \; b_0 ... b_T}{\pi'^0_i \, A'^0_{ij} ... A'^T_{ij} \; b'_0 ... b'_T}]}{\sum_{i,j \in E} \pi'^0_i \, A'^0_{ij} ... A'^T_{ij} \; b'_0 ... b'_T} . \tag{1.184}$$

where the quantity Q_T vanishes when the model λ coincides with the model λ'.

Maximizing Q_T with respect to the parameters (π, A, b) of the model λ therefore guarantees that the likelihood L'_T of the model λ' where the maximum of Q_T is reached will be greater than the likelihood L'_T of the model λ'. The Baum–Welch re-estimation formulas make it possible to explicitly find the parameters (π', A', b') of the model λ' as a function of the parameters (π', A', b') of the model λ'. By repeating this procedure, we construct a series of increasing likelihood models, and ideally this series converges to a model which reaches the maximum of the likelihood function.

The iterative algorithm for the maximum likelihood estimation of the model parameters in view of the observations $(o_0, ..., o_T)$, is given by the explicit re-estimation formulas:

$$\pi_i' = \frac{\alpha'_0(i)\,\beta'_0(i)}{L'_T} \tag{1.185}$$

$$A_{i,j}' = A_{i,j} \frac{\sum_{t=1}^{T} \alpha_{t-1}'(i)\, b_t'\, \beta_{t-1}'(j)}{\sum_{t=1}^{T} \alpha_{t-1}'(i)\; \beta_{t-1}'(i)} \tag{1.186}$$

$$b_t = \frac{\sum_{t=0 \cap y_t=\ell}^{T} \alpha_t'(i)\; \beta_t'(i)}{\sum_{t=0}^{T} \alpha_t'(i)\; \beta_t'(j)} \tag{1.187}$$

For each $i, j \in E$ where the two sequences α'_t and β'_t are the respective solutions of the forward and backward equations for the values (π', A', b') of the parameters.

1.5.6 Rainflow

The rainflow algorithm (also called Pagoda roof method) was introduced in T. Endo and M. Matsuishi in 1968 (Matsuichi and Endo [1968]). This method is often used to study material fatigue in association with the Miner's rule (Miner [1945]).

This algorithm extracts, from a loadtime profile/stress, the number of reversals. These last are the local minima and maxima where the load changes sign. Thus, it builds half-cycles. The rainflow algorithm relies to the flow of water down a series of pagoda roofs.

Let us consider a stress profile with peaks and valleys. The algorithm is detailed in Ariduru [2004]:

1. Extract a representative profile with peaks and valleys. This one begins with the point of the largest value and ends with a point with the same stress.
2. Consider this profile as rigid sheet. Turn the Figure clockwise 90° with the origin of time at the top. Thus we have a vertical time axis.

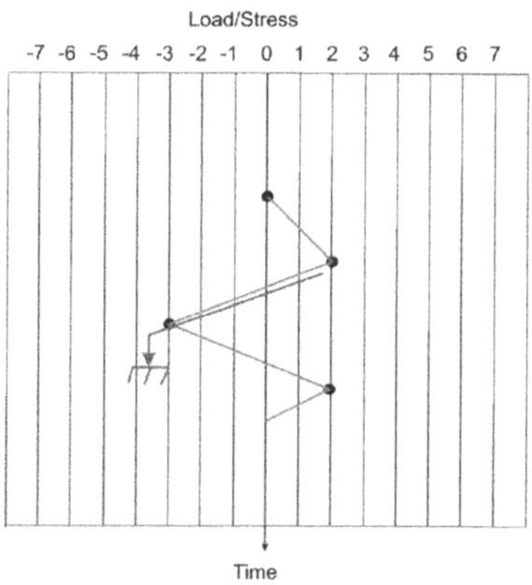

(a) Starting from a peak, the drop will stop if it meets an opposing peak larger than that of departure.

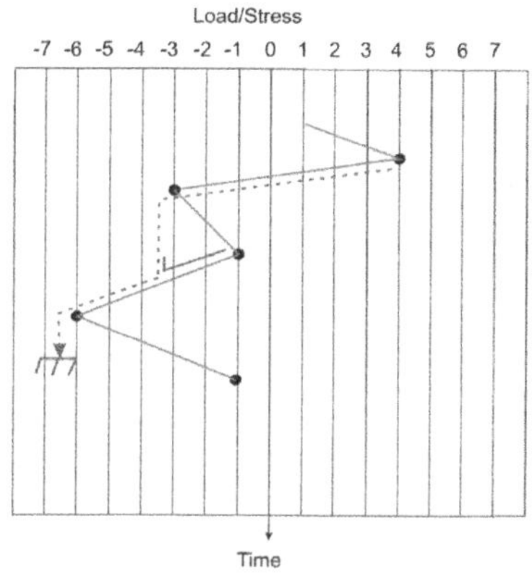

(b) Starting from a peak, the drop will stop if it crosses the path traversed by another drop, previously determined.

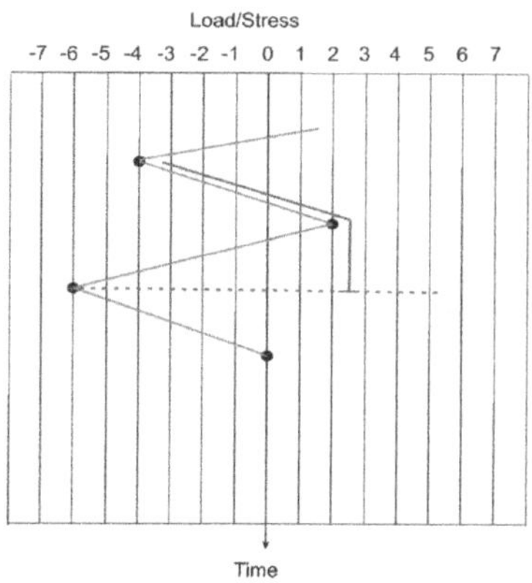

(c) Starting from a valley, the fall will stop if the drop meets a valley deeper than that of departure.

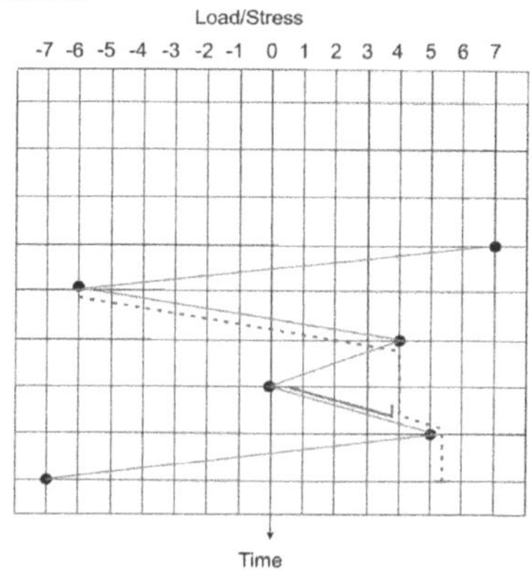

(d) Starting from a valley, the fall will stop if it crosses the path of a drop coming from a previous valley.

Figure 1.61 Different cases for the rainflow algorithm

3. Let us imagine a flow of rain starting at each successive extremum point.
4. If we start from a peak
   (a) the drop will stop if it meets an opposing peak larger than that of departure (Figure 1.61a).
   (b) the drop will stop if it crosses the path traversed by another drop, previously determined (Figure 1.61b).
   (c) the drop can fall on another roof and continue to slip according to rules 4.a and 4.b.
5. if we start from a valley
   (a) the fall will stop if the drop meets a valley deeper than that of departure (Figure 1.61c).

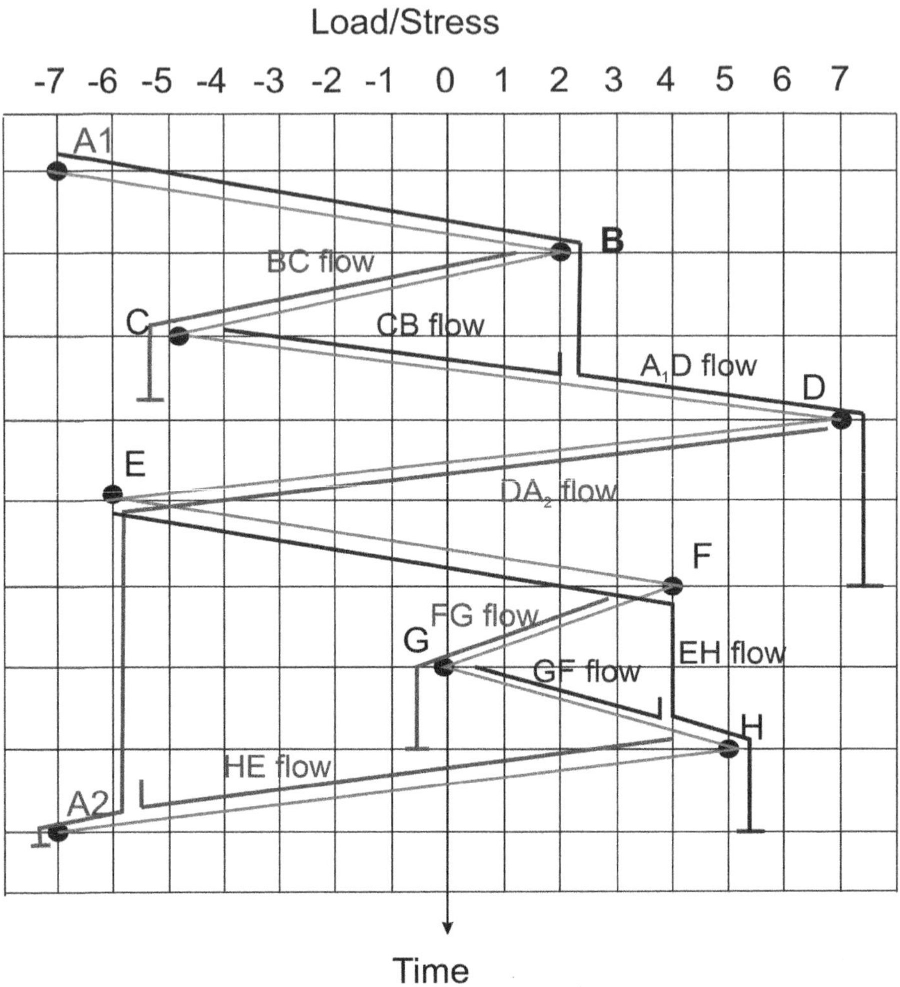

Figure 1.62 Example of the rainflow algorithm (Lee and Tjhung [2012])

   (b) the fall will stop if it crosses the path of a drop coming from a previous valley (Figure 1.61d).
   (c) the drop can fall on another roof and continue according to rules 5.a and 5.b.
6. Repeat from 3 until the end of the profile
7. Pair up half-cycles of identical magnitude (but opposite sense) to count the number of complete cycles. Typically, there are some residual half-cycles.

Let us consider the following example (Figure 1.62):
The different steps of the algorithm are described below:

1. Take point A1 as the starting point with the greater amplitude (largest peak). The end point is A2 with the same magnitude.
2. Identify the greater reversal A1 to D. We stop at the smallest valley D.
3. Identify the greater reversal D to A2. It starts from A and finishes at the end of the profile A2 (which is the same point as A1 as we describe a full cycle).
4. **Within the cycle A1–D**, identify the reversal BC that begins at B and ends at C (D is a larger maximum than B – case 4.a).
5. **Within the cycle A1–D**, identify the reversal CB that begins at C and ends at B (The flow meets a previous one at B – case 5.b).
6. **Within the cycle D–A2** (where A1 and A2 are merged), identify the reversal EH that starts at E and ends in the roof H.
7. **Within the cycle D–A2**, identify the reversal HE that begins at H and ends at E (The flow meets a previous one at E – case 5.b).
8. **Within the cycle D–A2**, identify the reversal FG that begins at F and ends at G (H is a larger maximum than F – case 5.a).
9. **Within the cycle D–A2**, identify the reversal GF that begins at G and ends at F (The flow meets a previous one at F – case 4.b).

Thus, we obtain the following reversals (half cycle) on Table 1.23 (Lee and Tjhung [2012]):

Table 1.23 Rainflow reversals

No of reversals	From point	To point	From load	To load	Range	Mean
1	A	D	−7	7	14	0
1	D	A	7	−7	14	0
1	B	C	2	−5	7	−1.5
1	C	B	−5	2	7	−1.5
1	E	H	−6	5	11	−0.5
1	H	E	5	−6	11	−0.5
1	F	G	4	0	4	2
1	G	F	0	4	4	2

Table 1.24 Rainflow cycles

No of reversals	From point	To point	From load	To load	Range	Mean
1	A	D	−7	7	14	0
1	B	C	2	−5	7	−1.5
1	E	H	−6	5	11	−0.5
1	F	G	4	0	4	2

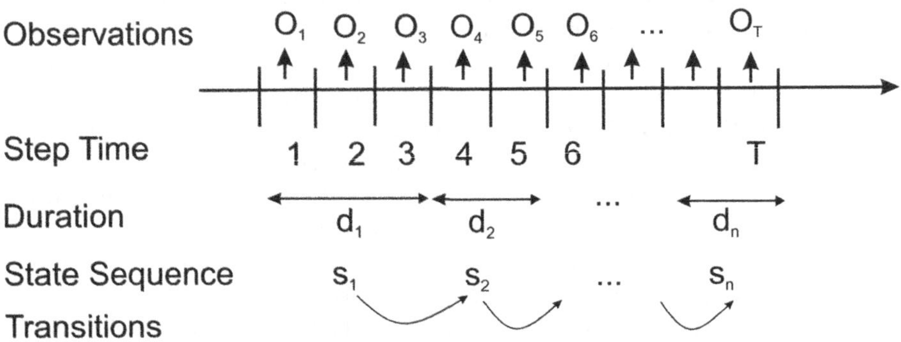

Figure 1.63 Hidden semi-Markov models

After pairing up these reversals, the cycles are given Table 1.24) (Lee and Tjhung [2012]):

1.5.6.1 Hidden Semi-Markov Models

Hidden semi-Markov models (Levy [1954]) are a generalization of hidden Markov model. In this case, each state can emit a sequence of observations. They allow a variable duration for each state and the number of observations emitted in a given state is determined by the duration spent in this state.

The detailed description of Hidden Semi Markov Model is beyond the scope of this book. The reader will find a very good description in Yu [2010].

References

Chris Aldrich and Lidia Auret. *Unsupervised Process Monitoring and Fault Diagnosis with Machine Learning Methods*. Springer Publishing Company, Incorporated, 2013. ISBN 1447151844.

A. Amrane, A. Larabi, and A. Aitouche. Fault detection and isolation based on nonlinear analytical redundancy applied to an induction machine. In *2017 6th International Conference on Systems and Control (ICSC)*, pages 255–260, 2017.

Jérôme Antoni. Cyclostationarity by examples. *Mechanical Systems and Signal Processing*, 23(4):987 – 1036, 2009. ISSN 0888-3270. doi: https://doi.org/10.1016/j.ymssp.2008.10.010. URL http://www.sciencedirect.com/science/article/pii/S0888327008002690.

Seçýl Ariduru. Fatigue life calculation by rainflow cycling method. Master thesis, Graduate school of natural and applied sciences of middle east technical university, 2004. URL http://www.vibrationdata.com/tutorials_alt/Ariduru_rainflow.pdf.

Georges Asch. *Les capteurs en instrumentation industrielle (8e édition)*. Technique et Ingénierie. Dunod, dunod edition, 2017. ISBN 2-10-076020-3.

Francois Auger and C. Doncarli. Quelques commentaires sur des représentations temps-fréquence proposées récemment. *Traitement du Signal*, 9:3–25, 01 1992.

Kevin Balanda and H. Macgillivray. Kurtosis: A critical review. *The Americal Statistician*, 42:111–119, 05 1988. doi: 10.1080/00031305.1988.10475539.

David Barber. *Bayesian Reasoning and Machine Learning*. Cambridge University Press, 2011.

G. Bartram and S. Mahadevan. Dynamic Bayesian networks for prognosis. *PHM 2013 - Proceedings of the Annual Conference of the Prognostics and Health Management Society 2013*, pages 167–184, January 2013.

Amy S. Beavers, John W. Lounsbury, Jennifer K. Richards, Schuyler W. Huck, Gary J. Skolits, and Shelley L. Esquivel. Practical considerations for using exploratory factor analysis in educational research. *PARE*, 18(6), 2013. URL https://scholarworks.umass.edu/pare/vol18/iss1/6/.

Mohamed Ben Marzoug. *Apport des cycles d'usage représentatifs synthétiques pour les études de vieillissement de batteries dans les nouveaux usages de mobilité*. PhD thesis, Université Claude Bernard Lyon 1, Villeurbanne, 2020.

H. Berriri, M. W. Naouar, and I. Slama-Belkhodja. Parity space approach for current sensor fault detection and isolation in electrical systems. In *Eighth International Multi-Conference on Systems, Signals Devices*, pages 1–7, 2011.

Christopher M. Bishop. *Pattern Recognition and Machine Learning*. Springer, New York, 2007. ISBN 978-0-387-31073-2.

Frédéric Bonnardot. *Comparaison entre les analyses angulaire et temporelle des signaux vibratoires de machines tournantes. Etude du concept de cyclostationnarité floue*. Theses, Institut National Polytechnique de Grenoble - INPG, December 2004. URL https://tel.archives-ouvertes.fr/tel-00103067.

Abdelhakim Boukar and Nacer Hamzaoui. Evaluation des indicateurs de surveillance par analyse vibratoire : Application aux engrenages et roulements. *International Journal of Innovation and Applied Studies*, 25(2):800–808, 2019.

ISSN 2028-9324. URL http://www.ijias.issr-journals.org/abstract.php?article=IJIAS-17-358-09.

Babak Nahid-Mobarakeh Breuneval R., Clerc G. and Badr Mansouri. Classification with automatic detection of unknown classes based on svm and fuzzy mbf: Application to motor diagnosis. *AIMS Electronics and Electrical Engineering*, 2(ElectronEng-02-03-059):59, 2018. ISSN 2578-1588. doi: http://dx.doi.org/10.3934/ElectrEng.2018.3.59.

T. Caliski and J. Harabasz. A dendrite method for cluster analysis. *Communications in Statistics*, 3(1):1–27, 1974. doi: 10.1080/03610927408827101. URL https://www.tandfonline.com/doi/abs/10.1080/03610927408827101.

C. Chatfield and Haipeng Xing. *The analysis of time series: An introduction, sixth edition*. Chapman & Hall/CRC texts in statistical science. CRC Press, 2019. ISBN 978-1-4987-9563-0.

E. Chow and A. Willsky. Analytical redundancy and the design of robust failure detection systems. *IEEE Transactions on Automatic Control*, 29(7):603–614, 1984.

G. Clerc and J.-C. Marques. Induction Machine Diagnosis Using Observers. In *Electrical machines diagnosis*, pages 93–129. Wiley edition, September 2011. ISBN 978-1-84821-263-3. URL https://doi.org/10.1002/9781118601662.ch4.

Guy Clerc. *Contribution à l'étude du contacteur statique autoprotégé et de sa stratégie de commande*. PhD thesis, Ecole Centrale de Lyon, Ecully, FRance, January 1989.

Leon Cohen. *Time-Frequency Analysis: Theory and Applications*. Prentice-Hall, Inc., USA, 1995. ISBN 0135945321.

M. Comanescu. A family of sensorless observers with speed estimate for rotor position estimation of im and pmsm drives. In *IECON 2012 - 38th Annual Conference on IEEE Industrial Electronics Society*, pages 3670–3675, 2012.

Antoine Cornuéjols, Laurent Miclet, and Vincent BARRA. *Apprentissage artificiel - 3e édition: Deep Learning, Concepts et algorithmes*. Eyrolles, May 2018. URL https://hal.archives-ouvertes.fr/hal-01829395.

Q. Cui, Z. Li, J. Yang, and B. Liang. Rolling bearing fault prognosis using recurrent neural network. In *2017 29th Chinese Control And Decision Conference (CCDC)*, pages 1196–1201, 2017.

D. L. Davies and D. W. Bouldin. A Cluster Separation Measure. *IEEE Transactions on Pattern Analysis and Machine Intelligence*, PAMI-1(2):224–227, 1979.

Bernard Desgraupes. Clustering indices, 2017. URL https://cran.r-project.org/web/packages/clusterCrit/vignettes/clusterCrit.pdf.

Gérard Dreyfus, J.M. Martinez, M. Samuelides, M. B. Gordon, Fouad Badran, Sylvie Thiria, and L. Herault. *Réseaux de neurones - Méthodologie et applications*. Eyrolles, January 2002. URL https://hal.archives-ouvertes.fr/hal-01125016.

Pierre Dumolard. Les réseaux de neurones. *Hermès*, 23(3):287–288, 1994. ISSN 0046-2497. URL https://www.persee.fr/doc/spgeo_0046-2497_1994_num_23_3_3327.

J. C. Dunn. Well-separated clusters and optimal fuzzy partitions. *Journal of Cybernetics*, 4(1):95–104, 1974. doi: 10.1080/01969727408546059. URL https://doi.org/10.1080/01969727408546059. Citation dunn1974 Publisher: Taylor & Francis tex.eprint: https://doi.org/10.1080/01969727408546059.

M.A. Efroymson. *Multiple Regression Analysis*. Mathematical Methods for Digital Computers. Wiley, ralston, a. and wilf, hs edition, 1960.

Samuel Eke. *Transformer condition assesment strategy : Outline solutions for aging transformers integrated management*. Theses, Université de Lyon, June 2018. URL https://tel.archives-ouvertes.fr/tel-02138989.

P. Ewert. Use of axial flux in the detection of electrical faults in induction motors. In *2017 International Symposium on Electrical Machines (SME)*, pages 1–6, 2017.

P. Flandrin. *Time-frequency/time-scale analysis*. ISSN. Elsevier Science, 1998. ISBN 978-0-08-054303-1. URL https://books.google.fr/books?id=FRwlO2rP5iIC.

E. Forgy. Cluster analysis of multivariate data: Efficiency versus interpretability of classification. *Biometrics*, 21(3):768–769, 1965.

Jerome H. Friedman. Regularized discriminant analysis. *Journal of the American Statistical Association*, 84(405):165–175, 1989.

J P Gauthier and I A K Kupka. Observability and observer for nonlinear systems. *SIAM J. Control Optim.,*, (32):975–994, 1994.

Pierre-Louis Gonzalez. L'analyse en composantes principales, 2020. URL http://maths.cnam.fr/IMG/pdf/A-C-P-.pdf.

Robert Gray, M. *Entropy and Information Theory*. Springer-Verlag, 2013. URL https://ee.stanford.edu/~gray/it.pdf.

L.I. Gudzenko. On periodic nonstationary processes. *Radio Engineering and Electronic Physics*, 4:220–224, 1959.

S. Hayashi, T. Asakura, and Sheng Zhang. Study of machine fault diagnosis system using neural networks. In *Proceedings of the 2002 International Joint Conference on Neural Networks. IJCNN'02 (Cat. No.02CH37290)*, volume 1, pages 956–961 vol.1, 2002. doi: 10.1109/IJCNN.2002.1005604.

Xiaofei He, Deng Cai, and Partha Niyogi. Laplacian Score for Feature Selection. In *Proceedings of the 18th International Conference on Neural Information Processing Systems*, NIPS'05, pages 507–514, Cambridge, MA, USA, 2005. MIT Press.

R.R. Herrera and D.S.L. Gac. *Initiation à l'analyse factorielle des données: Fondements mathématiques et interprétations, cours et exercices corrigés*. Ellipses, 2002. ISBN 9782729811198. URL https://books.google.fr/books?id=X_nYMgEACAAJ.

F. Hlawatsch and G. F. Boudreaux-Bartels. Linear and quadratic time–frequency signal representations. *IEEE Signal Processing Magazine*, 9(2):21–67, 1992. doi: 10.1109/79.127284.

H. Hotelling. Analysis of a complex of statistical variables into principal components. *Journal of Educational Psychology*, 24(6):417–441, 1933. doi: 10.1037/h0071325. URL https://doi.org/10.1037%2Fh0071325.

M. Hwang and J. H. Seinfeld. Observability of nonlinear systems. *Journal of Optimization Theory and Applications*, 10(2):67–77, August 1972. ISSN 1573-2878. doi: 10.1007/BF00934972. URL https://doi.org/10.1007/BF00934972.

Rob Hyndman. Time Series Data Library, 2005. URL http://www-personal.buseco.monash.edu.au/~hyndman/TSDL/.

TAHAR IFTENE and Habib Mahi. Approche de classification par rÉseaux de kohonen pour Établir des cartes d'occupation du sol de la rÉgion d'oran (algÉrie). *Teledetection*, 3:325–336, 01 2003.

J. J. A. Moors. The meaning of kurtosis: Darlington reexamined. *The American Statistician*, 40(4):283–284, 1986. doi: 10.1080/00031305.1986.10475415. URL https://www.tandfonline.com/doi/abs/10.1080/00031305.1986.10475415.

A. K. Jain, M. N. Murty, and P. J. Flynn. Data clustering: A review. *ACM Comput. Surv.*, 31(3):264–323, September 1999. ISSN 0360-0300.

Simon J. Julier and Jeffrey K. Uhlmann. New extension of the Kalman filter to nonlinear systems. In Ivan Kadar, editor, *Signal Processing, Sensor Fusion, and Target Recognition VI*, volume 3068 of *Society of Photo-Optical Instrumentation Engineers (SPIE) Conference Series*, pages 182–193, July 1997. doi: 10.1117/12.280797.

Henry F. Kaiser. The application of electronic computers to factor analysis. *Educational and Psychological Measurement*, 20(1):141–151, 1960. doi: 10.1177/001316446002000116. URL https://doi.org/10.1177/001316446002000116.

R. E. Kalman. A new approach to linear filtering and prediction problems. *Journal of Basic Engineering*, 82(1):35–45, March 1960. ISSN 00982202. URL http://dx.doi.org/10.1115/1.3662552.

R. Khelif, S. Malinowski, B. Chebel-Morello, and N. Zerhouni. Rul prediction based on a new similarity-instance based approach. In *2014 IEEE 23rd International Symposium on Industrial Electronics (ISIE)*, pages 2463–2468, 2014.

Thameur Kidar. *Diagnostic des défauts de fissures d'engrenages par l'analyse cyclostationnaire*. Theses, Université Jean Monnet - Saint-Etienne, March 2015. URL https://tel.archives-ouvertes.fr/tel-01538338.

T. Kohonen. *Self-Organization and Associative Memory: 3rd Edition*. Springer-Verlag, Berlin, Heidelberg, 1989. ISBN 0387513876.

J. Korbicz. Neural networks and their application in fault detection and diagnosis. In *IFAC Proceedings*, volume 30, pages 367–372, 1997. doi: 10.1016/S1474-6670(17)42429-3.

A. Lalami and R. Wamkeue. Parity-space approach for time-domain synchronous generator diagnosis. In *2012 25th IEEE Canadian Conference on Electrical and Computer Engineering (CCECE)*, pages 1–4, 2012.

Pat Langley, Wayne Iba, and, and Kevin Thompson. An analysis of bayesian classifiers. In *Proceedings of the Tenth National Conference on Artificial Intelligence*, page 223–228. AAAI Press, 1992. ISBN 0262510634.

Abdesselam Lebaroud. *Diagnostic des défauts des machines asynchrones : approche signal et système*. PhD thesis, 2007. URL http://www.theses.fr/2007LYO10288. Thèse de doctorat dirigée par Clerc, Guy Physique Lyon 1 2007.

V.L. Lebedev. On random processes having nonstationarity of periodic character. *Nauchnye Doklady Vysshchei Shchkoly. Seria Radiotekhnika i Elektronika*, 2: 32–34, 1959.

Yung-Li Lee and Tana Tjhung. Chapter 3 - rainflow cycle counting techniques. In Yung-Li Lee, Mark E. Barkey, and Hong-Tae Kang, editors, *Metal Fatigue Analysis Handbook*, pages 89 – 114. Butterworth-Heinemann, Boston, 2012. ISBN 978-0-12-385204-5. doi: https://doi-org-s.bibliopam.ec-lyon.fr/10.1016/B978-0-12-385204-5.00003-3. URL https://www-sciencedirect-com.bibliopam.ec-lyon.fr/science/article/pii/B9780123852045000033.

P. Levy. Syst èmes semi-Markoviens à au plus infinité d énombrable d'états possibles. volume 2, page 294, 1954.

Y. Liu, Z. Q. Zhu, and D. Howe. Instantaneous torque estimation in sensorless direct-torque-controlled brushless dc motors. *IEEE Transactions on Industry Applications*, 42(5):1275–1283, 2006.

P.J. Loughlin and B. Tacer. Instantaneous frequency and the conditional mean frequency of a signal. *Signal Processing*, 60(2):153 – 162, 1997. ISSN 0165-1684.

Elizabeth Ann Maharaj, Pierpaolo D'Urso, and Jorge Caido. *Time series clustering and classification*. Computer science and data analysis. CRC Press, Australia, 1st edition, 2019. ISBN 978-1-4987-7321-8. doi: 10.1201/9780429058264.

Masoud Malekzadeh, Mustafa Gul, Il-Bum Kwon, and Necati Catbas. An integrated approach for structural health monitoring using an in-house built fiber optic system and non-parametric data analysis. *Smart Structures and Systems*, 14:917–942, 11 2014. doi: 10.12989/sss.2014.14.5.917.

Francesca Mangili. *Development of advanced computational methods for prognostics and health management in energy components and systems*. PhD thesis, Politecnico di milano, Milan (Italie), March 2013. URL https://www.politesi.polimi.it/handle/10589/74461.

MATLAB. R2020a, 2020. URL https://www.mathworks.com/.

M. Matsuichi and T. Endo. Fatigue of metals subjected to varying stress. *Japan Society of Mechanical Engineering*, 1968.

K. Medjaher, D. A. Tobon-Mejia, and N. Zerhouni. Remaining useful life estimation of critical components with application to bearings. *IEEE Transactions on Reliability*, 61(2):292–302, 2012.

M.A. Miner. Cumulative Damage in Fatigue. *Journal of Applied Mechanics*, 12(3): 159–164, 1945.

Sidharth Mishra, Uttam Sarkar, Subhash Taraphder, Sanjoy Datta, Devi Swain, Reshma Saikhom, Sasmita Panda, and Menalsh Laishram. Principal Component Analysis. *International Journal of Livestock Research*, page 1, 2017. doi: 10.5455/ijlr.20170415115235.

A. Mohammed, J. I. Melecio, and S. Djurovi. Stator winding fault thermal signature monitoring and analysis by in situ fbg sensors. *IEEE Transactions on Industrial Electronics*, 66(10):8082–8092, 2019.

Sing Kiong Nguang, Ping Zhang, and Steven Ding. Parity relationbased fault estimationfor non linear systems: a lmi approach. *IFAC Proceedings Volumes*, 39 (13):366 – 371, 2006. ISSN 1474-6670. doi: https://doi-org-s.bibliopam.ec-lyon.fr/10.3182/20060829-4-CN-2909.00060. URL https://www-sciencedirect-com.bibliopam.ec-lyon.fr/science/article/pii/S1474667015344426. 6th IFAC Symposium on Fault Detection, Supervision and Safety of Technical Processes.

Olivier Ondel. *Diagnosis by Pattern Recognition: application on a set inverter – induction machine*. Theses, Ecole Centrale de Lyon, October 2006. URL https://tel.archives-ouvertes.fr/tel-00113102.

Karl Pearson. On lines and planes of closest fit to systems of points in space. *The London, Edinburgh, and Dublin Philosophical Magazine and Journal of Science*, 2(11):559–572, 1901. doi: 10.1080/14786440109462720. URL https://doi.org/10.1080/14786440109462720..

Karl Pearson and Francis Galton. Note on regression and inheritance in the case of two parents. *Proceedings of the Royal Society of London*, 58:240–242, 1895. doi: https://doi.org/10.1098/rspl.1895.0041.

Michael Pecht and Rubyca Jaai. A prognostics and health management roadmap for information and electronics-rich systems. *Microelectronics Reliability*, 50(3): 317 – 323, 2010. ISSN 0026-2714. doi: https://doi-org-s.bibliopam.ec-lyon.fr/10.1016/j.microrel.2010.01.006. URL https://www-sciencedirect-com.bibliopam.ec-lyon.fr/science/article/pii/S0026271410000181.

Franz Pernkopf. Bayesian network classifiers versus selective k-nn classifier. *Pattern Recognition*, 38(1):1 – 10, 2005. ISSN 0031-3203.

D. Alexander Poularikas. *The Transforms and Applications Handbook*. CRC, 02 2000. ISBN ISBN-10: 0849385954 ISBN-13: 978-0849385957.

R Core Team. R: A language and environment for statistical computing. manual, Vienna, Austria, 2020. URL https://www.R-project.org/. tex.organization: R Foundation for Statistical Computing.

C. Radhakrishna Rao. The use and interpretation of principal component analysis in applied research. *Sankhy: The Indian Journal of Statistics, Series A (1961-2002)*, 26(4):329–358, 1964. ISSN 0581572X. URL http://www.jstor.org/stable/25049339. Publisher: Springer.

S. Raschka and V. Mirjalili. *Python machine learning*. Packt Publishing, 2017. ISBN 978-1-78712-602-2. URL https://books.google.fr/books?id=_plGDwAAQBAJ.

M. Rayyam, M. Zazi, and Y. Hajji. Detection of broken bars in induction motor using the extended kalman filter (ekf). In *2015 Third World Conference on Complex Systems (WCCS)*, pages 1–5, 2015.

Paulo Roriz, Susana Silva, Orlando Frazão, and Susana Novais. Optical Fiber Temperature Sensors and Their Biomedical Applications. *Sensors (Basel, Switzerland)*, 20(7):2113, April 2020. ISSN 1424-8220. doi: 10.3390/s20072113. URL https://pubmed.ncbi.nlm.nih.gov/32283622. Publisher: MDPI.

Peter J. Rousseeuw. Silhouettes: A graphical aid to the interpretation and validation of cluster analysis. *Journal of Computational and Applied Mathematics*, 20:53 – 65, 1987. ISSN 0377-0427. doi: https://doi.org/10.1016/0377-0427(87)90125-7. URL http://www.sciencedirect.com/science/article/pii/0377042787901257.

Rumelhart, D., Hinton, and R. G., Williams. Learning representations by back-propagating errors. *Nature*, 323:533 – 536, 10 1986. doi: https://doi.org/10.1038/323533a0.

S. Russell, P. Norvig, and F. Popineau. *Intelligence artificielle: Avec plus de 500 exercices*. Pearson education. Pearson, 2010. ISBN 978-2-7440-7455-4. URL https://books.google.fr/books?id=DWTIFWSGxJMC.

Sachin Kumar, M. Torres, Y. C. Chan, and M. Pecht. A hybrid prognostics methodology for electronic products. In *2008 IEEE International Joint Conference on Neural Networks (IEEE World Congress on Computational Intelligence)*, pages 3479–3485, 2008.

S. A. Saleh and M. A. Rahman. Modeling and protection of and three-phase power transformer using wavelet packet transform. *IEEE Transactions on Power Delivery*, 20(2):1273–1282, Apr. 2005.

R. Satishkumar and V. Sugumaran. Estimation of remaining useful life of bearings based on support vector regression. *Indian journal of science and technology*, 9, 2016.

Claude Elwood Shannon. A Mathematical Theory of Communication. *The Bell System Technical Journal*, 27:379–423, October 1948.

A. Soualhi, M. Makdessi, R. German, F. R. Echeverría, H. Razik, A. Sari, P. Venet, and G. Clerc. Heath monitoring of capacitors and supercapacitors using the neo-fuzzy neural approach. *IEEE Transactions on Industrial Informatics*, 14(1): 24–34, 2018.

Abdenour Soualhi, Kamal Medjaher, Guy Celrc, and Hubert Razik. Prediction of bearing failures by the analysis of the time series. *Mechanical Systems and Signal Processing*, 139:106607, 2020. ISSN 0888-3270. doi: https://doi-org-s.bibliopam.ec-lyon.fr/10.1016/j.ymssp.2019.106607. URL https://www-sciencedirect-com.bibliopam.ec-lyon.fr/science/article/pii/S0888327019308283.

C. Spearman. The proof and measurement of association between two things. *The American Journal of Psychology*, 15(1):72–101, 1904. doi: 10.2307/1412159. URL http://www.jstor.org/stable/1412159.

Claasen TA and Wolfgang Mecklenbräuker. The wigner distribution - a tool for time–frequency signal analysis. iii. relations with other time–frequency signal transformations. *Philips Jl Research*, 35:372–389, 01 1980.

T. Tao and W. Zhao. A support vector regression-based prognostic method for li-ion batteries working in variable operating states. In *2016 Prognostics and System Health Management Conference (PHM-Chengdu)*, pages 1–5, 2016.

C. Torrence and G. P. Compo. A practival guide to wavelet analysis. *Bulletin of the American Metereological Society*, 79(1):61–78, Jan. 1998.

Stéphane Tufféry. *Data mining et statistique décisionnelle*. Technip, Paris, France, 4ème édition edition, 2012. ISBN 2-7108-1180-4.

C. H. van der Broeck, T. Polom, R. D. Lorenz, and R. W. De Doncker. Thermal monitoring of power electronic modules using device self-sensing. In *2018 IEEE Energy Conversion Congress and Exposition (ECCE)*, pages 4699–4706, 2018.

Vladimir N. Vapnik. *The Nature of Statistical Learning Theory*. Springer-Verlag, Berlin, Heidelberg, 1995. ISBN 0387945598.

J Ville. Théorie et applications de la notation de signal analytique. cables et transmission. In *Theory and applications of the notion of complex signal, Tech. Rept. T-92, The RAND Corporation*, volume 2A, pages 61–74, 1948.

E. A. Wan and R. Van Der Merwe. The unscented kalman filter for nonlinear estimation. In *Proceedings of the IEEE 2000 Adaptive Systems for Signal Processing, Communications, and Control Symposium (Cat. No.00EX373)*, pages 153–158, 2000. doi: 10.1109/ASSPCC.2000.882463.

Wang Chenchen and Li Yongdong. A novel speed sensorless field-oriented control scheme of im using extended kalman filter with load torque observer. In *2008 Twenty-Third Annual IEEE Applied Power Electronics Conference and Exposition*, pages 1796–1802, 2008.

Richard Weber. Time series, 2001. URL http://www.statslab.cam.ac.uk/~rrw1/timeseries/t.pdf..

John G. Webster. *Measurement, Instrumentation, and Sensors Handbook*. CRC Press, 1998. ISBN 0-8493-8347-1.

E. Wigner. On the Quantum Correction For Thermodynamic Equilibrium. *Physical Review*, 40(5):749–759, jun 1932. doi: 10.1103/PhysRev.40.749.

Wikipedia contributors. Microphone — Wikipedia, the free encyclopedia, 2020a. URL https://en.wikipedia.org/w/index.php?title=Microphone&oldid=973621749. [Online; accessed 28-August-2020].

Wikipedia contributors. Time series — Wikipedia, the free encyclopedia. https://en.wikipedia.org/w/index.php?title=Time_series&oldid=982786780, 2020b. [Online; accessed 13-October-2020].

Wikipédia. Silhouette (clustering) — wikipédia, l'encyclopédie libre, 2019. URL http://fr.wikipedia.org/w/index.php?title=Silhouette_(clustering)&oldid=162146683. [En ligne; Page disponible le 27-août-2019].

B. Wu, M. Wang, B. Wu, Y. Luo, and C. Feng. Cyclostationarity analysis and diagnosis method of bearing faults. In *2009 IEEE International Conference on Automation and Logistics*, pages 576–581, 2009.

X. Wu, J. Kumar, V. and Ross Quinlan, J. Ghosh, Q. Yang, H. Motoda, G. J. McLachlan, A. Ng, B. Liu, P.S. Yu, Z Zhou, M. Steinbach, D.J. Hand, and D. Steinberg. Top 10 algorithms in data mining. *Knowl Inf Syst*, 14:1–37, 01 2008. doi: https://doi.org/10.1007/s10115-007-0114-2.

K. Xue, Z. Wu, H. Li, and G. Yang. Research on identification method of aero-engine bearing fault using acoustic emission technique based on wavelet packet and rough set. In *2017 IEEE 2nd Advanced Information Technology, Electronic and Automation Control Conference (IAEAC)*, pages 1499–1503, 2017.

Yun Yang. *Temporal data mining via unsupervised ensemble learning*. Elsevier, July 2018. ISBN 978-0-12-811654-8.

Shun-Zheng Yu. Hidden semi-markov models. *Artificial Intelligence*, 174(2): 215–243, 2010. ISSN 0004-3702. doi: https://doi-org-s.bibliopam.ec-lyon.fr/10.1016/j.artint.2009.11.011. URL https://www-sciencedirect-com-s.bibliopam.ec-lyon.fr/science/article/pii/S0004370209001416. Special Review Issue. Special Review Issue.

Z. Zheng, Y. Li, and M. Fadel. Sensorless control of pmsm based on extended kalman filter. In *2007 European Conference on Power Electronics and Applications*, pages 1–8, 2007.

M. Zhong, S. X. Ding, Q. Han, and Q. Ding. Parity space-based fault estimation for linear discrete time-varying systems. *IEEE Transactions on Automatic Control*, 55(7):1726–1731, 2010.

Zhi-Jie Zhou, Chang-Hua Hu, Xiao-Xia Han, Hua-Feng He, Xiao-Dong Ling, and Bang-Cheng Zhang. A model for online failure prognosis subject to two failure modes based on belief rule base and semi-quantitative information. *Knowledge-Based Systems*, 70:221 – 230, 2014. ISSN 0950-7051. doi: https://doi.org/10.1016/j.knosys.2014.06.026. URL http://www.sciencedirect.com/science/article/pii/S0950705114002470.

H. Zhu, D. Kung, M. Cowell, and S. Cherukupalli. Acoustic monitoring of stator winding delaminations during thermal cycling testing. *IEEE Transactions on Dielectrics and Electrical Insulation*, 17(5):1405–1410, 2010.

D. Zurita, M. Delgado, J. A. Ortega, and L. Romeral. Intelligent sensor based on acoustic emission analysis applied to gear fault diagnosis. In *2013 9th IEEE International Symposium on Diagnostics for Electric Machines, Power Electronics and Drives (SDEMPED)*, pages 169–176, 2013.

2

Applications and Specifics

2.1 General Presentation of Motor Drives

Electric drives are widely used all over the world and occupy a growing place in industry, land transport and aeronautics. They offer a better power-to-weight ratio and easier maintenance than hydraulic or thermal drives. Therefore, their increasing complexity and their operating conditions lead to breakdowns, which must be detected as soon as possible in the incipient fault, and quickly to avoid production shutdowns or safety issues and, if possible, forecast to optimize maintenance.

They are composed of (see Figure 2.1):

(a) a DC source or an AC source and an AC rectifier;
(b) a storage capacitor with a current source or a storage inductor with a voltage one;
(c) an inverter;
(d) an AC motor such as an induction motor/generator, a synchronous machine (wound field or with permanent magnet).

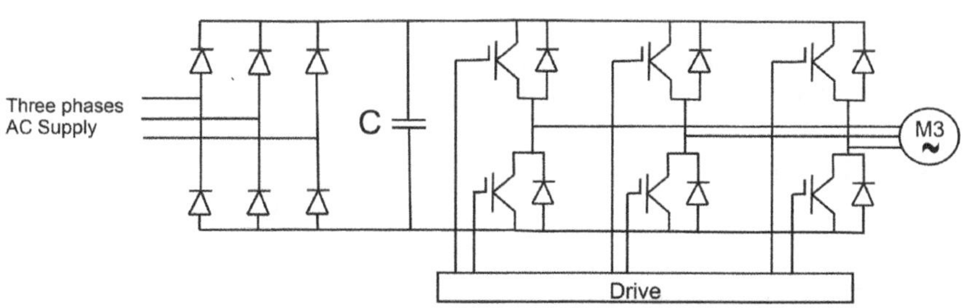

Figure 2.1 Motor Drive overview.

Fault Diagnosis, Prognosis, and Reliability for Electrical Machines and Drives. First Edition.
Elias G. Strangas, Guy Clerc, Hubert Razik, and Abdenour Soualhi.

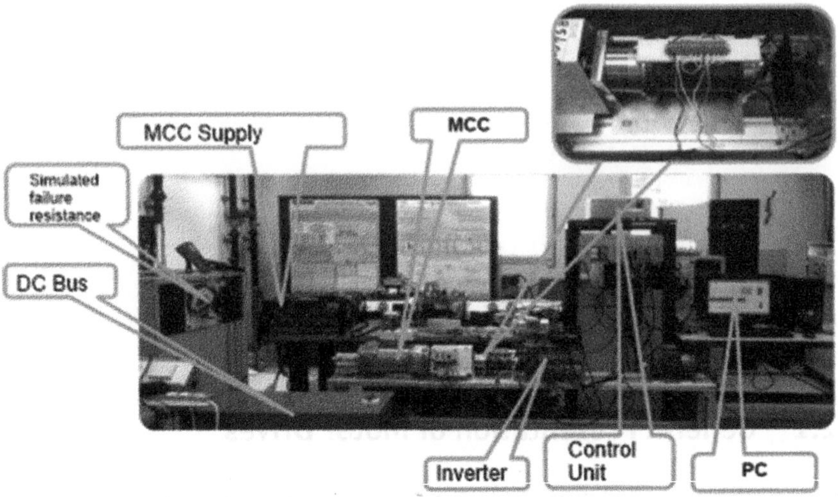

Figure 2.2 Diagnosis Test bench for synchronous machine in Green Laboratory (University of Lorraine).

All these components are subject to various electrical and thermal stresses during normal or accidental operation. These can lead to a reduction in performance or even an unplanned shutdown of the drive. Even more different failures interact between these different sub-systems and could lead to a total destruction of the full system.

Thus, it is important to detect early any failure and to prognose the loss of function. For that, different techniques presented in the previous paragraph have been applied in many industrial projects and research work (see Figure 2.2).

This section deals with the diagnosis and prognosis of:

(a) **AC electrical machines** such as induction motors, synchronous ones. The physics will be presented, the different type of failures (short-circuit, demagnetization, gear failures etc.),
(b) **the storage capacitors** with their different types and technologies, the applied health monitoring, by tracking the serial resistance or their capacitance and so on,
(c) **the inverter**: the monitoring of the static components (IGBT, Mosfet, diode) as well as that of the converter will be discussed.

2.2 Electrical Machines

The electrical machines, operating as motors or generators, are the main parts of electrical drive systems. There is a plethora of variations and options of both

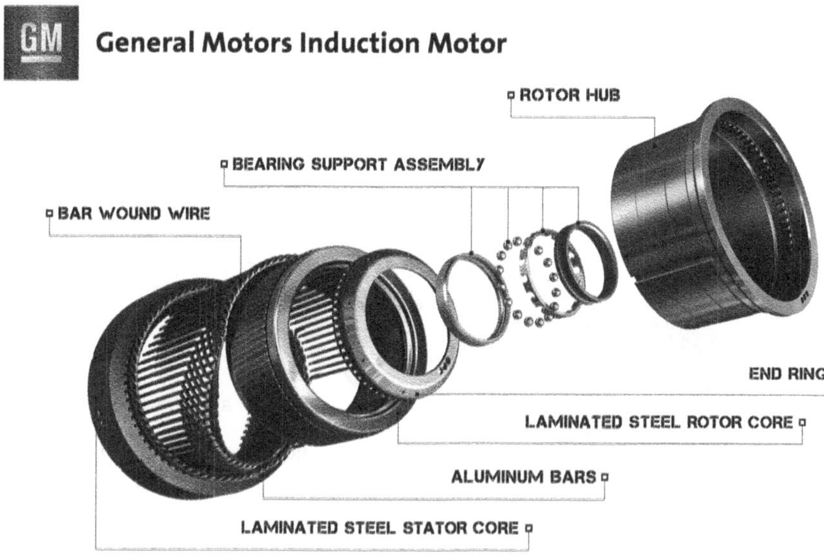

Figure 2.3 Chevrolet Spark Electric Vehicle Electric Induction Motor, GMS.

types, adapting to application requirements, and to the availability and cost of materials and manufacturing.

We will concentrate on two general types of machines, synchronous and asynchronous or induction ones. For the synchronous machines, we will study those with permanent magnets in the rotor, surface, or interior ones, as well as those that have an excitation coil. They are considered possible alternatives to PMAC machines in medium-sized drives. For asynchronous machines, we will study both cage and wound field rotors.

The basic parts of these machines are shown in the drawings below in applications for electric vehicles, 2.3 and 2.4.

In the induction machine, Figure 2.3, the parts that are clearly identifiable are the laminated stator with a three-phase winding and the rotor with aluminum bars and end rings. In the Permanent magnet machine shown in Figure 2.4, a three-phase winding is also installed in a laminated stator, while the rotor has a double layer of magnets.

Parts that are not seen clearly in these figures are the insulation and bearings. They require great attention in design and are susceptible to failures of a large percentage of electrical drives. Thus, they are addressed in detail later here.

Before we embark on discussing faults, their detection, diagnosis, and prognosis, we will cover quickly the basics of the operation of these machines types.

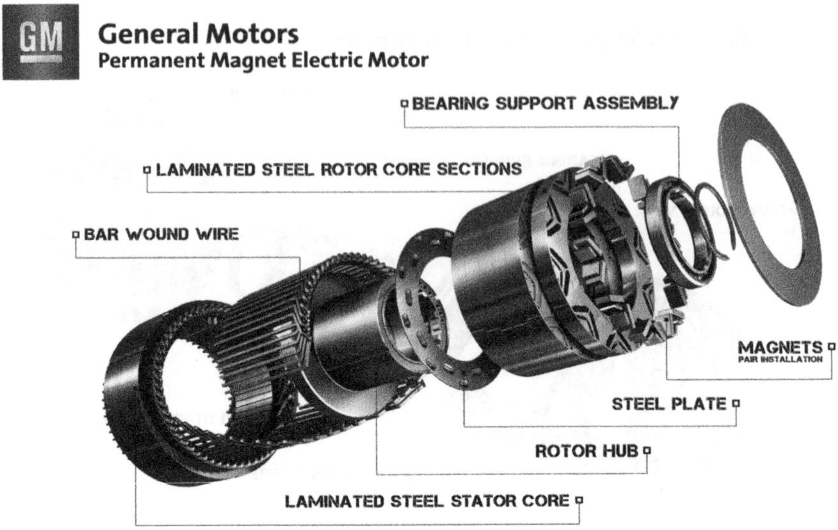

Figure 2.4 Chevrolet Spark Electric Vehicle Electric Permanent Magnet AC Motor, from GM, GMS.

2.2.1 Basics

In electrical machines, the primary variables are currents and magnetic flux. The currents flow in windings, and flux is established in magnetic steel, magnets, and air. Changing flux induces voltages and eddy currents, but also currents and changing flux cause losses and heat, which increase temperature that has detrimental effects on materials.

Flux is related to currents, according to Ampère's law.

$$\nabla \times \mathbf{H} = \mathbf{J} \ \ \mathbf{B} = \mu \mathbf{H}, \quad \text{or in integral form} \oint \mathbf{H} \cdot d\mathbf{l} = \int_s \mathbf{J} d\mathbf{S} \tag{2.1}$$

where **H** is the magnetic field intensity, **J** is the current density and **B** the flux density; μ is the material permeability. As flux changes, it induces an electric field, $\mathcal{E}$:

$$\nabla \times \mathcal{E} = -\frac{\partial B}{\partial t} \quad \text{or in integral form} \oint_{\partial S} \mathcal{E} \cdot d\mathbf{l} = -\frac{\partial \mathbf{B}}{\partial t}. \tag{2.2}$$

For a single turn coil, the flux through it, $\mathbf{\Phi}$, is defined as the integral of the flux density through it, and the resulting induced voltage in it is:

$$\mathbf{\Phi}_B = \int_A \mathbf{B}(t) \cdot d\mathbf{A} \tag{2.3}$$

$$\mathcal{E} = -\frac{d\Phi_B}{dt} \tag{2.4}$$

Induced electric fields, and hence voltages induced in windings are related to changing flux linkages λ in them:

$$\lambda = \sum_i \Phi_i \tag{2.5}$$

$$u = \frac{d\lambda}{dt} \tag{2.6}$$

Forces or torques are developed through the magnetomotive force $(N \cdot I)$ related to the magnetic field. They are the outputs of motors (or inputs to generators). They can be directly or indirectly determined using the Maxwell stress tensor, or derivative methods, e.g. co-energy. Besides their desired or basic components, forces or torques often include undesired torque pulsations, which can be exacerbated by faults.

2.2.2 Magnetic Steel and Magnets

Before discussing permanent magnets, let us present magnetic hysteresis in steel and its effects. We refer the interested reader to Bertotti (1988) for theory and details. For here, let us consider that the relation between the current and flux, or better, between the field intensity $\mathbf{H}$ and the resulting flux density, $\mathbf{B}$, in a homogeneous steel medium.

Let us take a simple homogeneous magnetic circuit, to which we place a winding in order to establish a uniform magnetic field in it, and assume that we can measure or calculate the field intensity H and flux density B. We start with a null field, i.e. $H = 0$, $B = 0$. In the idealized case, in the absence of hysteresis and eddy currents, the flux density B increases with field intensity H, close to linearly for low values of H, and then at a progressively lower rate. This nonlinear saturation characteristic affects the operation of the machine that uses these magnets, but as we will see, can also be used to evaluate the its health.

After reaching saturation, the current in our experiment, and hence H, is decreased. B decreases monotonically with it, but again nonlinearly; it does not follow the previous path, but rather, when H reaches zero, B reaches a non-zero value. This is the value of remanence, B_r. Further decrease of the current below zero and with H in the opposite direction, there still remains a positive B, until H reaches $-H_c$, the coercive field strength, which results in zero flux density. A complete cycle provides the loop that defines the hysteresis phenomenon. The integral of the surface $w = \int HdB$ gives the loss density in the circuit due to hysteresis. Of interest here are also the 'minor loops', which represent the

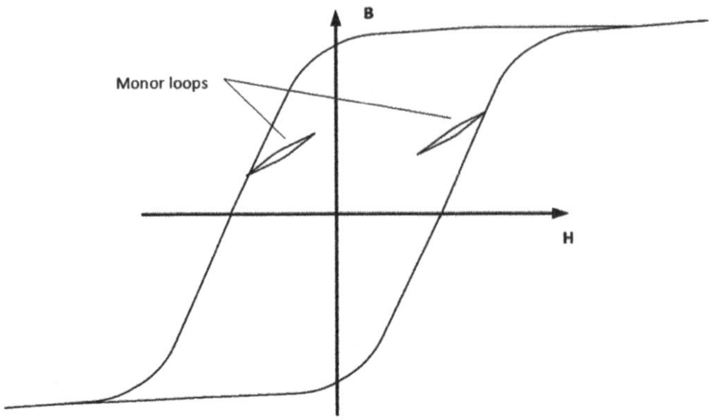

Figure 2.5 Hysteresis curve with minor loops.

change of the curve as the direction of the change of H is again reversed, but only for a short interval Figure 2.5.

In the second quadrant of the full B-H curve, the interest is the case of material categorized as permanent magnet. In this material, as the value of H becomes more negative and then increases again, a minor loop is established. In permanent magnets, the material has a well-established minor loop, shown in Figure 2.5. In Figure 2.6, load lines are shown, describing the presence of airgaps or coils currying current: **L1** (high permeability load), **L2** (high permeability load and demagnetizing current) and **L3** (low permeability load) result in operation at points P1, P3, and P2.

If the temperature of the magnet is increased, both B_r and H_c decrease, and the $B - H$ line, BH1, becomes BH2 as shown in Figure 2.7. The operating point with demagnetization of low permeance moves below the knee of the curve. Increasing the permeance by decreasing the air gap and/or removing the demagnetizating current, establishes a new recoil line, R2, and a new operating point, P3, at lower flux density. Decreasing the temperature does not always restore operation to the previous recoil line and operating point. When the operating point has moved below the knee of the curve, at a point on the new recoil line R1, this will result in a permanent decrease of flux density. This can happen accidentally and will lead to less flux in permanent magnet machines and decreased performance.

This demagnetization causes decreased induced voltage (electromotive force) and reduced torque. As a result, the controller increases the current, which in turn results in higher losses.

Magnet characteristics differ depending on their composition and production method.

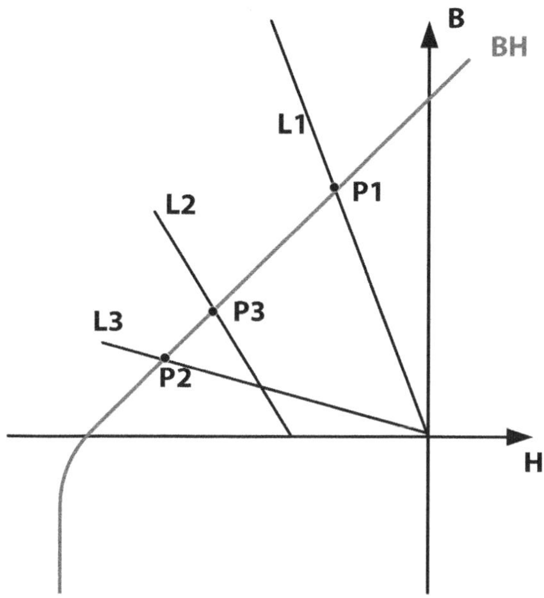

Figure 2.6 Demagnetization characteristics of rare earth magnets.

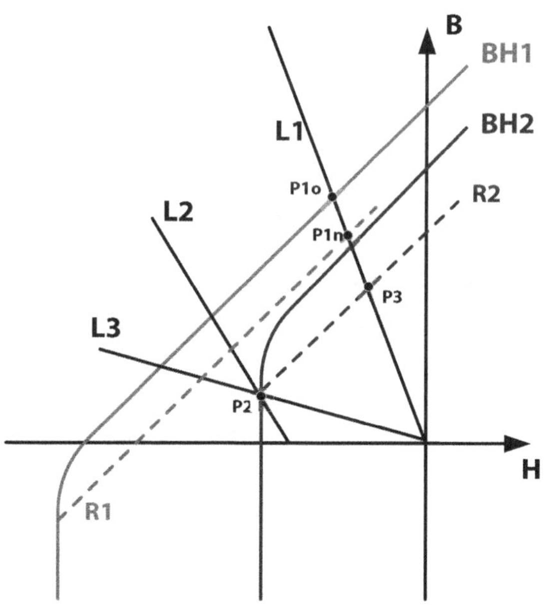

Figure 2.7 Demagnetization characteristics of NdFeB (Neodymium Iron Boron).

Rare earth magnets, made of NdFeB and SmCo, have the characteristics shown in Tables 2.1 and 2.2. The Nd-based magnetic materials are more susceptible to higher temperature than SmCo ones. Ferrite magnets have significant lower energy; they are sensitive to demagnetization but impervious to higher temperatures.

Table 2.1 Neodymium magnets characteristics: Arnold Magnetics.

Properties:	Br	HcB	HcJ	BHmax	Rev. Temp. Coef.		Temp.
Grade	Typical mT	min kA/m	min kA/m	Typical kJ/m3	Br %/C	HcJ %/ C	max C
N30H	1105	796	1353	235	−0.12	−0.605	120
N33H	1175	836	1353	267	−0.12	−0.605	120
N35H	1210	868	1353	283	−0.12	−0.605	120
N38H	1260	899	1353	307	−0.12	−0.605	120
N40H	1285	923	1353	322	−0.12	−0.605	120
N42H	1300	955	1353	330	−0.12	−0.605	120
N45H	1350	971	1353	354	−0.12	−0.605	120
N48H	1390	1011	1353	378	−0.12	−0.605	120
N50H	1415	1035	1274	390	−0.12	−0.605	120
N30SH	1125	811	1592	243	−0.12	−0.535	150
N33SH	1175	844	1592	267	−0.12	−0.535	150
N35SH	1210	876	1592	283	−0.12	−0.535	150
N38SH	1260	907	1592	307	−0.12	−0.535	150
N40SH	1285	939	1592	322	−0.12	−0.535	150
N42SH	1310	955	1592	330	−0.12	−0.535	150
N45SH	1350	979	1592	354	−0.12	−0.535	150
N48SH	1390	995	1512	374	−0.12	−0.535	150
N28UH	1075	764	1990	227	−0.12	−0.465	180
N30UH	1125	812	1990	243	−0.12	−0.465	180
N33UH	1175	852	1990	267	−0.12	−0.465	180
N35UH	1210	860	1990	283	−0.12	−0.465	180
N38UH	1260	876	1990	307	−0.12	−0.465	180
N40UH	1285	915	1990	318	−0.12	−0.465	180
N42UH	1310	955	1990	330	−0.12	−0.465	180

(Continued)

Table 2.1 *(Continued)*

Properties: Grade	Br Typical mT	HcB min kA/m	HcJ min kA/m	BHmax Typical kJ/m3	Rev. Temp. Coef. Br %/C	Rev. Temp. Coef. HcJ %/ C	Temp. max C
N45UH	1350	955	1910	358	−0.12	−0.465	180
N28EH	1085	780	2388	227	−0.12	−0.42	200
N30EH	1125	812	2388	243	−0.12	−0.42	200
N33EH	1165	820	2388	267	−0.12	−0.42	200
N35EH	1200	836	2388	279	−0.12	−0.42	200
N38EH	1235	899	2388	303	−0.12	−0.42	200
N40EH	1270	915	2388	314	−0.12	−0.42	200
N42EH	1310	971	2308	326	−0.12	−0.42	200
N28AH	1075	780	2706	223	−0.12	−0.393	220
N30AH	1120	812	2706	239	−0.12	−0.393	220
N33AH	1140	812	2706	231	−0.12	−0.393	220
N35AH	1195	883	2706	275	−0.12	−0.393	220
N38AH	1240	923	2626	299	−0.12	−0.393	220

Changing flux density in steel and magnets introduces eddy currents, resulting in losses and increased temperatures. These losses are proportional to the square of the flux and of the frequency:

$$w_{eddy} \propto \sum_{n=1}^{n=\infty} f_n^2 B_n^2. \tag{2.7}$$

Eddy losses can produce localized heating, decrease efficiency and, as a result, possibly magnets demagnetization.

2.2.3 Windings and Insulation

Windings embedded in stator slots are made of copper, and in the case of squirrel cage induction machines, the rotor windings are possibly made of aluminum. Stator windings are coated with insulating material, typically organic, and coils are insulated from each other and from the frame by insulating sheets.

At low temperatures, polyurethane insulation is used on the wires, up to 180^oC, polyester for up to 200^oC, and polyamide up to 250^oC.

Table 2.2 Samarium magnets characteristics: Arnold Magnetics.

Grade	BH max	Br	Hcb	Hcj	Rev. Temp. Coef. of Induction	Max Operating Temp
	kJ/m 3	T	k A/m	k A/m	%/K	C
18	143	0.87	650	2400	−0.045	250
20	160	.90	700	2400	−0.045	250
22	175	.94	730	2400	−0.045	250
25	200	1	775	2400	−0.050	250
24HE	195	1.02	765	2000	−0.035	350
26	205	1.04	765	2000	−0.035	350
26HE	215	1.07	800	2000	−0.035	350
28	225	1.1	800	2000	−0.035	350
28HE	225	1.1	805	2000	−0.035	350
30	230	1.12	820	1600	−0.035	250
30HE	230	1.12	830	2000	−0.035	350
30 S	235	1.12	845	2150	−0.035	350
32	240	1.15	835	1350	−0.035	250
32 S	245	1.15	850	1790	−0.035	250
33E	251	1.16	865	2100	−0.035	350
35E	265	1.19	880	1800	−0.035	300

Figure 2.8 shows rudimentary schematics of two basic windings in AC machines. The first, (a), shows a three-phase distributed winding, where each slot is occupied by one coil side; the windings are distributed in the stator to represent and approximate a spatially sinusoidal magnetomotive Force (MMF). The windings in the second Figure (2.8b) are further distributed in the slots, creating an MMF closer to a sinusoidal distribution. Each of these indicative winding distributions has distinct characteristics; there are many options in number of slots, coil pitch and distribution.

In Figure 2.9 similar rudimentary schematics are shown for concentrated windings. In (a) concentrated windings around one tooth are shown, although the same slot is occupied by two coil sides. In (b) only one coil side is occupying each slot. This last option separates coils mechanically and thermally.

Stator windings provide many different characteristics; please see many appropriate electrical machine design books, e.g. Pyrhönen et al. (2014). Among these:

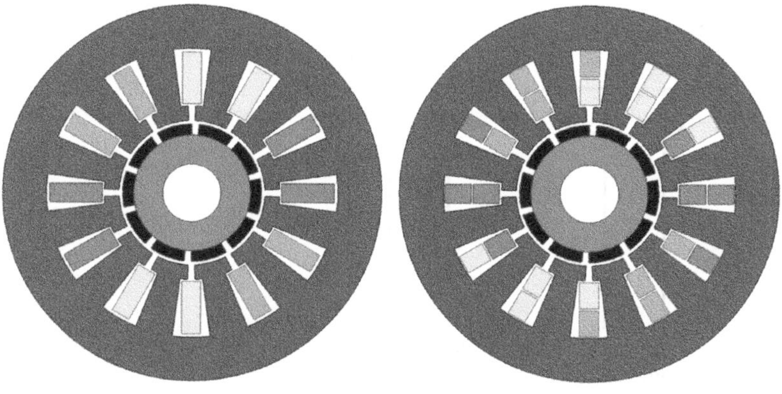

(a) (b)

Figure 2.8 Types of distributed windings. (a) Single layer distributed stator winding schematic. (b) Double layer distributed stator winding schematic.

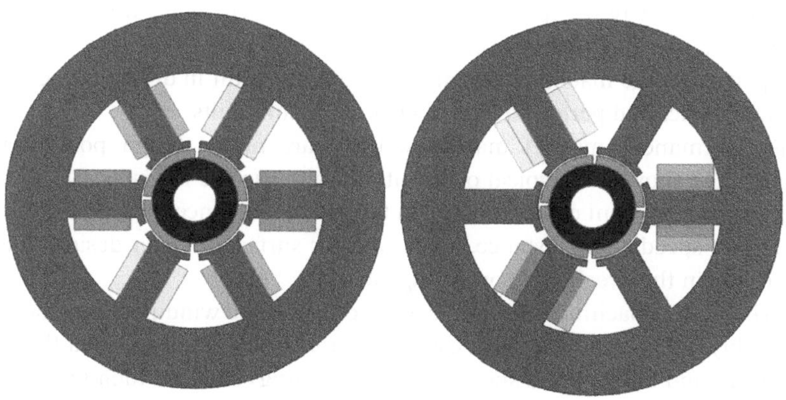

(a) (b)

Figure 2.9 Types of concentrated windings. (a) Double layer distributed stator winding schematic. (b) Single layer distributed stator winding schematic.

- Distributed windings provide smoother armature MMF and decreased torque pulsations, in particular, in induction machines.
- Concentrated windings result in torque pulsations, but much shorter end windings, hence lower losses, and allow the use of fractional slot/pole/phase windings.

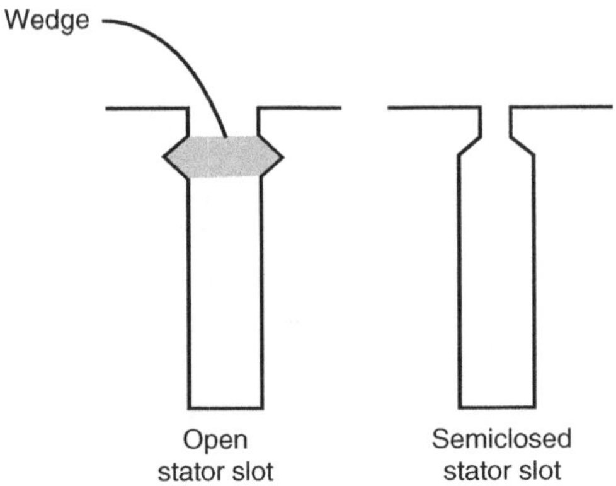

Figure 2.10 Stator slot designs.

- Windings with non-integer number of slots per pole per phase give the designer more freedom in terms of design topologies speed and torque, and again may result in higher torque pulsations.

Stator slots may be open or half closed, as shown in Figure 2.10. These designs matter in manufacturing ease, but also affect the magnetic flux density pattern in the machine, the inductances, and result in different operating characteristics and performance under short circuit faults.

For permanent magnet machines, there are many design possibilities. Figure 2.11 shows two typical ones: interior and surface magnets. These machines have different characteristics in terms of inductances and performance at higher speeds. For instance, some interior surface magnet designs allow operation in flux weakening operating mode.

In induction machines, the rotor can have distributed windings, as in the case of doubly fed machines. In that case, the rotor winding is connected through slip rings and brushes to a stationary (non-rotating) circuit. Much more common is the case of rotor bar windings, made of cast aluminum, cast copper, or copper rods brazed to end rings. Figure 2.12(b) shows this 'cage-type' winding. Rotor slots can also be of a great variety, including closed and semi-open type 2.12(b).

For all kinds of machines, **bearings** are selected to withstand the weight of the rotor, any axial forces, and the speed, as well as any possible vibrations arising from the machine itself or the equipment connected mechanically to it. They consist of inner and outer races, and rolling elements, as well as grease or lubricant. Bearings degrade due to operation in a contaminated environment (mechanically or chemical), high temperatures, and currents

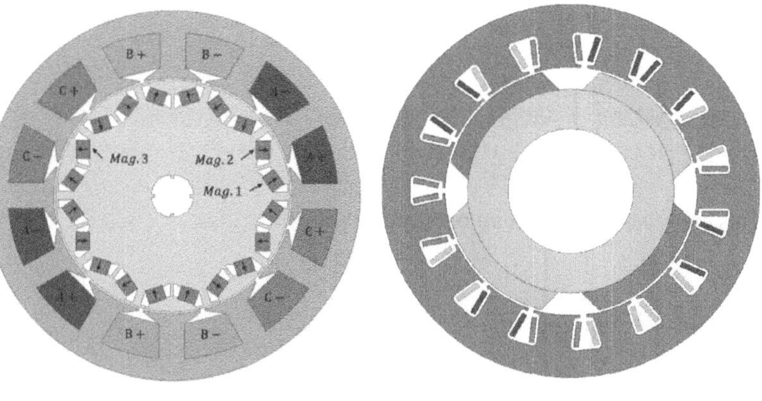

Figure 2.11 Interior and Surface Permanent Magnet Machines. (a) Single layer distributed stator winding schematic. (b) Double layer distributed stator winding schematic.

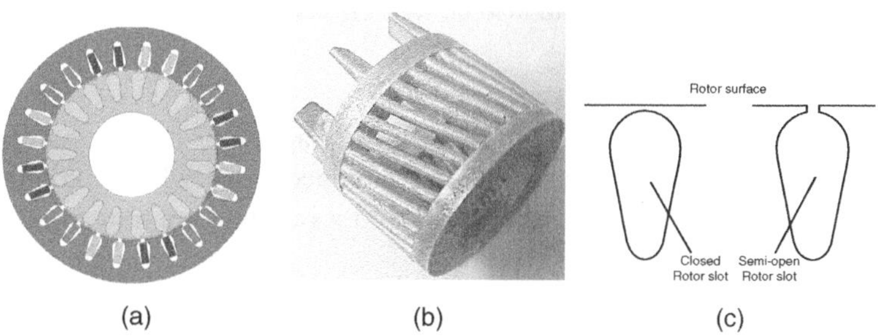

Figure 2.12 Induction machine rotor. (a) Induction machine with cage rotor. (b) Cage rotor with aluminum bars and end rings. (c) closed and semi-closed rotor slots in induction machines.

flowing in bearing. This degradation causes vibrations, noise, increased temperature, and failure.

2.3 Machine Models, Operation, and Control

2.3.1 Three-phase Windings

For three-phase machines, the variables assigned to each three-phase winding (currents, voltages, flux linkages) are transformed through the Clarke transform to two orthogonal ones. Phase currents $\mathbf{i}_{s,abc}$ are transformed to $\mathbf{i}_{\alpha\beta\gamma}$.

$$\mathbf{i}_{s,\alpha\beta\gamma}(t) = \frac{2}{3}\begin{bmatrix} 1 & -\frac{1}{2} & -\frac{1}{2} \\ 0 & \frac{\sqrt{3}}{2} & -\frac{\sqrt{3}}{2} \\ \frac{1}{2} & \frac{1}{2} & \frac{1}{2} \end{bmatrix}\mathbf{i}_{s,abc}. \tag{2.8}$$

This can also be used in an orthonormal form:

$$\mathbf{i}_{s,\alpha\beta\gamma}(t) = \sqrt{\frac{2}{3}}\begin{bmatrix} 1 & -\frac{1}{2} & -\frac{1}{2} \\ 0 & \frac{\sqrt{3}}{2} & -\frac{\sqrt{3}}{2} \\ \frac{1}{2} & \frac{1}{2} & \frac{1}{2} \end{bmatrix}\mathbf{i}_{s,abc}. \tag{2.9}$$

When the neutral of the winding is not connected, the third component (called "homopolar component") of the current is zero. This is a stationary frame of reference, attached to the physical location of the winding, and, in this reference, the currents can also be written in complex form as:

$$\mathbf{i}_{\alpha,\beta} = i_\alpha + ji_\beta \tag{2.10}$$

Yet another operation, the Park's transform, transforms these variables, e.g. currents $\mathbf{i}_{\alpha\beta\gamma}$, to a synchronously rotating reference frame, which is common to the stator and rotor. In this case, the α, β and γ variables are transformed, e.g. for the currents as below:

$$\mathbf{i}_{dqz} = \begin{bmatrix} \cos(\theta) & \sin(\theta) & 0 \\ -\sin(\theta) & \cos(\theta) & 0 \\ 0 & 0 & 1 \end{bmatrix}\mathbf{i}_{\alpha\beta\gamma}(t) \tag{2.11}$$

where θ is the relative position of the rotating frame of reference with respect to the stationary one.

2.3.2 Induction Machines

The stator and rotor flux linkages are caused by the currents in both stator and rotor windings. If the relative position of the rotor with respect to the stator is ϵ, the flux linkages are λ_S and λ_R respectively and the induced voltages in these stator and rotor windings are u_S and u_R in their natural frames of reference (Physical quantities):

$$\lambda_S(t) = L_S \underline{i}_S(t) + M\underline{i}_R(t)e^{j\epsilon(t)} \tag{2.12}$$

$$\lambda_R(t) = L_R \underline{i}_R(t) + M\underline{i}_S(t)e^{-j\epsilon(t)} \tag{2.13}$$

$$u_S(t) = R_S \underline{i}_S(t) + \frac{d}{dt}\lambda_S(t) = R_S \underline{i}_S + L_S \frac{d\underline{i}_S}{dt} + M \frac{d\underline{i}_R}{dt} e^{j\epsilon} + j\omega M \underline{i}_R e^{j\epsilon} \quad (2.14)$$

$$u_R(t) = R_R \underline{i}_R + L_R \frac{d\underline{i}_R}{dt} + M \frac{d\underline{i}_S}{dt} e^{-j\epsilon} - j\omega M \underline{i}_S e^{-j\epsilon} \quad (2.15)$$

$$T_e = \frac{2}{3} M \Im \left[\underline{i}_S(t) (\underline{i}_R(t) e^{j\epsilon})^* \right] \quad (2.16)$$

where $\omega = d\epsilon/dt$. At his point we have to define the various speeds and other para matters related to an induction machine:

- p number of pole pairs $p = P/2$.
- ω $d\epsilon/dt$, speed of the rotor for two-pair equivalent machine, electrical speed.
- ω_s speed of the stator variables for an equivalent two-pole machine, $\omega_s = 2\pi f_s$ with f_s the stator frequency.
- ω_m mechanical rotor speed, $\omega_s = \omega/p$.
- ω_r angular speed of rotor variables, for an equivalent 2-pole machine, $\omega_r = \omega_s - \omega$.
- s slip, $s = (\omega_s - \omega)/\omega_s$.

Here the variables are: $\boldsymbol{\lambda}_S$ is the d- and q- axis stator flux linkage, $\boldsymbol{\lambda}_S = \lambda_{dS} + j\lambda_{qS}$ and $\boldsymbol{\lambda}_R$ is the d- and q- axis stator flux component, $\boldsymbol{\lambda}_R = \lambda_{dR} + j\lambda_{qR}$, and $\mathbf{u}_S$ is d- and q-axis stator voltage complex vector, $\mathbf{u}_S = u_{ds} + ju_{qS}$. L_S is the stator inductance, L_R is the rotor inductance, M is the stator rotor mutual inductance, $\sigma = 1 - M^2/(L_S L_R)$ and ω_s the motor electrical angular velocity (p times the angular mechanical angular velocity, where p is the number of pole pairs).

In steady state with balanced, symmetric sinusoidal supply, the new variables (currents, voltages, flux linkages) are all constant. Using the transformed variables, the torque becomes in a non orthonormal reference frame:

$$T = \frac{3}{2} p \frac{M}{L_R} \left(\lambda_{dR} i_{qS} - \lambda_{qR} i_{dS} \right). \quad (2.17)$$

During operation in steady state, with sinusoidal supply and angular frequency of the stator voltages and currents f_s, the main magnetic field rotates at speed ω_s and the rotor rotates at speed ω, generally other than ω_s. This difference in speeds induces in the rotor voltages and currents of frequency $f_r = f_s - p\omega$, which leads to torque development. This is accomplished by induction, i.e the magnetic field is moving at a different speed than the rotor conductors, thus inducing voltages and currents in the rotor of frequency f_r. For a given flux, these currents increase with slip, hence they are high at starting, i.e. at $\omega = 0$, $\omega_r = \omega_s$.

In the case of a wound field induction machine, the stator (or the rotor) is connected to the power grid, operating at frequency f_s, while the rotor (or the stator) is fed from a source of controlled voltage and frequency. This makes it possible to control the stator power with a reduced power converter at the rotor (about one-third of the stator power).

Variables $\boldsymbol{\lambda}_S$ and $\boldsymbol{\lambda}_R$ consist each from the main flux and the stator and leakage fluxes correspondingly. The total flux can be considered as consisting of three parts, magnetizing flux the crosses the air gap, and leakage fluxes that link only the stator or only the rotor windings.

2.3.2.1 Induction Machine Rotor Field Orientation

In Rotor Field Oriented drives, the rotor flux is aligned to the d axis of the common synchronous reference frame (rotating at the speed ω_s), thus making zero the q-axis component of it, $\lambda_{Rq} = 0$. In this reference frame and for a squirrel cage machine ($\mathbf{u}_R = 0$), the equations of the machine become:

$$Mi_{ds} = \lambda_{dR} + \frac{L_R}{R_R}\frac{d}{dt}\lambda_{dR} \tag{2.18}$$

$$T = \frac{3}{2}p\frac{M}{L_R}\lambda_d i_{qs} \tag{2.19}$$

This leads to the idea of operating the drive so that the rotor flux is kept constant until base speed (or slow decreasing inversely proportional to the speed above the base speed), while the torque is primarily controlled through the d-axis stator current. This technique requires the accurate knowledge of the rotor flux in amplitude and position, and has given rise to two methods of implementation:

Direct Rotor Field Orientation requires the measurement of the rotor flux, or calculation from stator voltage and current measurements. Both versions are problematic, the first because of additional hardware requirements, the second because it collapses at low speed or requires rotor current measurements. The rotor flux is regulated on a closed loop.

Indirect Rotor Field Orientation uses an estimator or observer of the rotor flux amplitude and position, starting from an estimator of the rotor slip speed, and the use of a position sensor or an estimator. Its main drawback is that it depends heavily on the knowledge of the rotor time constant, which depends on saturation and temperature. Rotor flux is often computed in a open loop without feed-back.

The fluxes are then controlled through the applied voltages. It must be noted here that this is a very simplified model. In reality, the flux density distribution in the gap differs from sinusoidal, and this, along with saturation and other

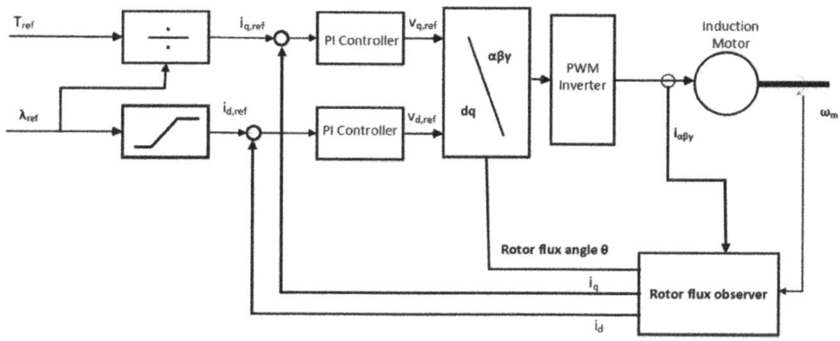

Figure 2.13 Schematic of Field Oriented torque control of an induction machine.

phenomena, leads to a developed torque that is not constant, but may have torque ripple.

Figure 2.13 is a conceptual drawing of the controller of an induction machine under Field Orientation mode.

2.3.2.2 Direct Torque Control

In the previously discussed torque in 2.3.2, we can substitute the currents for voltages as below:

$$\boldsymbol{\lambda}_S = L_S\mathbf{i}_S + M\mathbf{i}_R \tag{2.20}$$

$$\boldsymbol{\lambda}_R = L_R\mathbf{i}_R + M\mathbf{i}_S \tag{2.21}$$

Thus, obtaining an equation of torque:

$$T_e = \frac{3}{2}p\frac{M}{L_S L_R}|\boldsymbol{\lambda}_R||\boldsymbol{\lambda}_S|\sin\gamma \tag{2.22}$$

where γ is the angle between $\boldsymbol{\lambda}_R$ and $\boldsymbol{\lambda}_S$. The idea then is to control the stator and rotor fluxes and the angle between them, in order to produce directly the desired torque. This is done directly through the stator voltages. Since, due to the longer rotor time constant, the rotor flux is rather observed than directly controlled. At every step time, the stator voltage is adjusted to change the stator flux in amplitude or direction. There are two controlled variables: the stator flux amplitude and torque. They are controlled through hysteresis controllers. For each command, a hysteresis band is defined, with width for the torque H_T and for the stator flux $\Delta\lambda_S$.

Niu et al. (2016) utilize a switching table that consists of different voltage vectors. The rotor position sensing that is essential for field oriented control (FOC) is not necessary for DTC to operate properly, even if speed control loop is included in the direct oriented control (DTC) scheme. Although relatively simple, DTC achieves adequate torque control performance under both steady-state and dynamic conditions. On the plus side, DTC has low sensitivity to the accuracy of machine parameter estimation. Conversely, DTC has several disadvantages. Among them, the performance of DTC deteriorates at low speed due to reduced controllability, high torque ripple, and variable switching frequency.

The method uses the stator voltage directly to control the stator flux and the torque. The controller contains two PI regulators (one for flux and one for torque) and a sinusoidal voltage modulator (SVM) unit. It receives as inputs the stator flux and torque errors and generates the inverter drive signals (see eq. 2.3.2.2).

$$\delta T = \begin{cases} 1 & \text{for} \quad \Delta T > H_T \\ 0 & \text{for} \quad -H_T < \Delta T < H_T \\ -1 & \text{for} \quad \Delta T_e < -H_T \end{cases} \qquad (2.23)$$

$$\delta \lambda_S = \begin{cases} 1 & \text{for} \quad \Delta \lambda_S > H_\lambda \\ -1 & \text{for} \quad \Delta T_e < -H_\lambda \end{cases}$$

The selection of optimal voltage vector is implemented with reference to Table 2.3, which requires the desired stator flux variation, desired torque variation, and the sector that the actual stator flux located within (see Figure 2.14).

The voltage reference values are then obtained from the eq. 2.24:

$$u_{sd}^* = (K_{p\lambda} + K_{I\lambda}/s)(\lambda_s^* - \lambda_s) \qquad (2.24)$$

$$u_{sq}^* = (K_{pT} + K_{IT}/s)(T_e^* - T_s) + \lambda_s \omega_{\lambda s} \qquad (2.25)$$

Table 2.3 Voltage select vector for DTC of an induction machine.

$\Delta\lambda_S$	ΔT_e	Sector					
1	1	u_{s2}	u_{s3}	u_{s4}	u_{s5}	u_{s6}	u_{s1}
	0	u_{s0}	u_{s7}	u_{s0}	u_{s7}	u_{s0}	u_{s7}
	−1	u_{s6}	u_{s1}	u_{s2}	u_{s3}	u_{s4}	u_{s5}
-1	1	u_{s3}	u_{s4}	u_{s5}	u_{s6}	u_{s1}	u_{s2}
	0	u_{s7}	u_{s0}	u_{s7}	u_{s0}	u_{s7}	u_{s0}
	−1	u_{s5}	u_{s6}	u_{s1}	u_{s2}	u_{s3}	u_{s4}

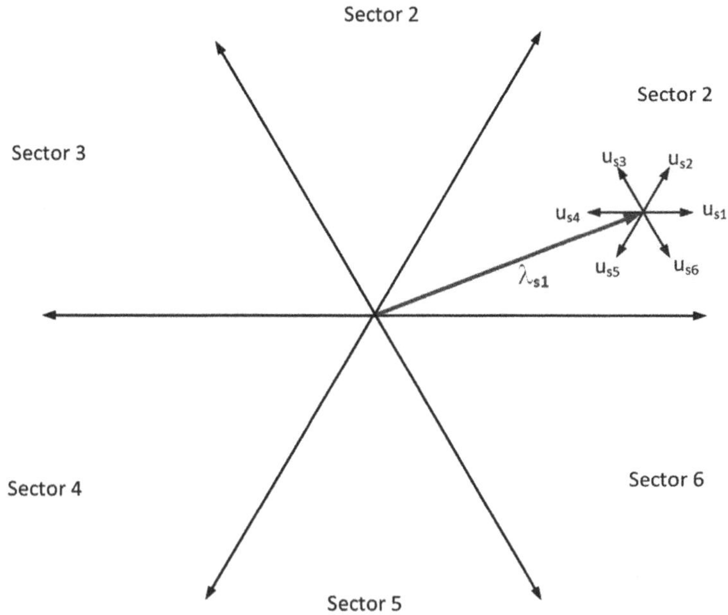

Figure 2.14 Schematic of flux and voltage sectors for Direct Torque Control.

while the torque is:

$$T_e = \frac{3}{2} p \frac{L_m}{L_\sigma^2} \lambda_s \lambda_r \sin(\theta) \tag{2.26}$$

where T_e is the developed torque, λ_s and λ_r are the stator and rotor flux magnitudes, θ is the angle between the stator and rotor flux vectors λ_s and λ_r, p is the number of pole pairs, $L_\sigma^2 = L_s L_r - L_m^2$, and L_s, L_r, and L_m are the stator, rotor, and magnetizing inductances, respectively.

Figure 2.15 gives the basic diagram of DTC control of an induction machine. The flux control is based on modifying the real component of the voltage, u_{sd}, i.e. the flux component of the voltage vector. This is accomplished by controlling the voltages applied to the stator through a sinusoidal voltage modulator (SVM) that produces the inverter control signals. It receives the reference voltages 2.24 in a stator flux reference frame. The SVM principle is based on the switching between two adjacent active vectors and a zero vector during one switching period.

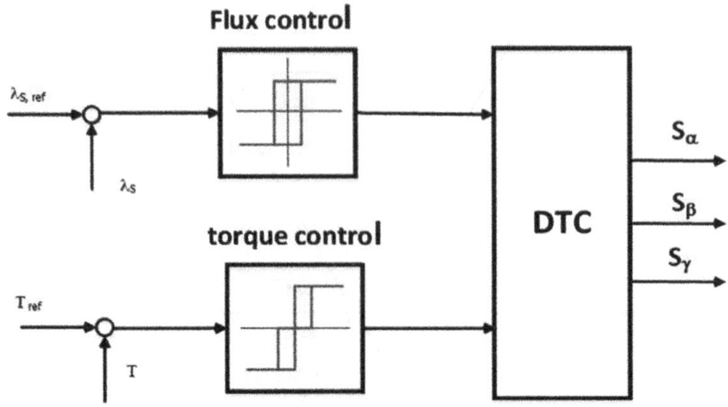

Figure 2.15 Schematic of Direct Torque Control Concept for Induction Machines.

2.3.3 Permanent Magnet AC Machines

Permanent magnet AC machines (PMAC) also called Permanent Magnet Synchronous Machines (PMSM) come in many varieties: one categorization is interior or surface magnets rotors, as shown in 2.16; others, defined by the stator windings, can be distributed or concentrated, and integral or fractional slots per pole per phase. Typically, the Park transformation uses as rotor reference the axis of the rotor magnets. In that frame of reference, the general form of the equations that describe the machine is below:

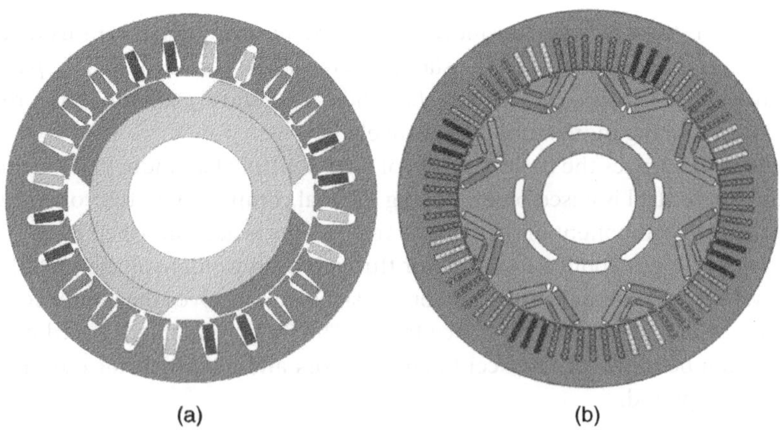

Figure 2.16 Typical cross sections of radial flux PMAC machines. (a) Surface Permanent Magnets. (b) Interior Permanent Magnets.

The equations that describe them are:

$$T = \lambda_f i_{Sq} + (L_{Sd} - L_{Sq}) i_{Sd} i_{Sq} \tag{2.27}$$

$$v_{dq} = v_d + j v_q = R_S i_{s,dq} + \frac{d\lambda_{S,dq}}{dt} + j\omega_s \lambda_{S,dq} \tag{2.28}$$

$$\lambda_{S,dq} = L_{Sd}(i_{Sq} + jL_{Sq}) + \lambda_f \tag{2.29}$$

where λ_f is the stator flux linkages due to the rotor magnets, and L_{Sd}, L_{Sq}, the stator d- and q-axis reactances.

Rotor Field Oriented Operation: A PMAC-based drive is conceptually similar to that of an induction machine. There are two differences: the rotor magnetic field is defined by the magnets and not the stator currents, except through saturation and flux-weakening mode, and the rotor in the case of the Interior Permanent Magnet machines is salient. Then, of the two torque components 2.28, one corresponds to magnets and armature currents, $\lambda_f i_{sq}$, and another, results from the machine variable reluctance and armature currents, $(L_{sd} - L_{sq}) i_{sd} i_{sq}$. In the interior PM machines similar to the one in Figure 2.16 (b), $L_{sd} > L_{sq}$, require negative i_{sd} for reluctance torque production. On the contrary, some PMSM have $L_{sd} < L_{sq}$ and could be used in Field Weakening mode under nominal speed. For surface PMAC machines, the second torque is minimal since $L_{Sd} \approx L_{Sq}$, while the first one is proportional to the stator d-axis currents.

Under limited available voltage, stator d-axis current is used at high speeds to weaken the field due to magnets and limit the required voltage at high speeds. For interior PMAC machines, the negative d-axis current is necessary to provide reluctance torque; in the interior PM machines, under both conditions, the demagnetizing stator current increases the risk of permanent demagnetization, in both ferrite and for high-temperature rare earth magnets. Therefore, for machines with inverse saliency ($L_{sd} < L_{sq}$), flux weakening can be done without exceeding electromagnetic constraints.

Direct Torque Control: Here, the rotor flux is defined by the magnets, and the torque control becomes simpler, as shown in Table 2.4.

Finite Element models provide detailed analysis of the magnetic field, losses, torque, and torque pulsations. For fault diagnosis and identification, the main use is to give a deeper understanding and insight of the underlying phenomena, rather than provide the characteristics of the variables and signals.

Table 2.4 Voltage select vector for DTC of a PMAC machine.

$\Delta\lambda_s$	ΔT_e	Sector					
1	1	u_{s2}	u_{s3}	u_{s4}	u_{s5}	u_{s6}	u_{s1}
	-1	u_{s6}	u_{s1}	u_{s2}	u_{s3}	u_{s4}	u_{s5}
-1	1	u_{s3}	u_{s4}	u_{s5}	u_{s6}	u_{s1}	u_{s2}
	-1	u_{s5}	u_{s6}	u_{s1}	u_{s2}	u_{s3}	u_{s4}

2.4 Faults in Electrical Machines

Detecting faults, identifying the type, determining the fault severity, and predicting the reaming useful life resulting from a developing fault are central to the safe operation of a drive system. Based on these, appropriate action can be taken to minimize any effects on health of personnel and cost.

These processes depend on the monitoring and processing of operating variables, which can either accessed directly through sensors, or estimated from sensed variables in Digital Signal Processor (DSP), microprocessor, etc. It is seldom that one or only a few faults are anticipated, and it is therefore important to have in place a supervisory system that can detect faults and classify them, even in a relatively general way, before exact categorization.

The faults that can appear in the machines within the electrical drives can be due to:

- Bearings,
- Eccentricity,
- Open or short circuit in inverters or windings,
- Insulation weakness,
- Broken bars in squirrel induction motor,
- Demagnetized magnets for PMSM.

Figures 2.17 and 2.18 give an overview of the different failures in AC machines. Some of these faults manifest themselves in similar ways. Many techniques discussed are applicable to more than one fault, and the discussion covers the separation and isolation of these faults, as well as the estimation of severity and failure prognosis. Developing faults in bearings and insulation have different characteristics than the rest, and the most important aspect of them is that they cannot be easily managed but instead should be monitored and their failure prognosis accurately estimated. Hence, they are discussed separately.

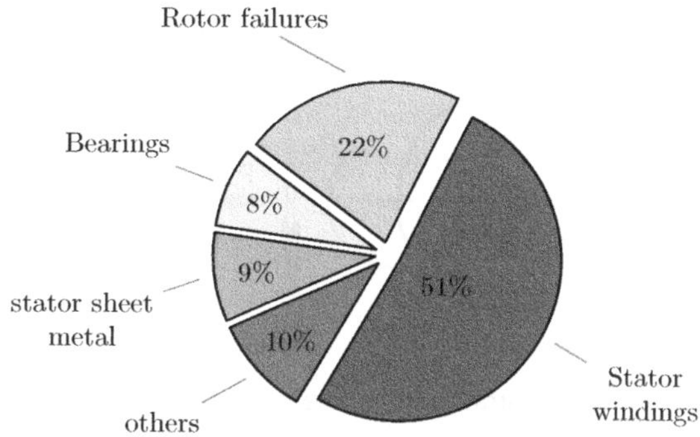

Figure 2.17 Failure distribution of Low and Medium Power Motors *Source:* France (1988).

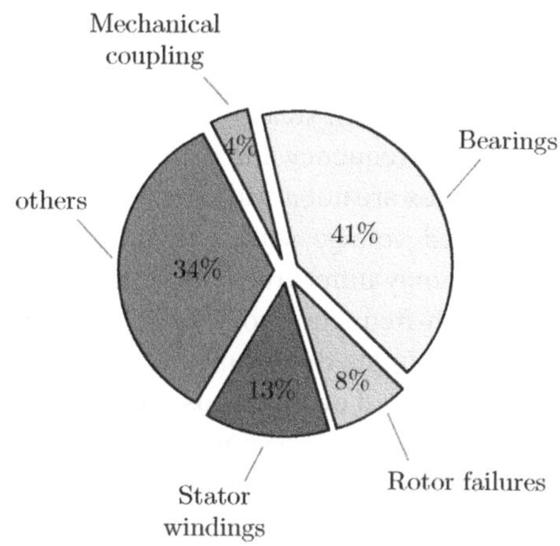

Figure 2.18 Failure distribution of High power motors for petrochemistry *Source:* Thorsen and Dalva (1995).

2.4.1 Operational Variables and Measurements

The faults in electrical machines operating in electrical drives manifest themselves in the variables of operation. It is always preferable to minimize the additional requirements of measurements beyond what is needed for the controller, both to keep the cost low and to limit the number of components and system complexity.

Currents are the easiest to use, as they are invariably measured as part of the controller. Still, the standard current sensors used in controllers are relatively low bandwidth, and they cannot always be used to handle the fast transients and high frequencies associated with some faults.

A number of techniques of fault diagnosis are based or various Motor Current Signature Analysis tools (MCSA). As many control schemes are based on voltage-source and current-controlled inverters, these methods are being eclipsed by methods using voltages. The reason is that the controller bandwidth

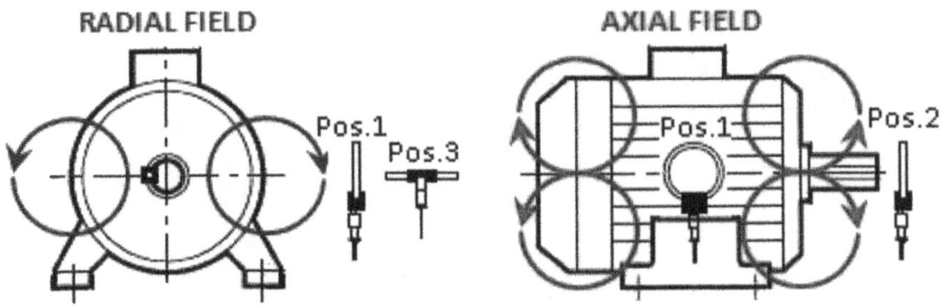

Figure 2.19 Stray flux and measurement positions Ceban et al. (2012).

is usually quite high in order to effectively impose the required current, thus masking the effects on the current that would be used. This is the case mostly when the signals of interest are of relatively low frequencies, e.g. eccentricity, demagnetization, steady-state winding faults, etc. There, faults are related to the power frequency and its harmonics, and/or slot-related flux pulsations.

Voltages are not always measured, although in many cases, when a current-controlled voltage source inverter is used, they are controlled. The actual voltages may differ from the expected ones, due to delays and switching, and have high-frequency components. In that case, the reference voltages can be used to estimate the actual ones.

Rotor speed or rotor position is commonly used, measured through an encoder or resolver. If acceptable, it can be omitted and replaced by a less accurate observer.

Power: real, reactive, or apparent is not directly measured, but at times, it may be preferable to either voltage or current, and it can be computed from these. In this case, particular attention must be paid to the synchronization of voltage and current measurements.

Torque, although very useful as a signal, is seldom measured, as torque sensors are bulky, expensive, and have relatively low bandwidth. Torque can be observed or estimated from direct measurements.

Vibrations cannot easily be estimated, and measurements are useful for determining the health of bearings or faults in other components. They are often used together with other measurements.

Flux densities and flux linkages are very important variables for determining the health of a machine; they cannot be directly measured. Flux linkages in windings can be computed from induced voltages in these winding. Flux density measurements of the leakage or stray field have been recently used as a promising measurement related to faults, see Figure 2.19, Ceban et al. (2012).

Jiang et al. (2017) give a review of the instruments that are used for measuring stray flux, namely search coils and Hall effect sensors, and, more recently, giant magnetoresistance sensors that provide higher performance in terms of sensitivity, linearity, and frequency response.

2.4.2 Supervision, Detection, and Fault Classification

Although well-developed methods have been proposed and are discussed in detail later here, there is a need to have an accurate supervision tool to detect a problem in a drive, identify the fault and, if possible, its severity. To this end, a number to tools have been developed, based on different approaches.

Lebaroud and Clerc (2008) proposed a classification technique based on the design of optimized time-frequency representation (TFR) from a time–frequency ambiguity plane in order to extract the feature vector. They use a decision criterion based on the Hidden Markov Model. The goal is the realization of an accurate diagnosis system of motor faults such as bearing faults, stator faults, and broken bars in induction machines (see Figure 2.20). λ_i is an HMM model specific of the failure i. Thus, each failure is associated to an HMM model.

For a given signal, a TFR can be uniquely mapped from a kernel designed for each classification task. The classification-optimal representation TFR is obtained by smoothing the ambiguity plane with an appropriate kernel, which is a classification optimal kernel. The problem of designing the TFR becomes equivalent to designing the classification-optimal kernel. This method, used to design kernels (and thus TFRs), optimizes the discrimination between predefined sets of classes.

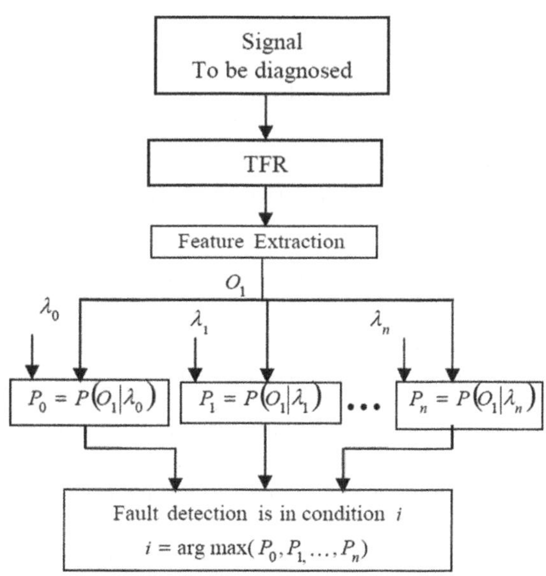

Figure 2.20 Classification based on TFR by A. Lebaroud, Lebaroud and Clerc (2008).

The resulting kernels are not restricted to any predefined function but are rather arbitrary in shape. This approach ascertains the necessary smoothing to achieve the best extraction features. The kernel determines the representation and its properties. A kernel function is a generating function that operates upon the signal to produce the TFR. In their classification procedure $C-1$ kernels were designed for a C-class classification system, in their case $C = 4$. The kernels were designed by training example signals.

Having trained the HMM, the first step is to extract the feature vector from the preprocessed signal under different load conditions. Next, the probability of the feature vector is calculated, given all the HMMs in the previously constructed database. The HMM, for which the probability is maximum, determines the type of the fault.

The method was tested for a bearing fault, broken bar, and imbalanced stator. It was trained at ten load current levels. They showed the superiority of the HMM to Mahalanobis distance to classify accurately the three faults independently of the load level.

In related work, Medoued et al. (2011) used a similar method to design optimal kernels, the classifier was designed using artificial neural networks, and the Levenberg-Marquardt algorithm was applied.

Zhang et al. (2020) designed an integrated supervisory system for fault isolation and prognosis for electric drive systems of battery electric vehicles. They based their method on an electric model. They use a hierarchical approach that first detects performance degradation in the drive system, and then goes down to the lower level component fault detection and isolation. In this approach, multiple health indicators are developed, and a rule-based strategy based on the calculated health indicators is presented to detect and isolate the faults through different fault signatures of different health indicators.

Considering that faults increase the losses and decrease the performance of the drive, the authors use the error between the calculated torque from the power balance and the estimated torque from the electromagnetic model, to represent the electric drive and determine system health. They address the following:

1. Influence of current sensor fault on the electric drive system-level health indicator.
2. Influence of resolver fault on the electric drive system-evel health indicator.
3. Influence of motor winding faults on the electric drive system level.

The detection and isolation method for a faulty sensor (any of the three-phase current sensors or rotor position sensor) is described in Figure 2.21, where a total number of four residual generators are computed in parallel, to detect and isolate a current sensor fault or position sensor (resolver) fault. RG1, RG2 and RG3 leverage three estimators (in the form of sliding mode observers) based on

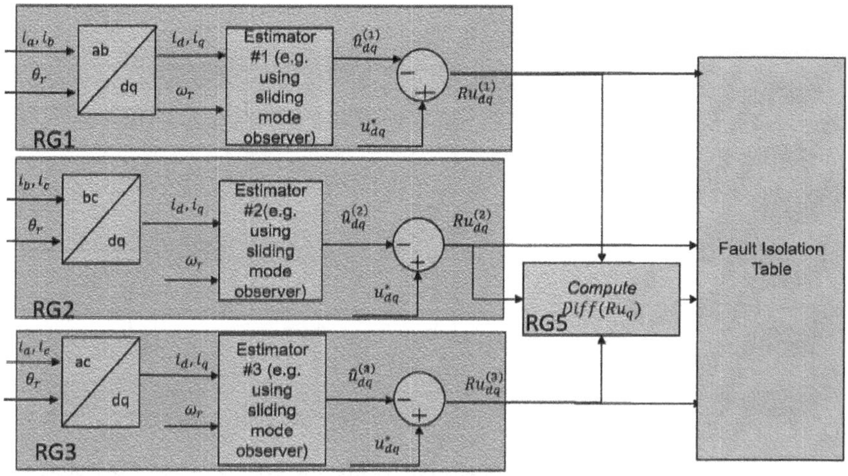

Figure 2.21 Sensor fault indicators, Zhang et al. (2020).

an inverse dq model of the electric motor to estimate the voltages in the dq reference frame. The three estimators use different phase current measurements, where RG1 uses Phase A and Phase B current measurements (i_a; i_b), RG2 uses the measured Phase B and Phase C currents (i_b; i_c), and RG3 relies on the current measurements from Phase A and Phase C (i_a; i_c). The concept is to use any two of the three current measurements to obtain the d and q axis currents, and then an estimator based on inverse model of the electric motor in dq reference frame is applied to estimate the motor input voltages (u_d, u_q). Because these estimators rely on different phase current measurements, they can be used to identify a particular faulty current sensor.

The hierarchical prognostic algorithm proposed is described in Figure 2.22. After an initial check on the sensor range/performance, if the performance of the sensors is within the pre-defined threshold (Th0), then the prognostic system will first perform the electric drive health evaluation by checking the performance of the electric drive system-level health indicator. If the State of Health (SoH) of the electric drive is below a certain threshold, the algorithm proceeds to the component level prognosis. A set of component health indicators are tested in parallel. A rule-based logic flowchart is implemented as the fault isolation algorithm to identify the faulty component, either due to a bias/drift or gain error in the current sensor, or an offset/internal imbalance in the resolver, or the motor degradation due to stator winding insulation fault. As each health indicator has a unique pattern with respect to faults, the fault isolation algorithm can determine what component is faulty by checking the performance of the various health indicators.

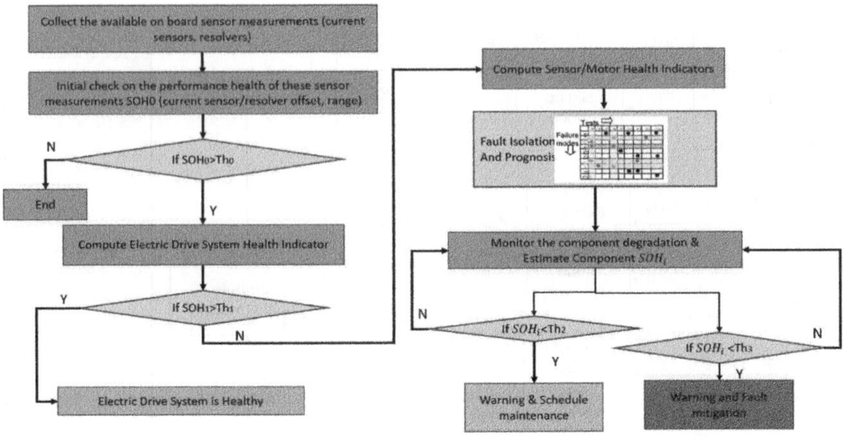

Figure 2.22 Integrated Prognosis Algorithm for Electric Drive Systems. SOH: state of health estimator, Zhang et al. (2020).

Ali et al. (2019) take an entirely different approach, using fault diagnosis method based on machine learning. It focuses on an induction machine and uses stator current and vibration data to detect faults. The feature extraction uses discrete wavelet transform (DWT) and matching pursuit (MP) and three classification algorithms, SVM, K-nearest neighbors (KNN), and ensemble. The faults tested were an unbalance shaft rotation (UNB), a bearing fault (BF), a combined BF and UNB fault, combined BF and one BRB faults, a combined BF, UNB, and unbalance voltage (UV) condition from the three-phase power supply, as well as one, two, and three broken bar (BRB) faults.

The authors used six steps to implement the method:

1. Conduct experiments for an induction motor under healthy, single- and multi-fault conditions.
2. Record stator current and vibration data simultaneously, where vibration sensors and a power quality analyzer, are required to take measurements.
3. Choose suitable signal processing methods, for features extraction.
4. Extract features for machine learning.
5. Conduct classification for electrical and mechanical faults using chosen classifiers.
6. Develop curve fitting equations to calculate features versus motor loading.

Rocchi et al. (2014) used a simpler algorithm and instrumentation. The amplitude of vibrations was the only measurement for the fault diagnosis and failure prognosis of electrical machines. They modeled through an exponential model:

$$y(t_k) = w_1 e^{w_2 t} \tag{2.30}$$

where t_k is the time step and w_1, w_2 are model parameters, estimated using an exponentially weighed Recursive Least Square algorithm with Forgetting Factor (RLSFF). The hypothesis under which the model is valid are that acceleration signal envelope will increase when wear and the deterioration model should reflect the vibration signals.

Faults and degradations were introduced in these motors, such as the degradation of the rubber piece that covers the bearing and forced overheating in order to experimentally test the proposed method.

2.4.3 Bearings

Faults in bearings are the most common in high-power electrical drives. Diagnosis and prognosis of bearing faults are of utmost importance, as very little can be done to mitigate them during operation, and redundancies require complete drive systems, often even including the load. Because of that and the interesting and challenging technical aspects, a large volume of research has been dedicated to this topic. To facilitate and enhance the results of research in these areas, test databases have been made available to researchers, e.g. from the Case Western University, see CWR; NASA, see NAS; and PRONOSTIA, see Nectoux et al. (2012).

Howard (1994), in an extensive review, discusses in detail many aspects of bearing faults and their manifestations in vibrations. The normal service life of a rolling element bearing rotating under load is determined by material fatigue and wear at the running surfaces. Premature bearing failures can be caused by a large number of factors, the most common of which are fatigue, wear, plastic deformation, corrosion. Most faults cannot be separated, except perhaps in their initiation, but the presence of one fault may accelerate the effects of another.

Environmental conditions, loading, and neglect can cause corrosion on the running surfaces. Lack of adequate and appropriate lubrication can lead to increased temperatures and related weakening of the steel.

A fault due to fatigue can manifest itself as pitting, spalling, or flaking of the bearing races or rolling elements. If the bearing continues in service, the damage spreads, since the localized stresses in the vicinity of the defect are increased. The surface damage disturbs the motion of the rolling elements and leads to the generation of short time impacts, repeated at the appropriate rolling element defect frequency. These early frequencies can cause pronounced effects through resonances and can be detected. As the damage spreads, but with distinct impulses, each response will follow another with a random irregular repetition rate. Instead, closely occurring impulses, each causing a response, lead to a general vibration, difficult to identify, and eventually to a failure. As the edges of the spall wear down, so the sharpness of the impacts will decrease, altering the vibration pattern.

Of the various causes of bearing degradation, temperature, lubricant condition, and contamination are the most difficult to assess, as there are no measurements or parameters to estimate. A major cause of faults in bearings, bearing currents, are increasingly caused by fast switching inverters. For this, some efforts have been made to estimate the occurrence and characteristics of the discharge phenomena that produce bearing currents and degradation, estimate the discharge energy and use them for predicting bearing faults. The manifestation of faults ranges from the appearance of small craters on the surface of the races and rolling elements, to increased general roughness, and excessive vibrations. An early review paper, Stack et al. (2004), discusses the basic forms of bearing faults.

When the faults are still single-point, resulting from a localized defect on the bearing surface, a spall, or a crater pit can be caused by a current discharge. Such a single-point (cyclic) defect produces an impact between the ball and the raceway and generates a detectable vibration. The frequencies at which these vibrations occur are predictable and depend on which surface of the bearing contains the fault, on the geometrical dimensions of the bearing, and on the rotational speed of the rotor f_r:

$$\begin{aligned}
\text{Outer race:} \quad f_o &= N/2 \cdot f_r(1 - d/D \cdot \cos\alpha) \\
\text{Inner race:} \quad f_i &= N/2 \cdot f_r(1 + d/D \cdot \cos\alpha) \\
\text{Ball:} \quad f_b &= D/2d \cdot f_r(1 - (d/D)^2 \cos^2\alpha) \\
\text{Cage:} \quad f_c &= 1/2 \cdot f_r(1 + d/D \cdot \cos\alpha),
\end{aligned} \tag{2.31}$$

where N is the number of balls, d is the ball diameter, D is the bearing pitch diameter, and α is the ball contact angle (see Figure 2.23).

When the number of these faults is small, they produce short changes in the relative radial rotor offsets from the stator, which, in turn, produce speed and/or torque variations and distinct vibrations, as well as characteristic acoustic noise. These cyclic faults can be identified by their expected frequencies, in eq. 2.31. They cause small variations in speed or torque, and they can appear in the stator currents or voltages. Attempts have been made to detect them through stray fields, e.g Frosini et al. (2019). Associated with each occurrence of such a fault due to a distinct imperfection, there is a transient phenomenon, shown in Figure 2.24, which could be detected through its frequency content.

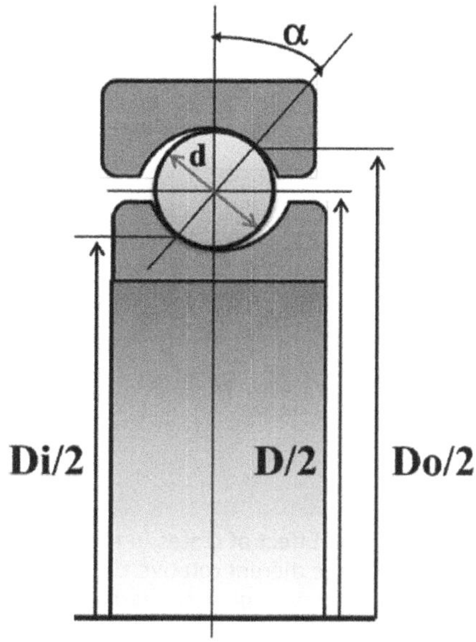

Figure 2.23 Bearing geometry.

Both the fundamental frequency given above and the short transients can be used to detect the fault.

The internal forces and stresses propagate through the structure, being attenuated at each interface, damped by the internal damping of the structure and modified by the frequency response of the system. Even when they remain distinct, the short impact duration and sharp rise times of the stress wave front mean that a wide range of natural frequencies and modes can be excited in the structure. The higher frequency stress waves (having smaller amplitudes) are attenuated the most. These transmission path effects ensure that the vibration which is finally measured by the transducer is a poor replica of the original internal forcing function, particularly for complex paths from source to transducer.

One of the major difficulties in detecting and diagnosing bearing faults through vibration monitoring is the rapidly changing nature of the vibration signal as the mechanical condition changes. Each new spall which is generated produces a whole family of impacts. As the edges of the spall wear down, so the sharpness of the impacts decrease, altering the vibration pattern. Each change in the rolling element or raceway geometrical profile can dramatically alter the vibration signal. Handling this difficulty has lead to techniques based on estimating the change of the spread in energy, and associating them with fault states, thus leading to more successful detection and prognosis methods.

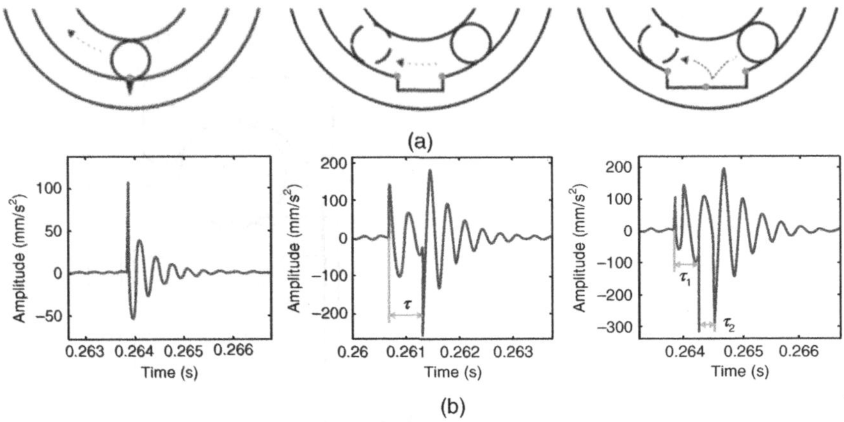

Figure 2.24 Effect of crates in the transient signal, Li et al. (2020). (a) Physical model of three dierent relative sizes of defects. (b) Simulated impulse pattern of the bearing defect with dierent widths.

Another complicating factor beyond the rapidly changing characteristics of the vibration is that the measured vibration signals collected from a defective bearing suffer from background noise. When the rotating machinery is in normal operation, its vibration signal is usually a Gaussian signal. Once operation deviates from normal, its vibration signal will turn into a non-Gaussian signal. A problem in detection is that superimposed to the vibrations is static random noise and discrete aperiodic noise. If the noise level is high, fault-induced impulse responses will be seriously corrupted, especially when the bearing fault is at its early stage. Background noise makes fault feature extraction more challenging. Therefore, effective denoising methods are required to highlight impulse responses. A lot of effort and increasingly complex techniques have been proposed, for example, by Zeng and Chen (2020), with good results as shown in Figure 2.25.

A voltage-source inverter presents a common-mode voltage source. The high-frequency (HF) components of this common-mode voltage interact with capacitances of the motor, thereby possibly generating inverter-induced bearing currents. Four types of inverter-induced bearing currents can be distinguished as described by Muetze et al. (2014). The first two are related to the influence of the common-mode voltage V_{com} on the bearing voltage; the last two are caused by ground currents that result from the interaction of V_{com} with high dv/dt and the capacitance between stator winding and motor frame C_{wf}.

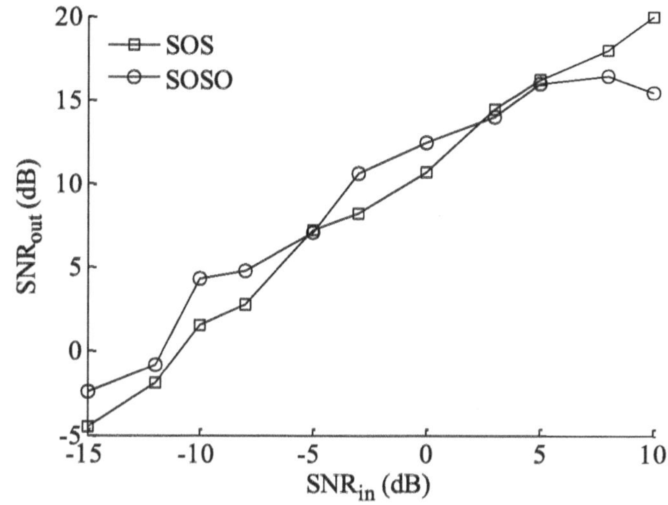

Figure 2.25 SNRs of the denoised signals from the SOS-based and SOSO based denoising algorithms for noisy simulated signals with different noise levels, Zeng and Chen (2020).

1. At low bearing temperature ($T_b \approx 25^oC$) and motor speed $n \geq 100rpm$, the dv/dt over the bearing, along with the bearing capacitance C_b, causes small capacitive $i_{\text{bcap max}} = 5 - 10mA$. At elevated bearing temperature T_b and/or low motor speed n, where the bearing acts electrically primarily as an ohmic resistance, the common-mode voltage v_{com} causes small bearing currents with amplitudes $i_{\text{bcap,max}} \leq 200mA$. This type of bearing current is of much smaller amplitude compared with the other types of bearing currents.
2. At intact lubrication film, the bearing voltage v_b mirrors the common mode voltage v_{com} at the stator terminals via the capacitive voltage divider bearing voltage ratio. If the bearing voltage v_b surpasses the threshold voltage of the oil film between the balls and running surfaces (approximately $5-30V$), the electrically loaded oil film breaks down. This breakdown causes an electric discharge machining (EDM)-current pulse. Peak amplitudes of the EDM-current pulse are $i_{\text{ibEDM max}} \leq 0.5 - 3A$.
3. The high dv/dt at the motor terminals, mainly because of the stator winding-to-frame capacitance C_{wf} causes an additional HF ground current i_g ('common-mode current'). This current excites a circular magnetic flux around the motor shaft. This flux induces a shaft voltage v_{sh} along the shaft, and between the two bearings, of the motor. If this voltage is large enough to puncture the lubricating film of the bearing and destroy its insulating

properties, it causes a circulating bearing current i_{bcir} in the 'stator frame–nondrive end–shaft–drive end' loop. Peak amplitudes vary depending on the motor size and are $i_{bcir\ max} \approx 0.5 - 20A$ for power less than $500kW$ ($Pr \leq 500kW$)

4. If the rotor is connected to earth potential (e.g., via the mechanical load) with a significantly lower impedance than the stator housing, part of the overall ground current i_g can pass the bearings as rotor ground current i_{rg}. Bearing currents due to rotor ground currents can reach considerable magnitudes with increasing motor size, and destroy bearings within short time of operation. When such currents flow, they add to potentially existing circulating bearing currents, where the latter generally have the much larger dominating amplitudes. In this case, however, the occurrence of EDM currents is impeded, because the lubricating oil film no longer has insulating properties. As a result, generally one of these types of inverter-induced bearing current is dominating.

Vibrations resulting from bearing degradation as discussed earlier are transmitted through the mechanical structure of the machine to its surface, although affected by the entire mechanical system consisting of frame, load, mounting, rotor, etc. They differ then from the original signals of the relative motion of rolling elements and the races because of resonances, damping, etc. Still they remain the closest to the source of the signal, and although their measurement requires additional sensors, they are by far the most widely used variable for bearing fault detection. Efforts have been made to use stator currents and stray fluxes, with limited success, although measurements of electromagnetic high-frequency signals have been used in to detect and identify the effects of bearing currents, (Muetze et al. (2014)).

Detection and identification of bearing faults and failure prognosis follows the standard processes. The features used are both in the time and time-frequency domain, and often advantage is taken of an understanding of the possible physical states mostly implicitly. Still, model-based approaches are not used; instead, both signal-based and artificial-intelligent approaches are favored. Unusual with respect to other faults is the use of entropy. Both unsupervised and supervised learning is used for the classification of faults, while prognosis tools include Hidden Markov Models, Hidden Semi-Markov models, Kalman filters, and variations.

Kim et al. (2012) propose a prognostic model that considers the discrete health state probability in the machine degradation process, which can effectively represent the dynamic and stochastic degradation of machine failure.

The health state probability estimation is carried out through exploring a full failure degradation process of the machine by selection of the optimum number health degradation states over time, from new to final failure stages. The cases of using too many health states in the degradation process can lead to the over-fitting problem of classification performance. On the other hand, insufficient number of health states may result in under-fitting. Therefore, both cases can significantly affect the performance of the classifier and will affect the remaining life prediction accuracy. In this work, the optimum number of health states was selected through the investigation of classification results with several health state cases.

The authors calculated ten statistical parameters using the time domain data. These feature parameters were mean, rms, shape factor, skewness, kurtosis, crest factor, entropy estimation, entropy estimation error, histogram lower and upper. In addition to these parameters, four parameters (rms frequency, frequency centre, root variance frequency, and peak) in the frequency domain were also calculated. The optimal features can be selected from the original feature sets according to the large distance evaluation criteria. In order to select the effective degradation features, the authors defined a value greater than 1.3 of normalised distance evaluation criteria. The ratio of 1.3 was selected based on past historical records for this particular application. From the results, the authors selected three features, Kurtosis, Entropy estimation value, and Entropy estimation error value for health state estimation. Their progression is shown in Figure 2.26.

To select the optimal number of health states to represent the bearing degradation process, several number of health stages were investigated using the data sets for training and prediction tests. As the basic kernel function of a Support Vector Machine, a polynomial function was used. Multi-class classification using 'one-against-one' was applied to perform the classification of bearing degradation. Several tests using different health states were conducted to verify the optimum number of health states, ranging from two states to ten states using the same test data arriving at five states. The probability distribution for each is shown in Figure 2.27.

After the estimation of current health state in terms of probability distribution of each health state, the RUL prediction was performed. For the prediction of machine remaining life, two parameters are used in this model, such as each health state probabilities at a certain time t and historical remaining life at each trained health state. The probabilities of each health state at certain time t provide a real time failure index in the machine failure process for RUL prediction. The RUL prediction of machine can be expressed as eq. 2.32 accordingly,

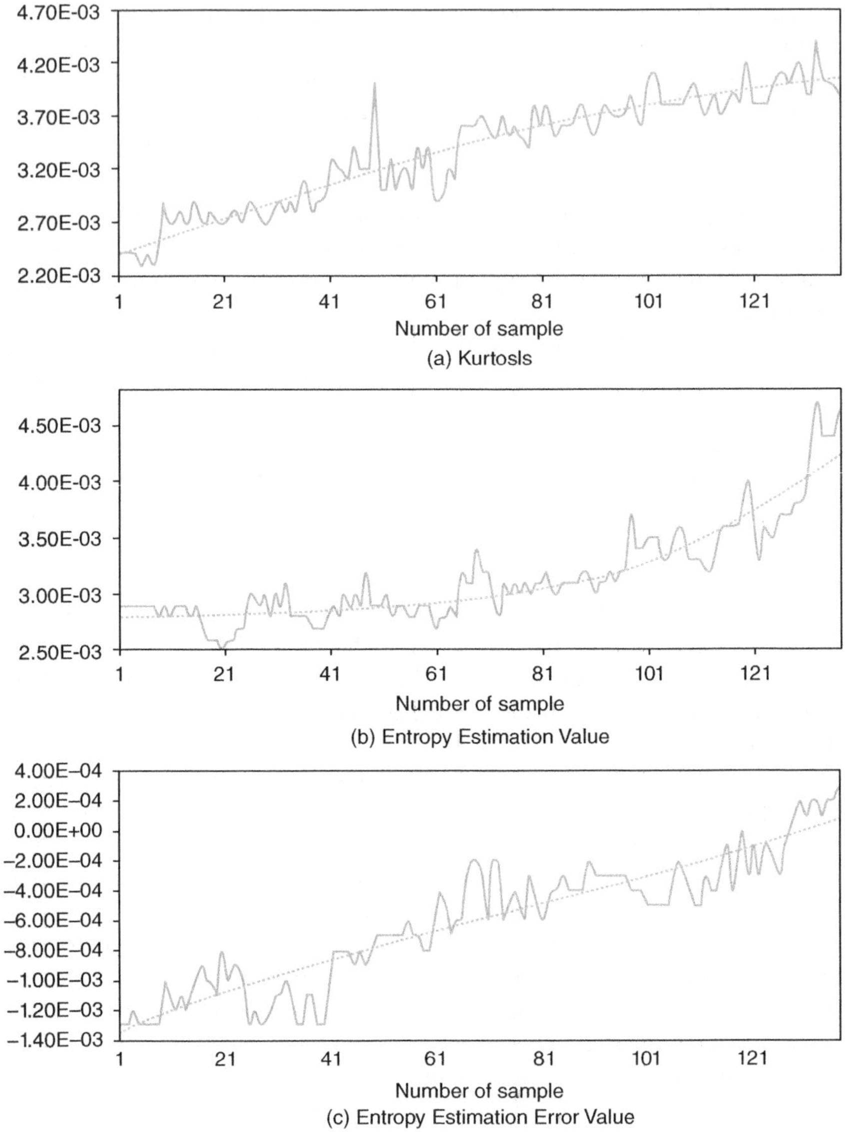

Figure 2.26 Feature trends of selected features Kim et al. (2012).

$$RUL(T_t) = \sum_{i=1}^{m} Pr(S_t = i \mid x_t, \dots, x_{t+u-1} \cdot \tau_i) \tag{2.32}$$

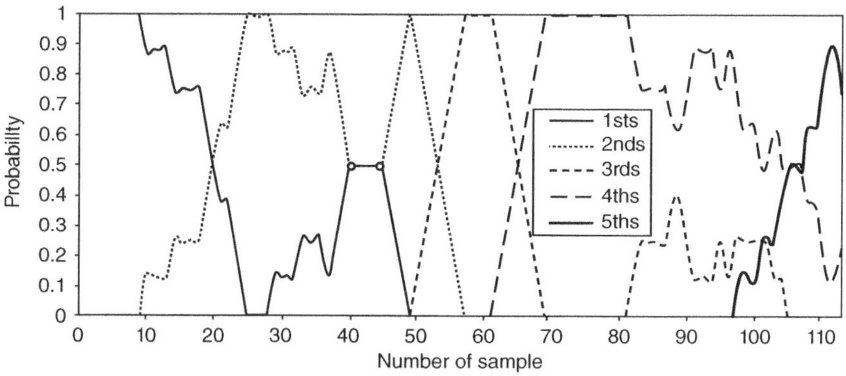

Figure 2.27 Probability distribution of each health state Kim et al. (2012).

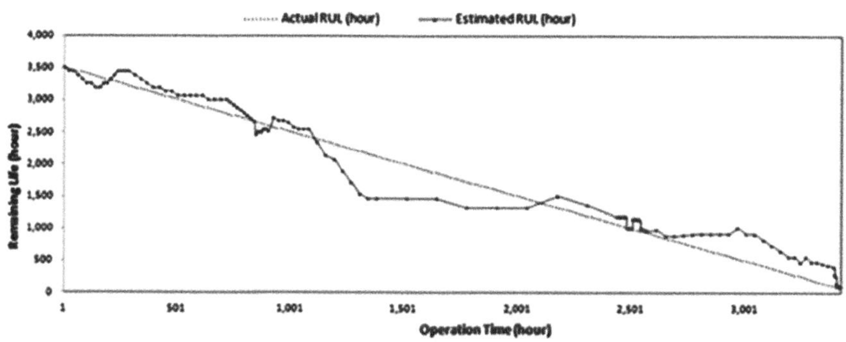

Figure 2.28 Comparison of actual RUL and estimated RUL Kim et al. (2012).

where S_t are current probabilities of each health state at time t, τ_i is historical remaining life at each trained health state i, m is number of health states, and x are the observations. Figure 2.28 shows the estimated RUL compared to the actual RUL.

Soualhi et al. (2014) combined a data-driven and experience-based approach: Time domain features extracted from the vibration signal were used as health indicators to track the degradation of bearings. The experience-based approach was used to provide the detection and the diagnosis of bearings by an unsupervised classification technique called the artificial ant clustering (AAC).

The data-driven approach was used to detect the 'imminence of the next degradation state ' (INDS) in bearings by Hidden Markov Model-based prognosis method. In addition to HMMs, a multistep time series prediction combined with the adaptive neuro-fuzzy inference system (ANFIS) was used to provide an estimation of the 'remaining time before the next degradation state' (RTNDS).

A classifier usually requires the input of identified, labeled data, a requirement that is often impossible to satisfy. The authors utilized an unsupervised classification technique, based on artificial ant clustering, so that similar data are grouped in the same classes, while dissimilar data are in separated classes. The number of classes is unknown.

For the diagnosis step, a similarity function is used, so that at any time the degradation state of the bearing was decided as the class that has the highest similarity. For the prognosis step, the results of the classifier step were utilized to adjust the HMM parameters. Applying the HMM, the imminence of the next probable state to the present one is determined to be as the state with the highest probability. That determination is based on the transition probabilities from the present state, the forward variable, and the emission probability in the new observation. Although the next probable (imminent) state can be determined, the time to arrive at this state is not determined. To accomplish that, a fuzzy inference system implemented in the framework of an adaptive neural network (see Figure 2.29). In this method, the authors used the rms, the mean value, and the average power of the vibration signal, recognizing that in a non-stationary operation time-frequency domain features would be preferable.

Entropy is an indicator to detect the dynamic change of time series, which has been widely applied in the fault diagnosis of rotating machinery. The vibration signal collected from a healthy bearing condition has a larger entropy value because of its high irregularity, while the vibration signal collected from a faulty bearing has a smaller entropy value due to its low irregularity. Therefore, entropy can be used to identify the bearing fault types or severities, and it has been used increasingly more in the fault diagnosis of bearings.

Singleton et al. (2015) introduced a stochastic data-driven approach, independent from fault severity diagnosis and continuously updates the RUL estimate as new data samples come in. For this purpose, they used an Extended-Kalman-Filtering based approach to first learn the degradation trend of the extracted features from the training data, then to apply this trend to testing data to predict RUL and, finally, to provide a confidence bound around the estimated RUL.

First, they consider both time and time–frequency domain features for tracking the degradation of the bearing. In the time domain, they use the variance feature as a reliable indicator of the bearing condition as it approaches

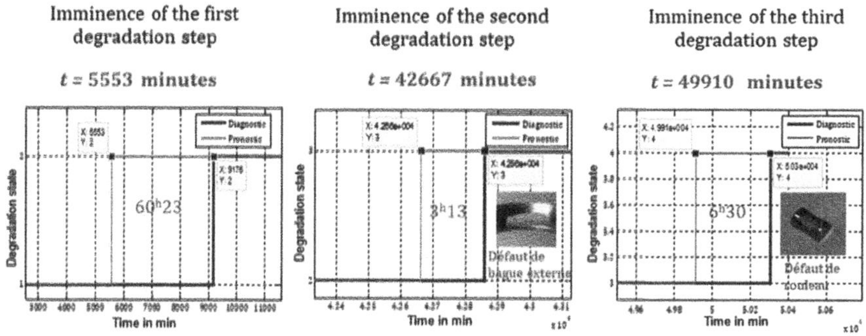

Figure 2.29 Neural network and HMM prognosis from Soualhi et al. (2014).

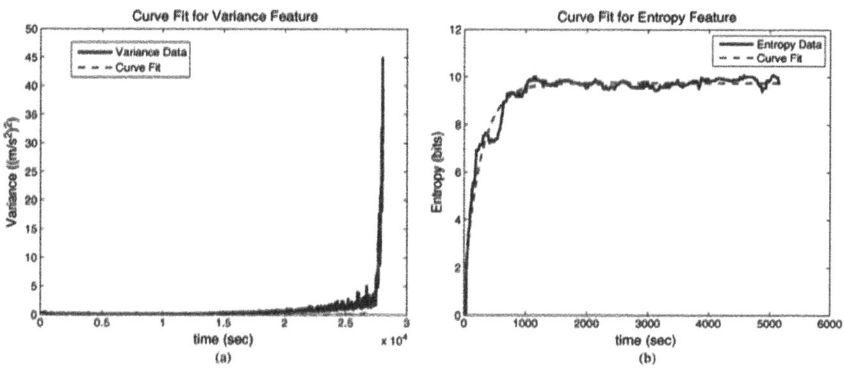

Figure 2.30 Curve fitting to variance and entropy features. (a) Variance curve. (b) Entropy curve fit Singleton et al. (2015).

failure. In the time-frequency domain, they use a novel entropy feature, which captures the complexity of the signal in both domains simultaneously. For the two different types of features, i.e., vibration and entropy, time-dependent degradation models are obtained through curve fitting. For the variance feature, an exponential of the form ae^{bt} was found to be the most suitable, whereas for the Time-Frequency (TF) entropy feature, a curve in the form of $a + be^{ct}$ was more suitable.

They notice that the entropy increases as soon as the first indications of fault develop, which relates well to the bearing physics of failure, where an initially localized fault (low entropy) becomes a general roughness with high entropy, see Figure 2.30.

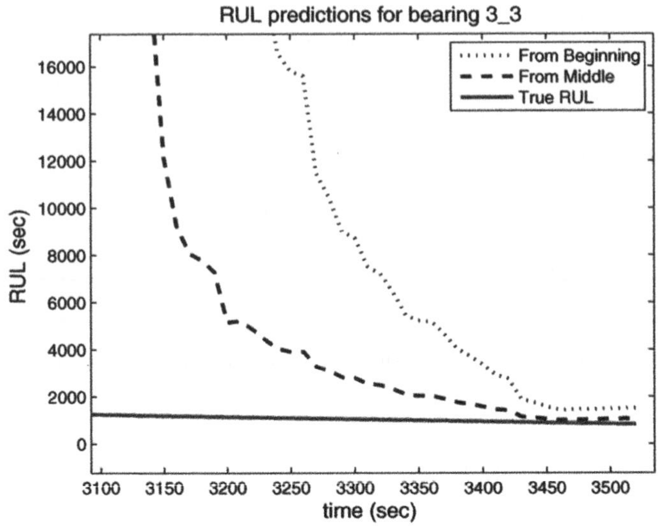

Figure 2.31 RUL estimation with different EKF tracking start times Singleton et al. (2015).

Second, they consider different analytic models for modeling the lifetime of the bearing and build the state vectors corresponding to each case. This consideration helps fully understand how different features evolve over the lifetime of the bearing and the effect of different model assumptions in the final RUL estimation. The main observation is that the starting point and duration of the process is an important factor in the accuracy of the estimation of the RUL, e.g. see Figure 2.31.

Third, they provide a confidence interval for the RUL estimates using the prediction errors calculated as part of Extended Kalman Filter (EKF), see Figure 2.32. An important aspect of prognosis and reliability is not just to predict the RUL but to add a confidence measure to it as well. In the testing phase, the confidence interval of the RUL prediction is calculated by using the error covariance matrix, $\mathbf{P}$. At the end of each step, the error covariance matrix is updated giving the uncertainty measurement between the predicted state and the true measurement. Since the first value in the state variable is the actual feature value, being either entropy or variance, the first element of the covariance matrix $\mathbf{P}(1,1)$ contains the uncertainty of the predicted state. Given this uncertainty, a 95% confidence interval can be placed around the estimate, and the upper and lower bounds. Similar

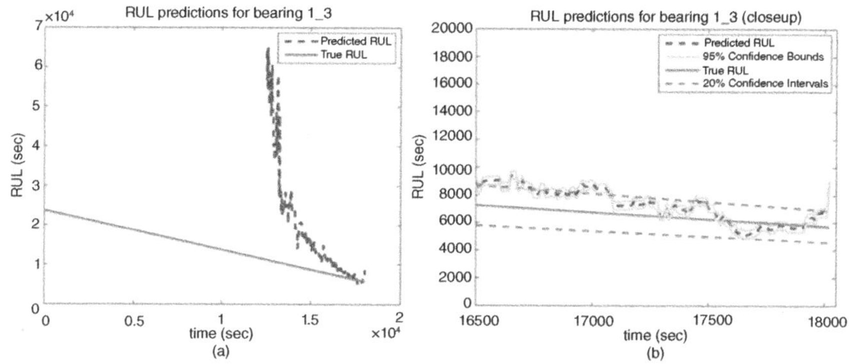

Figure 2.32 RUL estimation with the variance feature. (a) RUL prediction over time versus the actual RUL. (b) Confidence interval for the estimated RUL near the end of testing. Singleton et al. (2015).

to original RUL estimation, these values can also be extrapolated to the failure threshold to obtain upper and lower confidence bounds on the RUL predictions.

Ma et al. (2020) use the Hidden Semi-Markov Model (HSMM), which adds a temporal component to the HMM structure, to address the limitations of the HMM for the Markov property, and which can be directly used in prognostics. They recognize three problems in its application:

1. Determining the degradation states. Traditional works assume that the degradation states are known. If these states are predetermined, their number may not be accurate if there is a lack of corresponding prior knowledge.
2. Degradation feature selection. The wavelet packet decomposition, as well as many extensions are often used.
3. Prognostics using the HSMM. Combined with the HSMM, the particle filter (PF) and Monte Carlo (MC) methods are most commonly used for estimating the RUL.

The authors develop a new feature based on Hidden Semi Markov Models and sparse representation (SR). Here SR is used to find concise, high-level representations of Wavelet Packet Transform (WPT) subbands that match the structure of the degradation signal by using a learned dictionary.

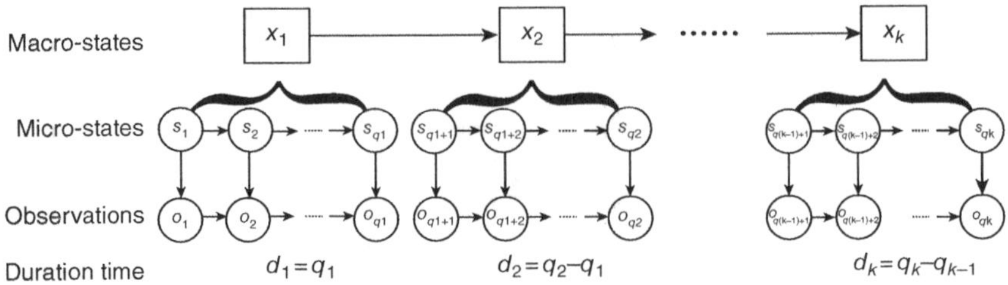

Figure 2.33 Observations, micro-states and macro states Ma et al. (2020).

The states in a segmental HSMM are called macro-states, and each macro-state consists of several micro-states. Only the transition between macro-states follows Markov principles. Figure 2.33 displays a basic framework of the HSMM. An HSMM can be described similarly to the Hidden Markov Models:

1. The number of macro-states is K, the segment at time t is s_t, and the end-point is q_i. All macro-states can be represented as $X = \{x_1, x_2, \dots, x_K\}$.
2. The initial state distribution is $\pi = \{\pi_i | \pi_i = P(x_i = i), i = 1, 2 \dots, K\}$, and the state transition probability distribution is:

$$A = \{a_{i,j} | a_{i,j} = P(s_t = x_j | s_{t-1} = x_i)\}. \tag{2.33}$$

3. The observation model $Y = \{y_1, y_2, \dots y_L\}$, the observation sequence with T elements $O = \{o_1, o_2, \dots o_T\}$ and the observation probability distribution in state I , $B = \{b_i(v)\}$, where:

$$B = \{b_i(v) \,|\, b_i(v) = P(o_v \,| s_t = x_i\,)\}. \tag{2.34}$$

The developed GMM-SR-HSMM Approach, shown in Figure 2.34, consists of three steps:

1. GMM clustering to determine the degradation states,
2. SR feature extraction to prepare the HSMM,
3. Degradation state recognition and RUL estimation based on the HSMM, using eq. 2.35 and shown in Figure 2.35

$$RL_i = \mathbf{k}U_i + \sum_{j=i+1}^{a} U_j, \tag{2.35}$$

where U_i is a unified variable of the duration values in state i, and $\mathbf{k}$ is a proportion value calculated adaptively.

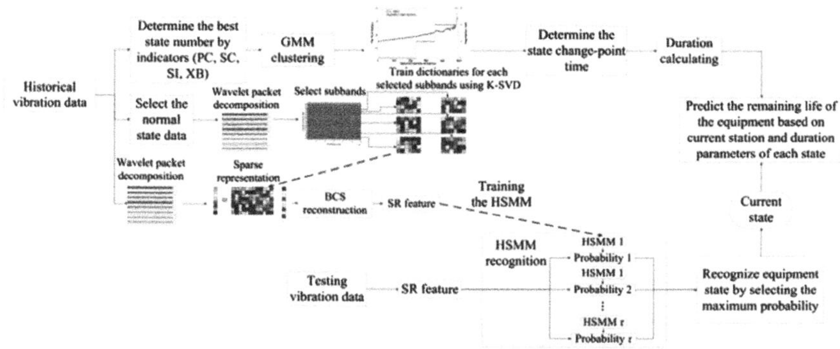

Figure 2.34 The framework of the proposed GMM-SR-HSMM approach. Ma et al. (2020).

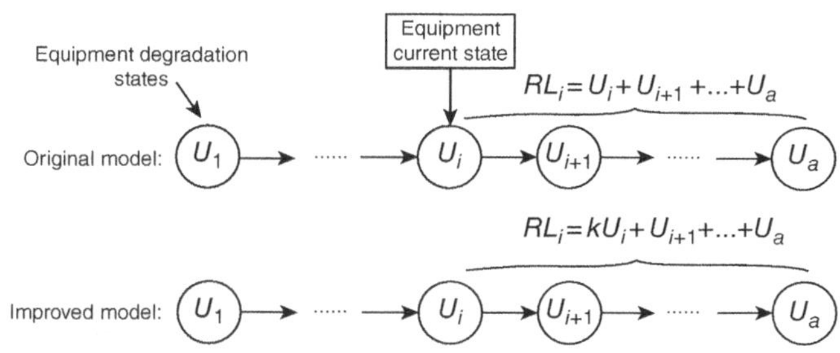

Figure 2.35 The calculation of remaining life for each state. Ma et al. (2020).

Following this method, the RUL was predicted using both SR and Renyi entropy (Rényi (1961)), with the SR method proving superior, see Figure 2.36.

More important for this type of failure than for other ones is calculating the bearing remaining useful life (RUL), as there is very little that can be done to mitigate a failure. Hence, only adequate warning will suffice. Using model-based, signal-based, or data-based methods, adequate training of any algorithm is required. The problem is significant, and hundreds of techniques have been proposed and hundreds of technical papers have been investigated. Three points of attention emerge:

- If a model is used, based on expected frequencies that will be encountered in the measured signals, Kim et al. (2017), Chen et al. (2020c), minimal training is required. As these frequencies result from localized faults, and as the number of these faults increases, the technique cannot be

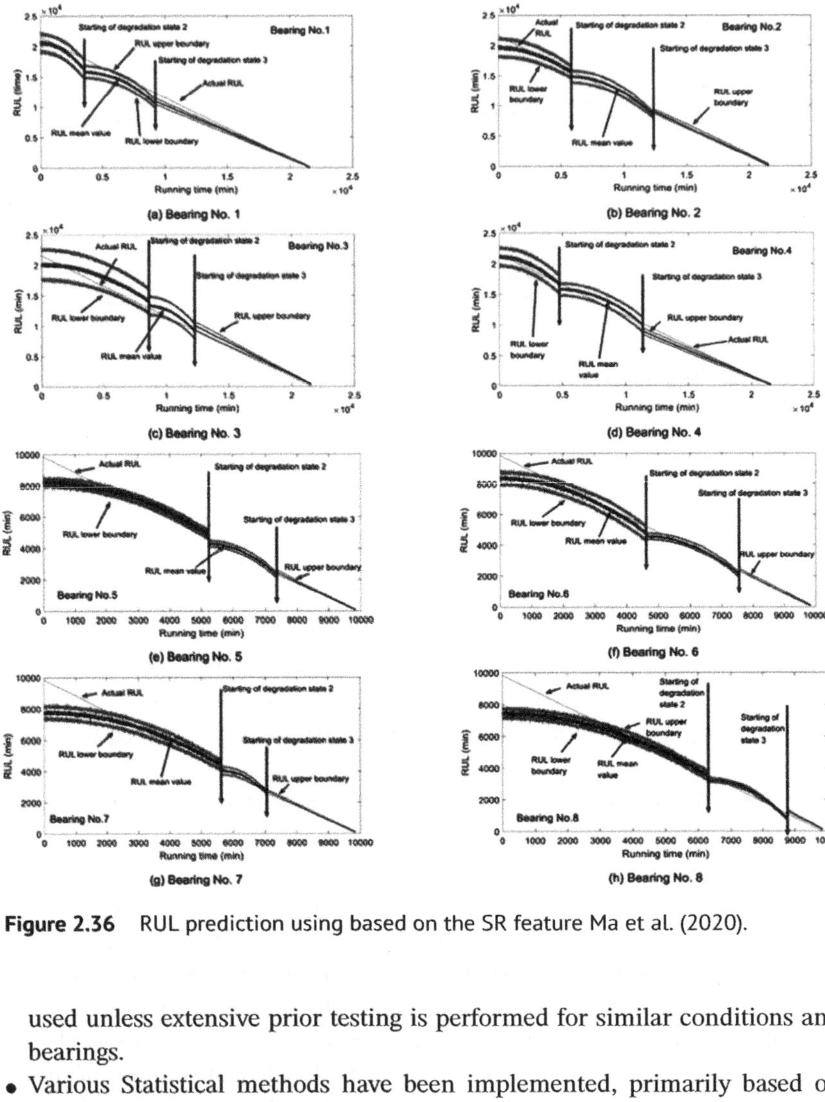

Figure 2.36 RUL prediction using based on the SR feature Ma et al. (2020).

used unless extensive prior testing is performed for similar conditions and bearings.

- Various Statistical methods have been implemented, primarily based on Hidden Markov Models, Ma et al. (2020) combined with AI tools Soualhi et al. (2014) or extended Kalman filter. There are complex and sometimes uncorrelated with physics.
- All these methods need a large amount of Data. However, training sets of aging data to apply and test proposed techniques Nectoux et al. (2012) have been developed, and they are useful for validating new methods, e.g. Singleton et al. (2015).

2.4.4 Insulation

Chapman et al. (2008) describe the insulating systems in a low-voltage electrical machine, identifying the characteristics and selection of each part, see Figure 2.37.

Maussion et al. (2015) discuss insulation characteristics for inverter-driven machines, standards, and testing. The International Standard Commission (IEC) published specific standards (IEC 60034–18-42 TS International Standard), respectively, considering type I machines, supposed to operate without any partial discharge (PD) and type II that can cope with moderate levels of PD. Type II are defined as 'system of stator/rotor windings of single or polyphase AC machines which are subjected to repetitive impulse voltages, such as pulse width modulation (PWM) converters and expected to withstand partial discharge activity during service'. These documents describe the required tests to qualify a material with regard to this standard, with several test voltages chosen to produce mean times to failure in the range from 100h to 3000h. It provides lifetime formulas between voltage and frequency in steady state. Despite its well-documented influence, temperature is not taken into account. Combining the frequency and voltage-dependent aging formulae leads to the general expression that gives a basic approach to insulation life expectancy.

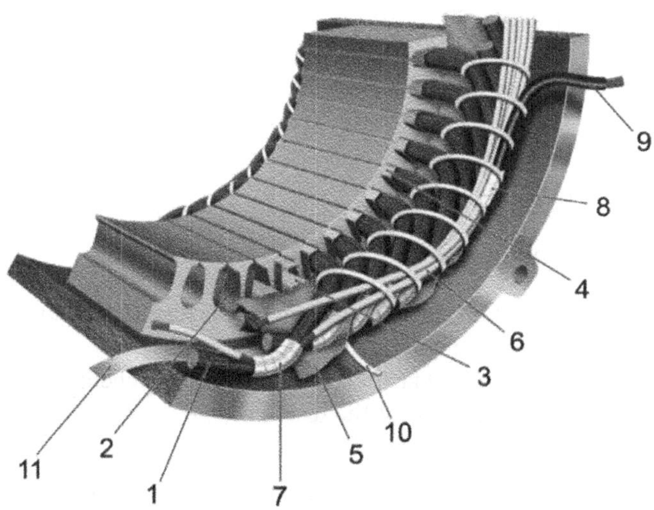

Figure 2.37 Overview of materials in a low-voltage insulation system: 1 turn insulation, 2 slot liner, 3 slot separator, 4 wedge, 5 phase separator, 6 lead sleeving, 7 coil-nose tape, 8 connection tape, 9 cable, 10 tie cord, and 11 bracing, Chapman et al. (2008).

In low-voltage electrical machines, the turn-to-turn is usually a thin layer of enamel, phase-phase insulation is a Nomex-type separator, and phase-ground insulation is also a Nomex-type liner. Each of these is subjected to thermal, mechanical, and electrical stresses, leading to aging, and decreased ability to handle the operating voltage, and leading to short circuit between turns, between phases or between winding and steel wall. Winding insulation is subject to four stress factors over its lifetime: material decomposition over time, the rate of which depends on operating temperature, electrically induced damage, contamination from the surrounding environment, and mechanical damage due to thermal cycling, opposing Lorentz force, and vibration. Small machines especially can be subjected to significant mechanical stress, however, little experimental evidence on the practical implication of these has been published. Maussion et al. (2015) realised a series of systematic experiments and using Design of Experiments and Response Surface identified voltage and temperature as the most influential factors affecting Type II insulation lifespan.

Many parameters have been used to detect insulation degradation other than partial discharges: as insulation ages, the resistance of the insulation decreases, hence a measurement of that resistance can lead to an understanding of the fault presence and severity estimation. The capacitance between turns and the grounded wall or other turns also depend on the insulation quality, and either directly or indirectly is a measure of this fault.

The thermal degradation of the can be considered a chemical reaction with effects on the electrical and mechanical characteristics. The Arrhenius Equation 2.36 describes the loss of life L due to the increased temperature:

$$L = L_0 e^{\frac{E_a}{R}\left(\frac{1}{T}-\frac{1}{T_0}\right)}, \tag{2.36}$$

where:

k is the rate constant (frequency of collisions resulting in a reaction),
T is the absolute temperature,
A is the pre-exponential factor, a constant for each chemical reaction,
E_a is the activation energy for the reaction,
R is the universal gas constant.

As insulation degrades, its ability to withstand mechanical stresses decreases, leading to breakdowns and possibly short circuits. Before that point is reached, its electrical strength is also diminished, i.e. its ability to withstand voltages and voltage transients. A possible indication is that, as the insulation degrades, its electrical resistance and capacitance decrease. The classic way to detect this is by measuring DC resistance, the polarization index (PI), and the dissipation factor (DF), the tangent of the angle δ between an AC voltage applied across the insulation (e.g. between the windings, or the windings and frame) and the current through it; the problem is that these measurements can only be taken with the machine off line. To address the problem of online detection of insulation health, a number of efforts have been made, based on measuring or estimating the leakage current from windings to insulation, and/or its characteristics.

Younsi et al. (2010) developed a testing scheme for online evaluation of health of insulation. Note that the equivalent capacitance, insulation resistance, and parameter DF are:

$$\begin{aligned} C_{eq,AC} &= C_{pg} + 2C_{pp}, \\ R_{eq,AC} &= R_{pg}R_{pp}/(2R_{pg} + R_{pp}), \\ DF &= \frac{|\tilde{I}_R|}{|\tilde{I}_c|} = \frac{1}{\omega R_{eq,AC}C_{eq,AC}} = \frac{2R_{pg} + R_{pp}}{\omega R_{pg}R_{pp}(C_{pg} + 2C_{pp})} \end{aligned} \tag{2.37}$$

with R_{pp} phase-to-phase resistance and R_{pg} phase to p ground resistance, C_{pp} phase to phase capacitance and C_{pg} phase to p ground capacitance.

The 'equivalent' insulation condition indicators, C_{eq}, R_{eq}, DF, for each of the three phases can be calculated online from the line-neutral voltage, $V_{abc,g}$, and insulation current, $_{Iabc,l}$ measurements from equations below:

$$\begin{aligned} C_{eq,abc} &= \left(2|\tilde{I}_{abc,l}| \cos \delta_{abc}\right) / \left(\omega |\tilde{V}_{abc,g}|\right) \\ R_{eq,abc} &= |\tilde{V}_{abc,g}| / \left(2|\tilde{I}_{abc,l}| \sin \delta_{abc}\right) \\ DF_{abc} &= \tan \delta_{abc} \times 100 = \tan\left[90^\circ - \angle\left(\tilde{V}_{abc,g}/\tilde{I}_{abc,l}\right)\right] \times 100 \end{aligned} \tag{2.38}$$

where δ_{abc} are the phase angles between $I_{abc,C}$ (capacitive current) and $I_{abc,l}$. The insulation condition indicators calculated from equations 2.37 are different from when the measurements are performed off-line using equations 2.38. There are three major differences between online and off-line measurements of C_{eq}, DF, R_{eq}, which are:

(a) the magnitude of the average voltage applied to the insulation,
(b) the distribution of the voltage applied from the line to neutral ends of the winding,
(c) the influence of the voltages in the adjacent phases on the phase-phase end winding insulation.

Their approach uses a developed High-Sensitivity Current Transformer (HSCT) to precisely and non-invasively measure the differential current (e.g., the Insulation current) of each phase winding from the motor junction box. Five stages were identified in the aging process:

(1) short-term behavior during initial heating,
(2) continued increase in the overall insulation current and at a constant loss angle,
(3) a peaking of the insulation current followed by a steep decrease. One interpretation for this peak is that the evaporation of volatiles and the beginning of resin weight loss competes with the thermally activated conduction and polarization,
(4) aging is the predominant process with loss of capacitance and increase in resistance, and,
(5) motor end of life with electrical breakdown.

An alternative to these measurements has been realized by measuring the response of the insulation to switching transients. Babel and Strangas (2014) proposed applying transients on the operating machine through a function generator, and measuring the leakage current through a low value resistor, to maintain the bandwidth of the current. They measured the response with external instruments and used the peak current at switching to determine the health of the insulation. They then used this overshoot in a model:

$$I_{leakage} = \alpha e^{\beta t} \tag{2.39}$$

$$\alpha = I_{healthy} \tag{2.40}$$

$$\dot{I}_{leakage} = \alpha e^{\beta t} + (\beta + t\dot{\beta}) + e^{\beta t}\dot{\alpha} \tag{2.41}$$

with slow changing parameters α and β. This dynamic model was used in an Extended Kalman Filter, to monitor aging.

Nussbaumer et al. (2015), Zoeller et al. (2017), Zoeller et al. (2016) utilized the fast transients due to the switching SiC Voltage Source Inverters in place of the imposed pulses. The time instant of the switching command is used as a pretrigger for the detection of the exact switching instant. Thus, the data collection starts before the actual switching instant. Although this reaction contains frequency components above the bandwidth of industrial current sensors (bandwidth between several tens to a few hundreds of kHz), no special sensors were necessary. The current sensors used in these investigations have a cutoff frequency of $300kHz$. The health indicator used the Fourier transformed signals of the healthy machine and the one under test to derive a health indicator called ISI. The root-mean-square deviation (RMSD) is chosen as a comparative value and serves as ISI for the respective phase:

$$ISI_{p,k} = RMSD_{p,k}(x_1, x_2)$$
$$= \frac{\sqrt{\sum_{g=n_{low}}^{n_{high}} \left(Y_{ref,p}(g) - \left|Y_{con,p,k}(g)\right|\right)^2}}{n_{high} - n_{low}}. \quad (2.42)$$

The Fourier transformed signals Y_{ref} and Y_{con} refer to healthy machine condition (reference) and a later condition assessment, respectively. The index p defines the investigated phase (U,V,W). The variables n_{high} and n_{low} define the compared frequency range and depend on the sampling rate and investigated window length. The definition of the evaluated frequency range allows separating other influences like cabling or grounding that is also leading to a change in the drive's high-frequency behavior. Due to the fact that a single measurement's duration is in the range of $100\mu s$, the procedure can be repeated m times to increase the accuracy of the method. Thus, the index $k(1, 2, 3, \ldots)$ denotes the number of the consecutive measurements. g is the discrete frequency, calculated using the sample frequency, number of samples, and window length.

Jensen et al. (2018) and Guedes and Silva (2020) studied the response of the impulse testing associated with fast switching caused by the power electronics, which themselves, as discussed already, may be the cause of insulation failure. This response can be monitored online and depends on the parameters of a distributed RC circuit that describes the insulation.

Jensen et al. (2018) designed and proposed a simple peak-sampling circuit of the current at switching. They used elevated temperature for accelerated aging, and determined failure when insulation had visibly deteriorated. Figure 2.38 shows the estimation the RUL during accelerated aging.

As noticed in the papers discussed earlier, degradation affects the dielectric properties of the insulating material, as well as its resistance. Tsyokhla et al. (2019) go beyond the measurement of the current transient resulting from the switching impulse of PWM. They developed a method based on the spectral analysis of the common-mode impedance of the Winding- Ground Wall, exploiting again the intrinsic common-mode excitation due to space-vector pulse width modulation (SV-PWM) in standard two-level converters. They base their method on the common mode voltage and leakage current using measurements of all three-phase voltages and leakage current, Figure 2.39.

Finally, the capacitance and resistance of the insulation material are shown to depend on frequency (Figure 2.40), and hence the detection and prognosis scheme dependent on the switch characteristics.

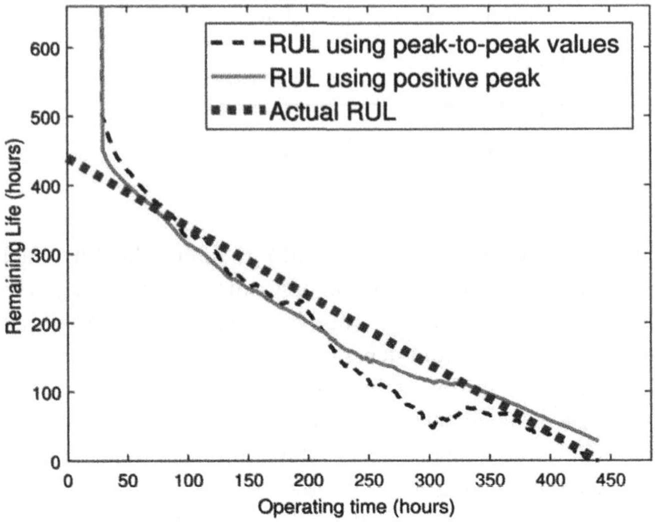

Figure 2.38 RUL estimate using peak-to-peak value of the switching current, Jensen et al. (2018).

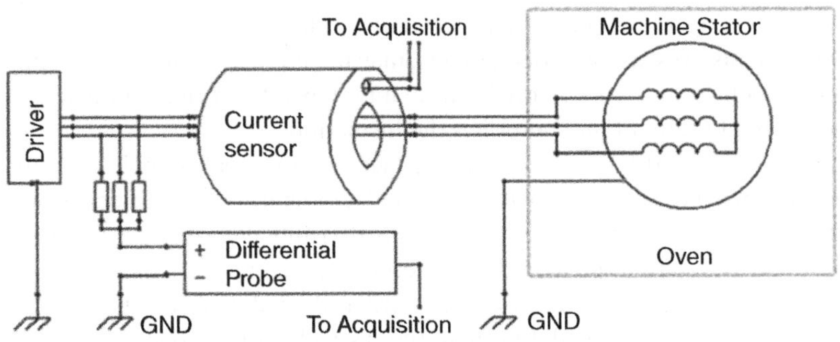

Figure 2.39 Experiment equipment setup, Tsyokhla et al. (2019).

Then, the method is based on the spectral analysis of the common-mode impedance of the Ground-to-Winding insulation, The technique has been successfully applied to the insulation of medium-voltage drives, with accurate RUL estimates based on the capacitance trend, as shown in Figure 2.41.

Guedes and Silva (2020) proposed a system based on the use of a single high-sensitive current transformer (HSCT) and a conventional voltage transformer (VT) for the online evaluation of the insulation condition. An HSCT was developed to allow the measurement of leakage currents with magnitudes in the

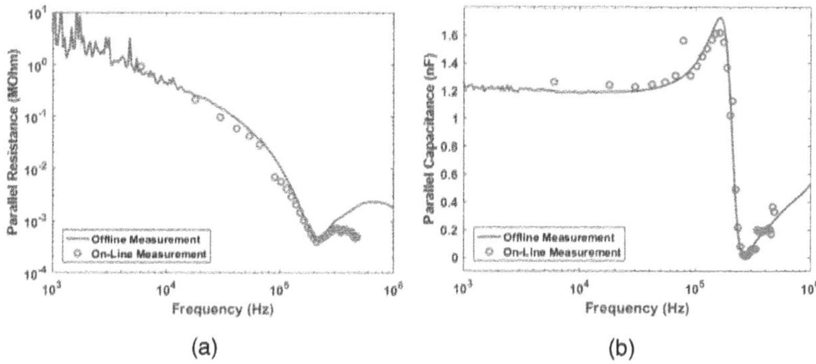

Figure 2.40 Insulation frequency-dependent characteristics, Tsyokhla et al. (2019). (a) Normalized Ceq over time. (b) Fitting of normalized Ceq over time.

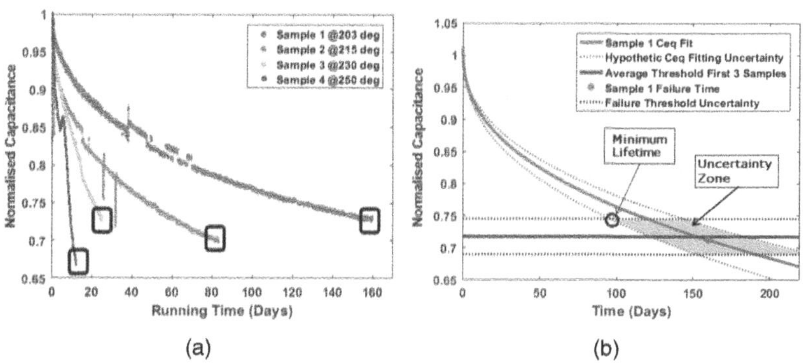

Figure 2.41 Equivalent capacitance and RUL estimation, Tsyokhla et al. (2019). (a) Normalized C_{eq} over time. (b) Fitting of normalized C_{eq} over time.

order of μA. Advantages of this proposed system are: it can be applied to low- and medium-voltage machines with three or six accessible terminals; it can be used for in-service or new installations, since no special requirements or prerequisites are necessary, besides the installation of the HSCT and the VT. The current measured by the HSCT is the sum of the phase currents of the machine and of the insulation leakage currents. Since the machine does not have any neutral connection, the sum of the phase currents is zero. For this reason, the current measured by the HSCT is the resultant leakage current. From the evolution of these quantities over time, compared to their values when the machine was considered adequate, an ANN was used to detect and identify the stress agent affecting the insulation. The insulation failure forecast and identification of the stress agent are obtained from the analysis of the evolution of the parameters R_{eq}, C_{eq}, and DF_{eq} over time, starting from the base values. From the evolution of these quantities over time, compared to their values when the machine insulation was considered adequate, an ANN is used to detect and identify the stress agent affecting the insulation.

The data set was assembled from test reports of 115 machines. From this total, 100 motors had 2 test results and 15 had only one. Thus, the data set had a total of 215 entries for RI, CI and DF. The tested machines have been in operation for 5–15 years in thermoelectric power plants and refineries. The resilient backpropagation algorithm (RPROP) was used to perform the training of the ANN. The RPROP algorithm uses the sign of the partial derivative, $\partial e(t)/\partial w_{ij}$ (derivative of the error in relation to the weight w_{ij}) instead of the value of the derivative, to adjust the network weights.

The forecast of the behavior of the resistive component (IR) of the insulation current of the machine was developed as a means to estimate the time-to-failure for the insulation. Both a linear (ARMA) and a nonlinear (ANN) models were evaluated.

Testing was in a thermal chamber to control temperature; moisture and oil contamination were introduced to the sample through mist. Two errors were used to evaluate the forecast, root mean square error (RMSE) and the time-to-failure (FETF). The results show that as the time (lag) increases, both models converge with the reduction of the estimation errors for the resistive component of the current, $\widehat{I}_R$ (RMSE), and $\widehat{T}_F$ (FETF). However, it was noticed that the neural model presents smaller prediction errors in all cases.

Partial discharge is a localized dielectric breakdown in insulation, that does not completely bridge the two conductors. Partial discharges can start in air spaces within the insulation (i.e., voids), leading to polymeric bonds breaking and resulting in conductor short-circuits (see Madonna et al. (2019)). As degradation progresses, these short-circuits result in changes of the insulating material makeup and create pockets or voids in it; discharges in these voids are initiated, a precursor of arcing and breakdown.

Partial discharges are generated in insulation due to high voltage power frequencies and high-voltage spikes. Partial discharges within an insulating material are usually initiated within gas-filled voids within the dielectric. Because the dielectric constant of the void is considerably less than the surrounding dielectric, the electric field across the void is significantly higher than that across an equivalent distance of dielectric. Figure 2.42 shows a schematic of the capacitances between insulation surfaces. Due to the different dielectric characteristics in the void and the insulation, a voltage spike containing high-frequency components, can lead to the breakdown in a void. The breakdown is accompanied by a short fast current pulse. Once partial

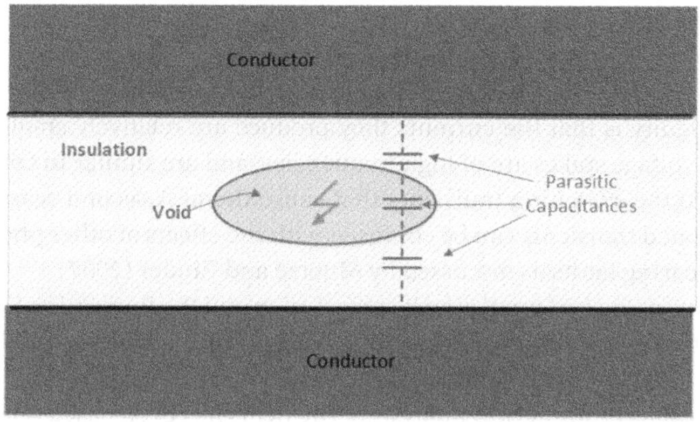

Figure 2.42 Partial discharges in voids.

discharges start, they lead to progressive deterioration of the insulating materials, and eventually breakdown.

Detection of these discharges is therefore useful, although difficult to conduct online. The high values of the rate of voltage rise, dv/dt, resulting from high-frequency PWM, both increase the Partial Discharge occurrence and make it more difficult to distinguish the high frequencies of the signal and of the discharge (see Hammarström (2018)). Partial discharges during operation have high-frequency components, which can be similar to noise and those resulting from the operation of healthy and faulty power electronic switches.

Unlike power frequency sinusoidal voltages, fast voltage impulses create an uneven distribution of the voltage in a winding. This unevenness increases with the rise time; with rise times less than 200*ns*, over 50% of the voltage drop can occur between the first and second turns of the coil, thus increasing the stress between these first turns. Little has been done to alleviate the negative effects.

In low- and medium-voltage electrical drives, the cause of PDs is primarily the fast voltage transients caused by the inverter PWM, exacerbated by the length of the cable between the inverter and motor. Overstress on winding can be due to uneven voltage distribution across the windings arising from high-voltage levels and fast rise times of impulsive waveforms. It has been reported that for rise times less than 200*ns*, over 50% of the voltage drop can occur between the first and second turns of the coil (see Stranges et al. (2009)). If PDs are initiated or enhanced in defects within the insulation or on insulation/air interface as a result of this overstress, this will lead to accelerated degradation of the insulation. Type I insulations must be PD-free for all its operating life, and design and qualification tests are described in to achieve this goal with reference to rotating machine insulation.

Detecting partial discharges on line is difficult, for a number of reasons: A main difficulty is that the currents they produce are relatively small, and the resulting voltage spikes are of high frequencies, and are similar in content and location to the switching transients that cause them. A second reason is that the associated transients can be confused with the effects of other phenomena, like the bearing faults as discussed by Muetze and Binder (2007).

Detecting aging and predicting its progression and the Remaining Useful Life of insulation is a daunting task, and it is made more complex by the use of fast switching inverters. Modelling the insulation materials through analytical expressions or numerical analysis of the fields has been based on extensive testing.

Four separate methods have been used to evaluate for Partial Discharges: separate testing of materials or semi constructed motors (motorettes), off line testing, usually at the end of the production line, before commissioning, simulations, and online testing, an emerging area of research.

In an entirely experimental approach, Hammarström (2018) used multilevel inverters feeding twisted conductor pairs to investigate the inverter effects on the PD magnitude. The experimental results show that the total summed PD magnitude (exposure) drops considerably when applying the three- or higher level inverters, and that the rise time of the pulse is significant.

They presented results that provide additional evidence on the relation between rise time and PD characteristics, also in cases when the number of voltage levels in PWM inverters is increased from two to at least five. It is further shown that the total PD exposure drops considerably, already when using a three-level inverter.

Neacsu et al. (2000) and Bidan et al. (2003) proposed offline testing through impulse generators and measurements of terminal voltages and currents. Bhure et al. (2017) used an antenna to measure the emitted radiation during PD, and processed the results using Discrete Wavelet Transform and classifiers.

Nyanteh et al. (2011) discuss simulation models as alternatives to extended experiments to determine the feasibility of estimation of the remaining useful life of insulation. The two most important equations to be solved in a numerical simulation are the continuity equation for the flow of discharged ions and the Poisson equation for the distribution of potential at each simulation step.

$$\frac{\partial N_e}{\partial t} = \frac{\partial (N_e W_e)}{\partial x} + \frac{\partial}{\partial x}\left(\frac{D \partial N_e}{\partial x}\right) \tag{2.43}$$

$$gradV = \frac{\partial}{\partial x}\left(\frac{\partial V}{\partial x}\right) + \frac{\partial}{\partial y}\left(\frac{\partial V}{\partial y}\right) = -\frac{\rho}{\varepsilon}, \tag{2.44}$$

where N_e is the electron density, W_e are the velocity of the electrons, and D is the diffusion coefficient. These equations couple the temporal and spatial charge distribution in a 1D dielectric material undergoing PD. V is the potential at given x and y coordinates for a two-dimensional PD simulation. The authors consider the finite element approach of calculating the field flux of the dielectric during discharge and the formation of electrical trees to be accurate for most research purposes. They also propose a new approach to fault diagnosis on shipboard power systems using measurements of PD and AI to correlate the extent of insulation damage due to electrical treeing.

In a series of articles, Ghosh et al. (2020) address the issue of online detection of partial discharges. Their objective is to develop an automated method for the detection of PD in the presence of commutation disturbance, which occurs during the rising front of impulsive voltages. This work demonstrates that PD detection can be facilitated by the repetitive and deterministic nature of such disturbances, which can serve as a reasonable basis to determine the presence of PD in a group of traces recorded from the detector. Successive commutation disturbances corresponding to successive switching events will trigger the acquisition device to acquire signal frames, which may consist of two parts: the deterministic commutation disturbance and the stochastic PD. The deterministic parts of the signal correspond to the switching characteristics, depending on the type of switching device used, the rise time, and applied voltage. As long as these factors remain unchanged, which is indeed the case in practical applications, the commutation disturbance corresponding to successive switching events remains largely similar in magnitude, frequency, and shape. This will mean that the instantaneous magnitude of the deterministic part of the signal at any time instant will have a relatively small variation across successive switching events. On the other hand, if there is any PD, the stochastic and nonrepetitive nature of the PD will definitely result in significant differences between the acquired signals with respect to the amplitude and time of occurrence of pulses. This will cause a significant variation in instantaneous magnitude of the signal at any time when a PD is initiated at any switching event. A measure of the time variation of signal magnitudes, therefore, can be used to track whether PD is present in a group of acquisitions and to identify the instant of occurrence of PD.

Once a PD is detected, the next step is to separate the commutation disturbance from each frame so that only PD part of the signal is left, thus identifying PD–pulse characteristics as well as distribution and concentration of the PD around the switching instant. PDs can appear progressively closer to the commutation instant with faster switching and higher supply voltage levels, and it is a fundamental parameter for insulation design and acceptance/rejection criteria, being related to the (PDIV).

To separate the commutation disturbance, the instant in the standard deviation time plot when there is an abrupt change in the standard deviation magnitude must be identified accurately. This is achieved using the mean value of the standard deviation. More precisely, if we track the mean value of standard deviation in successive overlapping short time windows of a few nanoseconds (windowed or sliding mean), we should be able to identify the approximate region of PD onset by determining the point around which the windowed mean exceeds the overall mean value of the standard deviation.

The instant corresponding to the PD inception splits each frame into two halves: the first half of the signal is, therefore, the repetitive signal duration in each frame and represents the commutation disturbance, while the second half is the stochastic duration which pertains to PD, extracted using the above method.

The method was tested on a $400V$ three-phase motor electric motor. The extraction of PD from the commutation transient signal is shown in Figure 2.43, for an applied voltage of $2.5kV$ peak-to-peak (Repetitive Partial Discharge Inception Voltage RPDIV defined in the IEC 61934 standard) having a rise time of $60ns$, at which a number of PD can be observed. It has to be noted that, in this case, the PD pulses are clearly separated from the commutation disturbance.

2.5 Open and Short Faults, Eccentricity, Broken Magnets and Rotor Bars

All these faults can be detected, diagnosed, and their failure predicted using primarily operating signals, currents, voltages, fluxes. The methods used can and often have to differentiate between relatively similar manifestations of different faults.

Following the discussion in many works, e.g. Hu et al. (2020b), the different methods can be classified in:

Model-based methods that utilize a number of models: The switching state function model, state-space model, and mixed logical dynamic model diagnose the open-switch faults with the help of observers. These methods claim fast diagnosis, often within one fundamental power period. Nevertheless, they are sensitive to the model accuracy. An inaccurate model may lead to poor diagnosis performance, e.g., false positive and negatives.

Signal-based methods utilize the variables already in the control system to infer the health state of insulated gate bipolar transistors (IGBTs) without

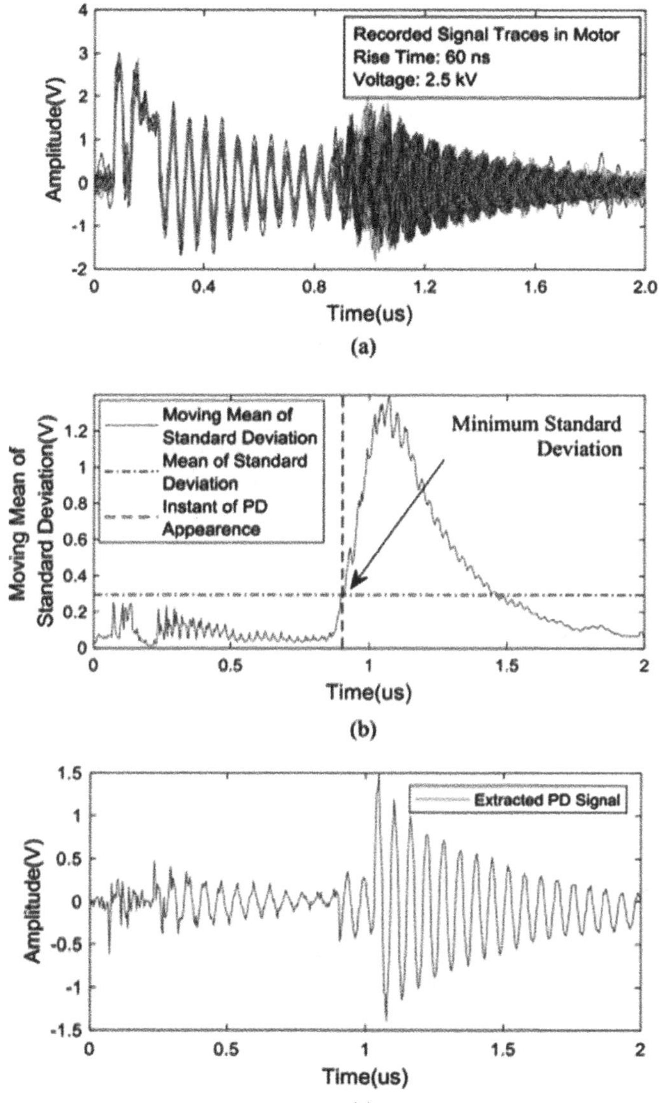

Figure 2.43 Separation of PD from commutation disturbance in motors using moving mean of standard deviation for $60ns$ rise time and applied voltage of $2.5kV$ (RPDIV). (a) Group of acquired signals. (b) Standard deviation time plot. (c) Extracted PD signal, Ghosh et al. (2020).

additional hardware. The current trajectory method differentiates between the trajectories of the current, between healthy operation and the different fault conditions in order to identify the faulty components. It can diagnose the single and double switch failure faults within at least one fundamental period and can be normalized to remove load dependence. Since trajectory tracking is sensitive to transients, this method can be inaccurate, although filtering can improve its characteristic. Filtering, though, or mean value-based methods require many points, stressing memory requirements and increasing time of detection.

Data-driven methods require two stages before diagnosing the open-switch fault, namely, feature extraction and training. The panoply of tools presented in other section have been used: wavelet decomposition, e.g. detail part coefficients and detail part energy. Moreover, principal component analysis has been used to reduce the feature dimension and improve training efficiency. Ensemble learning has also been used to improve accuracy, however, the performance has not always been satisfying. Deep learning, has been attempted, but it suffers from limited data availability and long execution times.

2.5.1 Induction Machines

The faults in induction machines that are considered here include primarily electrical faults (open and short windings, broken rotor bars, and end rings), and eccentricity. For diagnosis of faults the whole set of measurements discussed in subsection 2.4.1 are used: measurement of stator voltages and currents, real and reactive power, vibrations, and fluxes, both inside the machine and stray, or leakage.

Although an increasing portion of induction machines is supplied by inverters, and thus are central to the topic of this book, many large machines are operating directly connected to the grid. Research on fault diagnosis and failure prognosis on these machines produced significant understanding and results. Some are discussed here, as the developed methods can be transferred to induction motors operating in electrical drives.

Most of the techniques discussed in this section, detect and identify more than one faults, and separate them in terms of diagnosis and severity estimation. Failure prognosis is discussed in a limited subset of these works, either because the faults progress quickly and hence immediate decision and action is required, or the fault is stable and independent of operating conditions, and the operator has to wait till the next scheduled maintenance.

The selection of sample papers presented here aims to demonstrate the wide breadth of signals and techniques used for the diagnosis on Induction Machine faults.

2.5.1.1 Stator Fault Diagnosis

Briz et al. (2004) aimed to detect asymmetries in the stator. They injected a high-frequency carrier voltage signal in an induction motor operating under indirect field orientation.

This allows the measurement of the spatial saliences and different q- and d- axes stator transient inductances in a stationary reference frame. When there is an asymmetry or saliency in a machine's stator transient inductance, the interaction between the carrier-signal voltage, and the saliency will produce a carrier-signal current consisting of both positive- and negative-sequence components. The high-frequency machine model can be written as:

$$v^s_{qds\_c} = V_c e^{j\omega_c t} \tag{2.45}$$

$$v^s_{qds\_c} \cong j\omega_c L_{\sigma s} i^s_{qds\_c}. \tag{2.46}$$

When there is an asymmetry or saliency in the stator transient inductance, the interaction between the carrier-signal voltage and the saliency will produce a carrier-signal current, consisting of both positive- and negative-sequence components, as in Equation 2.47, where θ_e is the angular position of the saliency in electrical radians, h is the harmonic number of the saliency, and ω_c is the carrier frequency in radians per second.

$$i^s_{qds\_c} = i^s_{qds\_cp} + i^s_{qds\_cn} = -jI_{cp}e^{j\omega_c t} - jI_{cn}e^{j(h\theta_e - \omega_c t)}. \tag{2.47}$$

They use this injection of high frequencies to detect short circuits through the unbalance. Shorts give rise to a detectable noticeable variations of the stationary saliency and consequently of the dc component of the negative-sequence carrier-signal current spectrum. The magnitude of this component is large enough, even for a relatively small number of turn-to-turn faults, that it is distinguishable from parasitic components of the negative-sequence carrier-signal current, i.e., those that exist even in a healthy machine. Not only can the presence of a fault be detected, but also the phase in which it is located can be identified, due to the difference in the phase angle of the dc component complex vector.

In a $380V$ machine, the carrier signal used was $10V$ peak at $535.7Hz$. The dc component of the negative-sequence carrier-signal current spectrum corresponds to the turn-turn short circuit.

A short circuit poses a major problem in operation, since it can create high currents, increased temperatures, forces, and quickly catastrophic failures. A short circuit in the inverter, leaving a switch continuously in the conducting stage, e.g. due to latch-up, can be determined from the inverter monitoring scheme. A short circuit in the windings, though, turn-to-turn, phase-to-phase and phase-to-frame cannot be as easily and quickly detected, although only a

Figure 2.44 Open circuit faults: a. Parallel coils in a phase, b. One coil in a phase, c. Open in an inverter switch.

few ms or 1 to $2s$ are available for mitigation. In induction machines the fault cannot easily be directed using Field Orientation, since the fault disrupts the flux and the field oriented control of induction machine drives.

In Rotor Field Oriented induction drives, it is therefore necessary to use directly the signals of the stator, and the detection techniques that have been used, include the signature analysis of the stator real and reactive power, Drif and Cardoso (2014). They discuss how, when the fault affects a number of turns rather than all of a phase, the twice frequency component in the instantaneous reactive power is directly related to the number of open turns.

In a healthy induction machine the real and reactive power can be calculated from the space vector of currents and voltages:

$$[s] = u_s(t)i_{s,0}(t) = p_0(t) + jq_0(t). \tag{2.48}$$

With a stator fault, the instantaneous active and reactive power become:

$$p_d(t) = p(t) + \Delta p_d(t) \tag{2.49}$$

$$q_d(t) = q(t) + \Delta q_d(t), \tag{2.50}$$

where $p(t)$ and $q(t)$ are the mean values of the faulty motor active and reactive powers, and $\Delta p(t)$ and $\Delta q(t)$ are, respectively, the active and reactive power pulsations due to the fault at the disturbance frequency $f_d - 2f$ and f is the voltage supply frequency.

The presence of the stator fault manifests itself in the instantaneous active and reactive power by the presence of a spectral component at twice the fundamental supply frequency. This additional frequency component in the two quantities spectra, subsequently called characteristic component, provides an extra diagnostic information about the machine condition in the presence of stator faults in operating three-phase induction motors both fed from the grid and supplied by an inverter.

Ghosh et al. (2017) model a squirrel-cage aluminum motor with a stator-flux based control method. The space harmonics of the rotating magnetic flux can be presented as in eq. 2.51 where ω_{sh} is the electrical speed of the space harmonics of the angular MMF and k is the positive order of the flux space harmonics.

$$\omega_{sh} = \left(\frac{z}{p}\omega_r \pm k\omega_o\right) \tag{2.51}$$

The harmonic slip frequency is in terms of the time harmonics. Eq. 2.52 gives the window for space harmonics as well as time harmonics tracking.

$$f_{sh} = f_{sh0} - \frac{z}{p}\left(\frac{1}{s_k} - 1\right)f_r. \tag{2.52}$$

In the stator flux control method, again the torque is given by:

$$T_e = \frac{3}{2}\frac{P}{2} \cdot \lambda_{ds} \cdot i_{qs} \tag{2.53}$$

but for i_{ds} to be aligned with the d axis component of the stator flux λ_s, a decoupling compensation is used.

The proposed drive system is a stator flux-oriented control using current, voltage, flux, and speed as feedback. The stator inter-turn fault is detected in control logic by measuring the flux using flux sensor and time and space harmonic tracking.

The authors used particle swarm optimization to track the change in parameters of the stator winding, thus detecting the short circuit fault.

Liu et al. (2015) take a stator inter-turn fault detection approach for the inverter-fed Induction Machine based on the Teager - Kaiser energy operator (TKEO).

The continuous form of the TKEO, applied to a signal s(t) is given by eq. 2.54:

$$\psi(x(t)) \triangleq \left(\frac{dx}{dt}\right)^2 - x(t)\left(\frac{d^2x}{dt^2}\right) \tag{2.54}$$

By making the most of the characteristic of the switching modulated signal, additional high-frequency injection is not required, and the approach is more suitable for continuous operation. Considering the fact that high-frequency response is insensitive to the parameter variation, the zero sequence voltage (ZSV) signal under the excitation of the inverter PWM wave is analyzed.

They show that the characteristic spectra of the Common Mode Voltage (CMV) around the carrier band, which contains the information of ZSV that indicates the fault, has the same sideband features as those of currents under the rotor related faults. As a result, the TKEO is applied to realize the cancellation of the undesired inverter ZSV noises. After the processing of the original common mode voltage signals, the fault severities can be extracted briefly by using the reliable spectral analysis. Under faulty conditions, the common-mode voltage becomes:

$$v_{sz}^f = V_c \cos(\omega_c t) + V_{fl} \cos(\omega_l t + \phi'_{fl})$$
$$+ V_{fr} \cos(\omega_r t + \phi'_{fr}), \tag{2.55}$$

where $\omega_r = \omega_c + 2\omega_1$ and $\omega_l = \omega_c - 2\omega_1$. ω_c, ω_1 are the carrier frequency and fundamental frequency, respectively.

For the healthy case, the common-mode voltage is:

$$v_{sz}^h = V_c \cos(\omega_c t) \tag{2.56}$$

There are rich voltage harmonics in a VSI drive system, which are mostly due to the modulation strategy. When an inter-turn fault arises in the windings, those harmonic can also be detected in the Common Mode Voltage (CMV) of the Voltage source inverter (VSI). The zero sequence model of the Induction Machine is:

$$v_{os} = -\frac{1}{3}\mu r_s i_f - \frac{1}{3}\mu l_{\sigma s}\frac{d}{dt}i_f, \tag{2.57}$$

where v_{os} is the zero sequence voltage (ZSV) of the induction machine, i_f is the current in the short-circuit path and μ is the quantity defined as the relative fraction of the short-circuit turns in a phase. r_s and $l_{\sigma s}$ are the stator resistance and leakage inductance.

Analyzing the TKEO for the zero sequence voltage, the two harmonic amplitudes identified as the fault indicator can be derived:

$$\eta_1 = \psi[v_{sz}^f(t)]|_{2\omega_1}$$
$$\eta_2 = \psi[v_{sz}^f(t)]|_{4\omega_1} \tag{2.58}$$

After the processing of the original CMV, the information of two sidebands can be combined and the fault severities can be extracted and used to detect fault severity.

In large induction motors, one can afford to make extensive use of sensors, both because of the importance of continuous operation and because they represent a small portion of the motor cost. Mirzaeva and Saad (2018b) present a method based on the use of miniature Hall effect sensors in the airgap for early diagnosis of stator turn-turn faults and static eccentricity. Time harmonic magnitude are plotted against angular position relative to stator in Figure 2.45. As expected, the most dominant time harmonic is the fundamental frequency harmonic. The sawtooth-like variation of the fundamental frequency harmonic, evident in Figure 2.46, is not a simulation artifact but a real effect. It is due to the presence of higher-order space harmonics in the stator flux density.

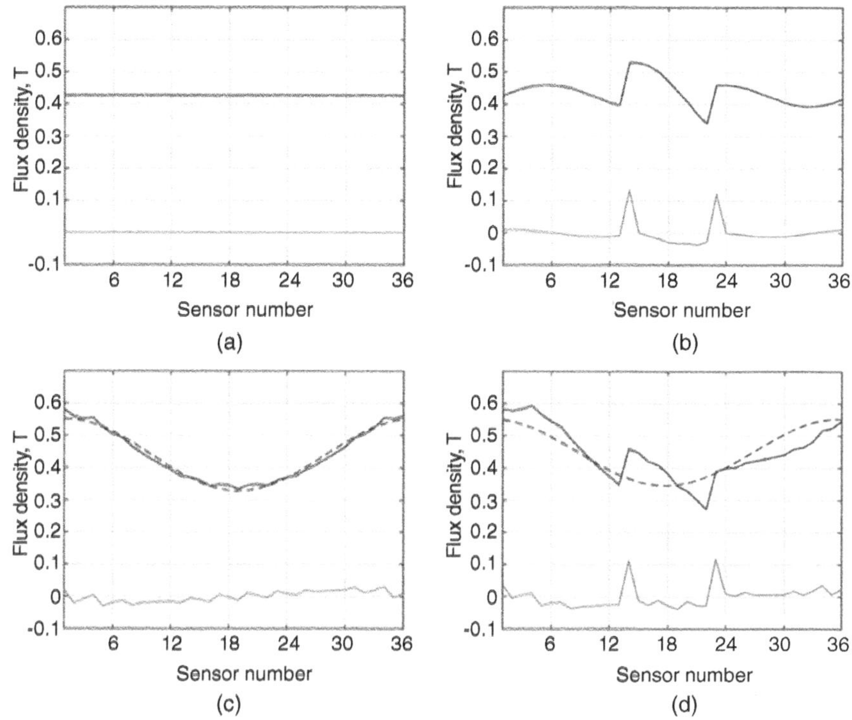

Figure 2.45 First harmonic magnitude analysis after data regularization. Top plot: first harmonic magnitude; bottom plot: magnitude difference between adjacent sensors. (a) Healthy motor. (b) Stator short. (c) Static eccentricity. (d) Short + static eccentricity, Mirzaeva and Saad (2018b).

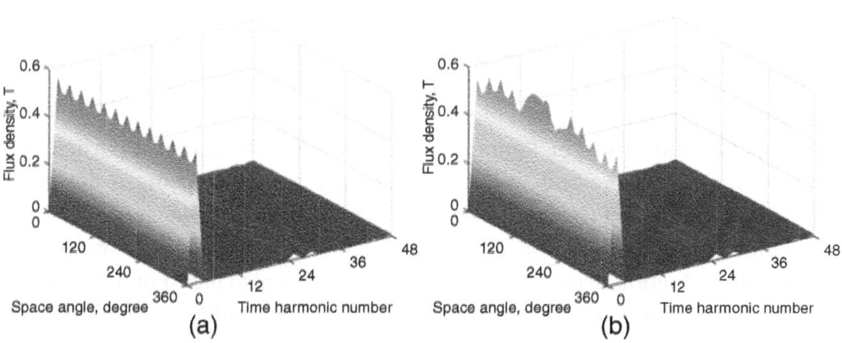

Figure 2.46 Analysis of time harmonic magnitudes in space. (a) Healthy motor. (b) Motor with stator short Mirzaeva and Saad (2018b).

The magnitudes of the stator fault and static eccentricity effects can be comparable. Therefore, it is important to be able to discriminate between them, or to identify the stator fault in the presence of static eccentricity , and vice versa. The regular sawtooth component can be estimated and removed from all the flux plots. After such a processing, the magnitude variations due to the higher-order space harmonics of flux density are mostly eliminated from the metric signal. However, the question of differentiation between the static eccentricity and the stator short fault still has to be addressed. Eccentricity can be estimated from the fundamental harmonic magnitude by calculating its first space harmonic and its angle.

A model-based diagnosis and prognosis of a interturn fault in induction machines was proposed by Nguyen et al. (2017). The process follows the classic progression: it uses multiple fault indicators for diagnosis and particle filters for prognosis. The fault is considered to be represented primarily by two parameters:

- the fault loop resistance r_f,
- the fraction of shorted turns $\mu = N_{asf}/N_{as}$, where N_{asf} and N_{as} are the number of turns in the faulted portion of the winding, and the total turns in it.

A state-space model of an induction machine under stator winding interturn short-circuit fault is the basis of this work. A number of indicators are used to start prognosis. Sequence component-based indicators are selected, generated from the model-based approach, which can capture the fault characteristics, so that the fault effect can be discriminated from that of voltage imbalance and load variation. They are implemented requiring no sensors other than voltage and current measurements.

Figure 2.47 shows the framework for diagnosis and prognosis. Fault progression is considered as step-wise change of parameters, and the parameter estimation method is shown in Figure 2.48.

The estimation of the of Remaining Useful Life is based on a particle filter technique. The method was experimentally validated using a designed degradation process of the fault resistance, with exponential decay, $r_f = ae^{-bt}$.

A technique for diagnosis of open circuits in fault-tolerant and operation after the fault in induction drives by Hu et al. (2020b) was also based on the voltage residuals. The phase voltage for open-switch fault in field oriented IM drives and their residuals are estimated. The voltage residuals, r_k, i.e. the difference in stator voltages between healthy and faulty-switch case depend on the switching states:

$$r_k = u_{kN} - u_{hF}, \tag{2.59}$$

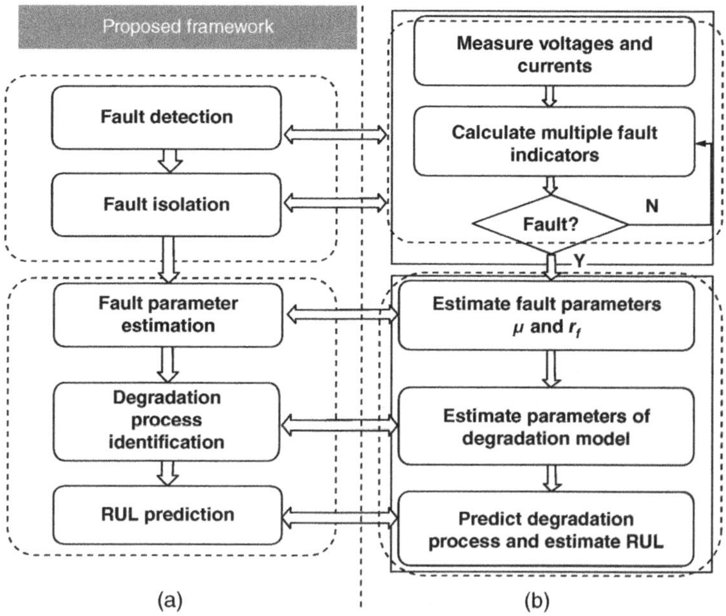

Figure 2.47 Framework for diagnosis and prognosis of electrical machines. (a) General framework. (b) Specific framework for interturn fault, Nguyen et al. (2017).

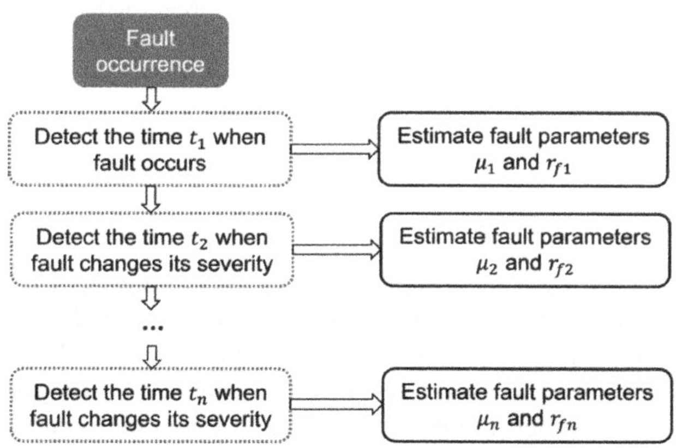

Figure 2.48 Triggering of fault parameter estimation when fault occurs or changes in its severity, Nguyen et al. (2017).

where the stator voltages were estimated from flux linkages:

$$\hat{u}_a = \frac{d\lambda_a}{dt} + i_a R_s, \tag{2.60}$$

and the flux linkages are estimated in the rotating frame of reference as

$$\lambda_{dr} = \frac{L_m}{T_r p} i_d. \tag{2.61}$$

A table relating faulty switches to fault indicators is used to identify the faulted switch and the threshold of the residuals is used to account for noises and modeling errors.

Stator faults and direct torque control: Direct Torque Controlled drives are not based on an estimation of the rotor field. Since the current is not controlled or imposed, once a turn-turn-fault occurs, it creates its own magnetomotive force (MMF). Therefore, the air-gap flux will carry fault signatures. Those signatures will be reflected in the motor current, air-gap magnetic flux, stray flux, and induced emf.

The motor current signatures could be analyzed directly or through investigating their reflection on the instantaneous power, or the phase-shift between phase current and phase voltage, or through the investigation of the system impedance.

If the stator currents are expressed first in a clockwise synchronous reference frame and the constant terms are removed from each one of the d-q current components, and subsequently transformed to a counterclockwise synchronous reference frame, the resulting d-q current components in this frame of reference, denoted by the superscript +, can be regarded as the sum of the value they would have in the absence of faults, plus an additional component due to the presence of the stator fault. Cruz et al. (2005):

$$i_{ds}^{+} = i_{ds}^{h} + i_{ds}^{f} \tag{2.62}$$

$$i_{qs}^{+} = i_{qs}^{h} + i_{qs}^{f}, \tag{2.63}$$

where the superscripts h and f stand for the healthy condition of the motor and additional current components due to the presence of the fault, respectively.

In order to calculate the fault indicator, which is defined by the vector $\Delta \mathbf{i}^f = i_{ds}^f + j i_{qs}^f$, it is necessary to calculate i_{ds}^f and i_{qs}^f, for the specific value of rotor speed, at which the motor is running. Using the motor parameters determined by the drive during the commissioning phase, the calculation of these quantities is straightforward. The calculation of the severity factor involves the determination of the quantities i_{ds}^h and i_{qs}^h. This can be done using the motor parameters determined by the drive during the commissioning phase in conjunction with the measured or estimated rotor speed.

Eldeeb et al. (2018) used a Discrete Wavelet Transform of the normalized current of a Direct Controlled Drive, and a new thresholding tool to detect fault location and severity. Contrary to methods using Motor Current Signature Analysis and Park's Vector Analysis sequence components, this method detects the controller's response to the changes due to a short introduces currents changes in the faulted phase. The short circuit transforms the IM from a balanced to an unbalanced three-phase system. The existence of the $2f$ component in the developed torque result in two terms added together: the first is the static error in the developed torque (constant term), that introduces an offset in the developed torque, while the second term is the dynamic error, that is oscillatory term at the double fundamental frequency.

The stator fault introduces inter-harmonics in the stator current at frequencies f_{ITSC}:

$$f_{ITSC} = f\left[\frac{m}{p}(1-s) \pm l\right], \qquad m = 0, 1, 2, \ldots, l = 0, 1, 3, 5, \ldots \tag{2.64}$$

through the use of Discrete Wavelet Transform, normalized and used for detection through a thresholding method.

Direct torque control with a multi-phase drive has been studied under healthy and faulty conditions, and it has been shown that it may behave better under fault compared to RFOC, but not necessarily under healthy conditions, see Bermudez et al. (2018).

2.5.1.2 Eccentricity

In both induction and synchronous machines, eccentricity is initiated at the manufacturing and assembly state. It causes vibrations and noise, magnetic pull of the rotor to one side, and bearing wear-out. It worsens progressively over time.

As the eccentricity worsens and airgap becomes nonuniform, the flux through the airgap, through the teeth, and in the back iron change (Fig 2.49), as well as the flux linkages. The frequencies of the fluxes, flux linkages, voltages and currents of the windings, are relatively low, relating to those of rotor and stator slots, harmonics of power frequency and speed. Since other faults (e.g. stator winding shorts, broken rotor bars) can produce signals in the same frequency range, it becomes complex to separate these faults.

Fundamental analysis is based on work by Alger; early work by Nandi et al. (2001) studied the harmonic content of the stator current as defined by the rotor slot harmonics. The frequency of the induction motor stator variables (currents, but also voltages) is:

$$f_h = \left[(kR \pm n_d)\frac{(1-s)}{p} \pm \nu\right] f \tag{2.65}$$

where R is the number of rotor slots, f_s the stator supply frequency, S is slip, ν the stator time harmonics of the supplied voltage, and p the number of fundamental pole pairs; $n_d = 0$ for static eccentricity and $n_d = 1, 2, ...$ for dynamic eccentricity. In the case of static eccentricity the permeance function of the gap becomes:

$$P \approx P_0 + P_1 \cos x, \tag{2.66}$$

where x in the mechanical position in the rotor frame of reference.

The permeance variation in eq. 2.66, the slot harmonics and the stator MMF produce airgap flux components varying sinusoidally with the position of the rotor, the slip, and the electrical angle $\omega_s t$, modifying the flux harmonics compared to the healthy machine.

The effects of faults on the principal slot harmonics (PSH) were studied by Climente-Alarcon et al. (2015a), who proposed a general method for the diagnosis of induction motor faults, including broken rotor bars, stator shorts, and eccentricity.

In the case of eccentricity, the permeability of the circuit modulated both by the PSH and eccentricity. The authors used Wigner-Ville Distribution for Time-Frequency Distribution of the energy of the PSH under different loads, and Particle Filter feature extraction. The motor will show PSH in the current spectrum. The effect of faults on the PSH were:

1. The effect in the air gap of the stator short-circuit current can be modeled as a square-shaped MMF with a spatial distribution corresponding to the shorted turn's pitch superimposed to the one caused by a healthy winding.
2. A condition of mixed eccentricity changes the permeability to constant to the following approximate expression:

   $$\Lambda \approx \Lambda_0 + \Lambda_1 \cdot \cos\theta + \Lambda_2 \cdot \cos(\theta - \omega_r t). \tag{2.67}$$

   the modulation of the fundamental flux with the space-dependent terms in 2.67 introduce low-frequency sidebands.
3. A broken rotor bar disturbs the symmetry of the field in the air gap, increasing the saturation of the rotor and stator teeth. This defect is modeled assuming a current across the faulty bar equal in magnitude to the healthy one but opposite in direction. A two-pole magnetic field in the rotor is created, which increases the reluctances of the stator and rotor teeth due to saturation, having a constant direction whose magnitude pulsates at the slip frequency. This can be decomposed, according to the doubly revolving field theory, in two rotational components of constant magnitude moving in opposite directions at a frequency equal to the pulsating field. The negative frequency one produces the known harmonics at twice the slip frequency

that is used to diagnose the fault, whereas the positive sequence one rotates at the speed of the fundamental wave, thus modifying its value.

Ceban et al. (2012) discuss the axial and radial stray fluxes as in Figure 2.19. In case of static eccentricity the axial flux is found to be of the supply frequency, while in the case of dynamic eccentricity, this flux moves with the rotor.

Jiang et al. (2017) review the condition monitoring of induction motors based on stray flux. They recognize that the analytical or numerical modeling of the external (stray or leakage) field is very difficult, but also not necessary. Still, analysis can provide a basic overview and understanding of the phenomena. they identify the following faults:

- Insulation Failure Detection is in itself complex and inaccurate, although for specific sensors it is possible to detect a short circuit fault.
- Bearing faults detection through flux probes has been promising but has not yet been widely verified and accepted.
- Dynamic eccentricity has been shown to be detected accurately through stray flux.

Zamudio-Ramirez et al. (2021) review recent contributions related to the monitoring of electrical machines dealing with flux-based methods. The potentials of this approach are highlighted and comparisons are made. Moreover, many references are given for several kind of machines.

2.5.1.3 Multi-fault Diagnosis with Stray Flux and Flux Sensor

Capolino et al. (2019) provide a detailed well referenced discussion of the faults that can be detected in both induction and synchronous machines through stray flux measurements.

Gyftakis and Marques Cardoso (2020) recognize that the short circuit diagnosis in induction machines through stray flux measurements has been shown to be successful under high severity faults, and attempted to detect this fault at very low severity. They carefully placed three flux sensors close to the machine core in a cylindrical fan. They noticed a significant increase of the third harmonic from -45.58dB for healthy to -37.83dB for a low severity 0.25% fault.

Zamudio-Ramirez et al. (2020) explore the effect of rotor winding asymmetries in wound rotor induction motors using again measurements of external field. The coil flux sensor was installed in the vicinity of the motor frame at three different positions to measure the axial and radial flux components according to their placement. A normalized indicator γ_{DWT}, given in eq. 2.68, relies on the fact that energies of the faulty band wavelet component increase as the fault severity does, and for the case in which a magnetic field is monitored, it relates the emf induced in the coil sensor to that of the wavelet signal

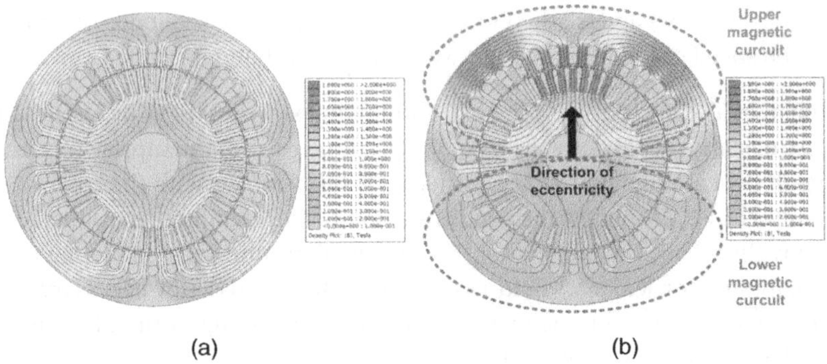

Figure 2.49 DC Simulation of the field in healthy (a) and eccentric (b) induction motor Hong et al. (2013).

containing most of the fault component evolution, d_n.

$$\gamma_{\mathrm{DWT}}(dB) = 10 \cdot \log \left[\frac{\sum_{j=Nb}^{N_s} emf_j^{\,2}}{\sum_{j=Nb}^{N_s} [d_n(j)]^2} \right], \tag{2.68}$$

where N_s is the number of samples of the signal until the fault-related oscillations disappear in the wavelet signal and N_b the sample number at the origin of the considered time interval.

For the time–frequency analysis, two different tools were applied to the different coil-sensor emf signals captured and stored during transient. First, the Discrete Wavelet Transform (STFT) was used to track fault components evolution under starting. In healthy conditions, only the fundamental component was expected to be present. Under the presence of rotor asymmetry, the fault component was expected to be accompanied by the characteristic time-frequency patterns caused by the evolution of the fault components. Alternatively, signal processing technique used is the Discrete Wavelet Transform (DWT).

The method proved able to detect the asymmetry level, but is also very sensitive to the position of the flux sensors.

Hong et al. (2013) propose an accurate, but off-line method to detect eccentricity through the measurement of the inductance slope, i.e. incremental or differential inductance, with accurate results.

Along the same line, detecting the change in flux density due to eccentricity, Mirzaeva and Saad (2018a) measure the main air gap flux density, in both time and space, with the help of an array of miniature Hall effect flux sensors (HEFS) installed around the circumference of the stator inside the motor air gap. They compare the signals of the sensors for different poles, and when synchronizing with the rotor position they can differentiate between broken rotor bars, stator short circuits (see Mirzaeva and Saad (2018b)), as well as static and dynamic eccentricity.

2.5.1.4 Open Faults in Windings and Inverter

Faults in windings, along with faults with bearings, are the most common ones. They can compromise the operation of the drive, leading to catastrophic failures. Although short circuits are caused by insulation degradation, mechanical faults, and faults in power electronics switches, are equally severe.

Open circuits in the windings or the inverter are important, not only because they can cause significant performance issues, torque pulsations, torque reduction, and indirectly over-currents and overheating, but also because many of the effects can be mitigated through control and employment of redundancies.

The faults can be located in windings or in the electronic switches. In this section we address the diagnosis and effects on the machine, rather than the switches themselves, which are covered in other sections. The work on the detection and managing open circuit faults in electrical drives has been going on for decades, and there are different aspects of the fault location:

- A switch is staying open, and the antiparallel diode is also damaged or it is conducting.
- The whole winding of one phase is disconnected or broken, either because of the switch malfunctioning, or because each phase consists of a single coil that is open, or, when the phase winding consists of parallel segments, one of coils connected in parallel is open, while the rest remain connected.
- The fault is intermittent, lasting a short time, not allowing the standard schemes to detect it.

The techniques that have been used to detect open circuits in both Permanent Magnet AC machines and Induction machines, whether based on Rotor Field Orientation or on Direct Torque Control, have been numerous and have attempted to satisfy three criteria: Accuracy even at low speeds and loads, independence of parameters, and fast fault detection and location identification.

Eickhoff et al. (2018) discuss how a single open switch fault leads to a current control deviation. Immediately after the fault occurs, the deviation increases and the angle of the deviation indicates the faulty switch. In their proposed method the current control deviation in the stator reference frame α, β is monitored and used as a fault indicator. If the trigger criterion is met, an additional voltage test is enabled to further validate the fault. This test increases the desired voltage in the deviation direction to counteract the deviation. The current trajectory of a fault is shown in Figure 2.50.

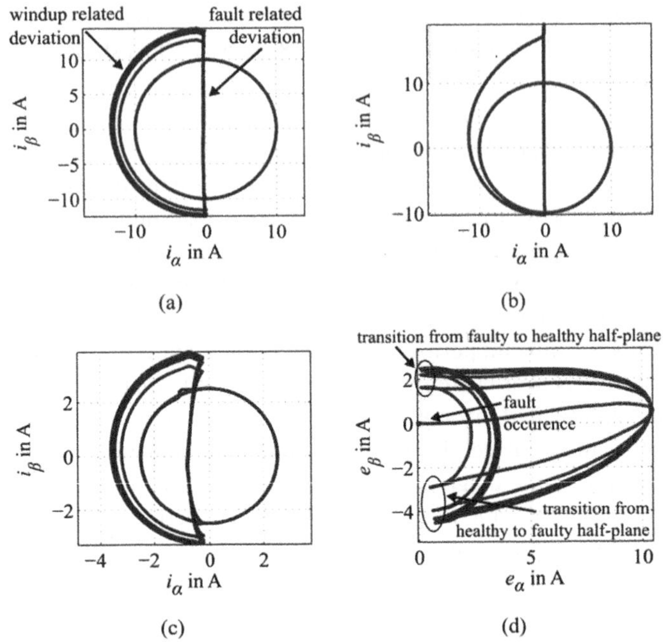

Figure 2.50 Simulated current in the stator frame before and after the SSF (S1) occurrence at different operating points of the drive: (a) 1000 r/min and $i_q^* = 10A$; (b) 200 r/min and $i_q^* = 10$; (c) 1000 r/min and $i_q^* = 2.5A$ and (d) control deviation after the fault occurrence of the operating point of case (a) 1000 r/min and $i_q^* = 10A$. Note the different scaling of the α-axis and the β-axis in (d). Eickhoff et al. (2018).

2.5.1.5 Broken Rotor Bars

Many effots have historically been made to detect broken bars. Fault diagnosis of induction machine began with such failure, even if this one is not the most common.

Rotor bar breakages in induction machines occur possibly due to underlying manufacturing problems but are caused by increased currents in the bars under heavy loading. This occurs more often in line starting induction machines, and less frequently during operation through inverters. The fault in line-operated machines is detected through analysis of the stator currents, but in a machine operated through a current-controlled inverter, these signatures are difficult to detect. Alternatives include the detection of stray flux and complex signals (e.g., real, reactive and apparent power).

The topic has been addressed primarily through the measurement of the stator current under steady state conditions. The stator current develops harmonics of the order:

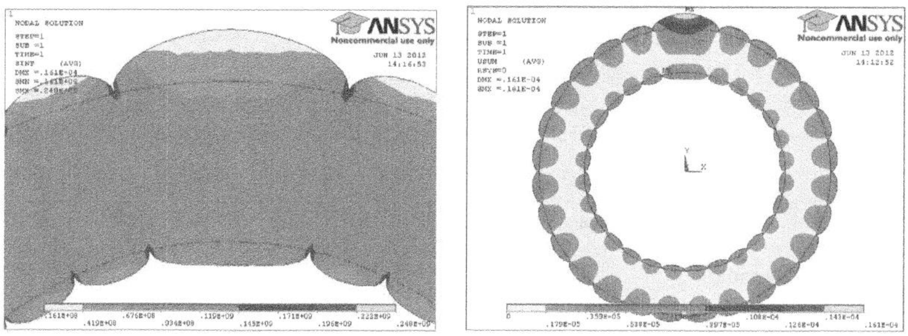

(a) (b)

Figure 2.51 Numerical analysis of stresses and displacements after a broken bar in an induction machine. (a) Von Mises stresses in end ring segments after a broken bar. (b) Displacement in a ring segments resulting from broken bar.

$$f_{bb} = f_s(1 \pm 2ks), \tag{2.69}$$

where f_s is the supply frequency, s is the slip, and k is an integer. Generally, as long as machine is not heavily damaged, faulty frequency sidebands amplitude are small. As the number of broken bars increases, the amplitudes of the harmonics increase. As the load is decreased, thus decreasing slip, the harmonics due to the fault approach the fundamental frequency and become smaller.

When the machine is fed from a variable frequency drive, the stator current contains harmonics related to the switching frequency. In high-performance drives, these are far apart, but if the current is controlled as in a Field Oriented Induction drive, the controller action masks the fault, by commanding the current. It has been proposed then to detect the fault through measurements of the flux, e.g by measuring the effects on the Principal Slot Harmonics (see Climente-Alarcon et al. (2015a)). An additional advantage in this method is its applicability to other faults, e.g. eccentricity as discussed earlier, and line-line shorts. Similar use of stay flux has been extensively reported, e.g. Panagiotou et al. (2018).

A broken rotor bar results in an uneven current distribution, with neighboring bars being more stressed. This leads to further linear expansion and stresses to the end ring. Since rotor bar breakages develop gradually and do not immediately affect the operation of the drive until they become severe, it is useful to predict the Remaining Useful life or the bars. Some attempts have been made in that direction such as in Climente-Alarcon et al. (2015b). Figure 2.51 shows the stresses and displacement leading to fault progression.

Many attempts have been made to diagnose rotor broken bars with models. For instance, Boumegoura et al. (2004) has proposed a method based on a high-gain extended observer applied on a dq model of an asynchronous machine in a

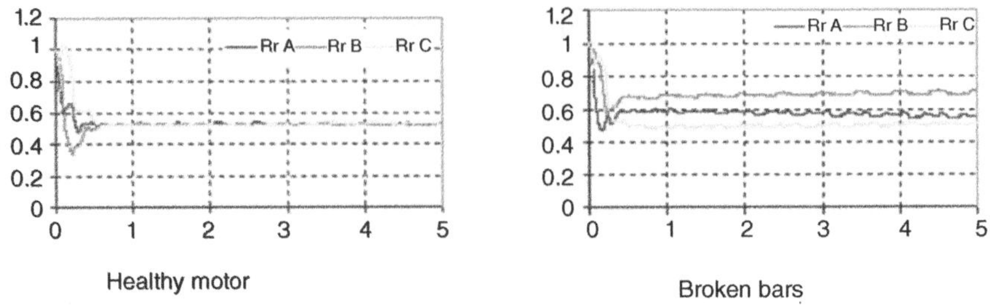

Figure 2.52 Three-phase rotor resistance with and without broken bars. Boumegoura et al. (2004).

synchronous frame and an extended Kalman observer applied on a three-phase stator frame model of the induction machine.

2.5.2 Permanent Magnet AC Machines

2.5.2.1 Demagnetization of Permanent Magnets

In permanent magnet AC machines demagnetization can be due to several reasons: Aging of the magnets, or operation of the machine under field weakening and high temperatures, often accidentally due to a fault in field orientation or in the inverter. Both uniform and partial demagnetization can have detrimental effects in the operation of the drive, decrease of average torque, and efficiency, and increase in the torque ripple. The effects of demagnetization on measured variables, direct and quadrature axis voltages and harmonics are similar to that of eccentricity as both affect the airgap flux and flux linkages.

Detecting demagnetization is important, and as already pointed out, it is also important to separate it from other faults with similar manifestations, eccentricity, and short circuit. All three faults have a defining effect of the stator voltages. Haddad et al. (2017) studied these voltages as they vary with the fault severity, and observed different trajectories with different faults. Figure 2.53 shows three trajectories. To avoid dependence of operating point and temperature, compensation and interpolation were used, along with standard classification tools.

Goktas et al. (2016) base their technique on that under broken or demagnetized magnets, the reluctances of the magnetic circuit remain the same, but the values of back-emf are reduced. Conversely, they expect that under static eccentricity, the flux amplitudes remain similar to the healthy machine, but the stator inductances change depending on eccentricity. They conclude that the 0.25th and 0.5th harmonics point out the broken magnet failure, whereas 0.75th harmonic is related to static eccentricity fault.

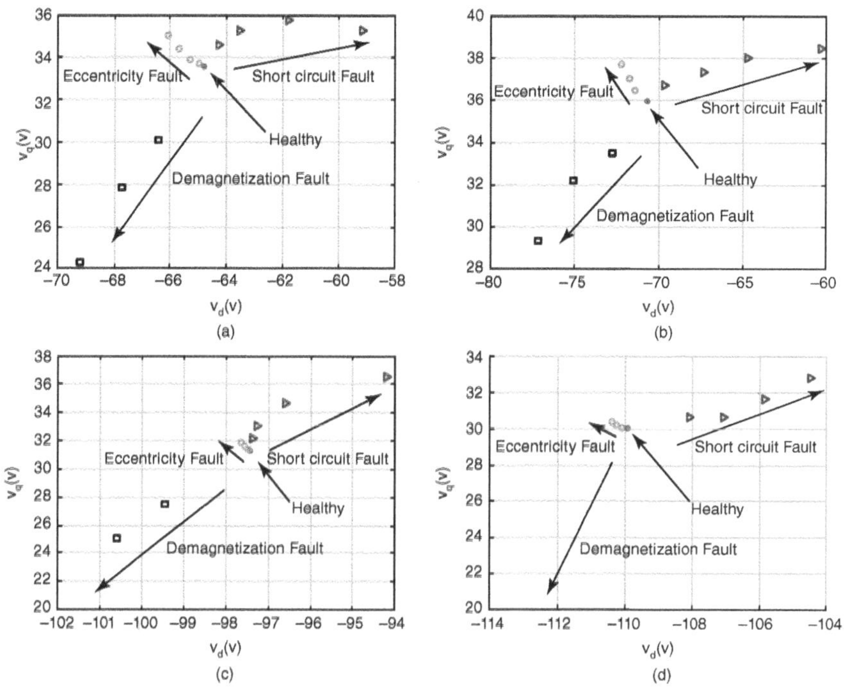

Figure 2.53 Comparison between simulation and experimental results for the change in V_d and V_q for healthy and different faults for $I = 5A$, $\delta = 120^o$ and $I = 10A$, $\delta = 120^o$. (a) V_d versus V_q simulation results $I = 5A$, $\delta = 120^o$ b) V_d versus V_q experimental results $I = 5A$, $\delta = 120^o$. (c) V_d versus V_q simulation results $I = 10A$, $\delta = 120^o$. (d) V_d versus V_q experimental results $I = 10A$, $\delta = 120^o$. Haddad et al. (2017).

To overcome the limitations of off-line testing and the need to use spectrum-analysis-based detection, Park et al. (2019) utilized the analog Hall sensors installed in PMSMs for initial position estimation, for the purpose of identifying the cause of rotating asymmetry. They investigated interior permanent magnet synchronous motor (IPMSM) under eccentricity, local demagnetization, and load unbalance conditions during transients. They placed three sensors accurately in the end shield, in locations shown in Figure 2.54. Results for the three investigated faults are shown in Figure 2.55. They show the effectiveness of the technique.

Addressing the prognosis of the irreversible demagnetization (IDF) of the permanent magnets, Ullah and Hur (2020) determine the operating point of the machine using machine learning to monitor the variation in the operating point of the machine and predict the remaining margin of temperature that a machine can endure before the occurrence of IDF.

In the first step, the magnetic flux densities of PM at different temperatures, loads, and field weakening levels are obtained using the finite element method

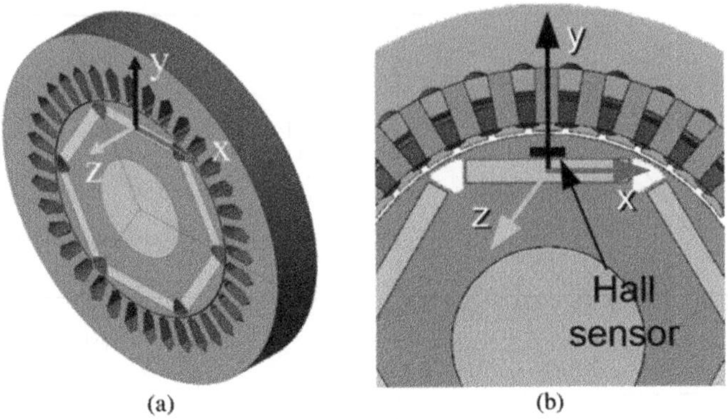

Figure 2.54 Location of Hall sensor installed for FE and experimental verification of proposed method Park et al. (2019).

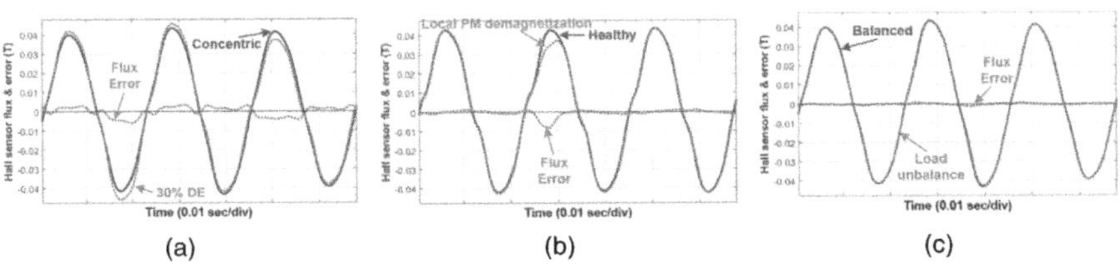

Figure 2.55 Experimental results for three IPMSM faults, Park et al. (2019). (a) IPMSM with 30% eccentricity. (b) IPMSM with with local demagnetization (3% of one PM chipped). (c) IPMSM under load unbalance that results in 0% and 8.6% torque oscillations.

(FEM) simulation and high priority are given to the edges, while calculating magnetic flux density. The aim is to find the weakest point or the area of the PM with the lowest magnetic flux density because the IDF is going to affect that portion first. Through Finite Element analysis and machine learning, they estimate the flux density under load, B_L at each operating point during operation. Since the B_L depends upon three factors, temperature, stator current, and the field weakening, which are not linear, they use a fifth degree multivariate polynomial regression (MPR) to model B_L. The predicted value of the magnetic flux density ($\widehat{B}_L$) is used to estimate the margin between the irreversible demagnetization point B_k and the actual flux density given in eq. 2.70. The value of B_k can be obtained using a simple two-dimensional lookup table of the magnetization curve of the permanent magnet.

$$B_{mar} = B_L - B_k \tag{2.70}$$

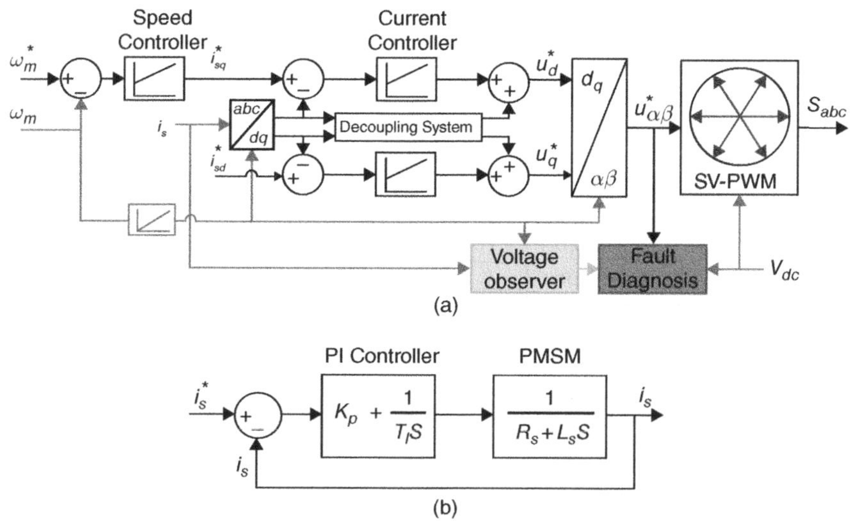

Figure 2.56 RFOC strategy with SVM applied to the PMSM-side converter. (a) Control strategy block diagram, including the proposed fault diagnostic technique. (b) Current control loop, Freire et al. (2014).

2.5.2.2 Open and Short Circuit

Detection alone in a Permanent Magnet AC machine is not enough, since invariably, the inertia of the system will continue the motion of the magnets, thus inducing a voltage and feeding the short-circuit. The case is not as severe in induction machines, where the field is weaker and decays quickly.

Of the voltage estimation techniques to detect an open circuit, Freire et al. (2014) used the average of the difference between commanded and estimated voltages d', as the diagnostic variable.

$$d'_n = \left\langle \frac{(u_n^* - \hat{u}_n)}{V_{dc}} \right\rangle, \tag{2.71}$$

where $\hat{u}_n$ stands for the estimated phase voltage, and u_n^* for the commanded one. The voltage estimate is calculated from the flux linkages:

$$\hat{u}_a = \hat{u}_{sa} = \frac{d}{dt}\hat{\lambda}_{sa} = \frac{d}{dt}\left(L_d i_{sa} + \lambda_{PM}\cos\theta_r\right) \tag{2.72}$$

The diagnostic technique was successfully applied to both Rotor Field Oriented and to Direct Torque Controlled electrical drives. The diagnosis technique is shown in Figure 2.56, while the control strategy is shown in Figure 2.57.

Following the detection of an inter-turn short circuit in interior Permanent Magnet machines, Qi et al. (2019) proposed a method to determine the number

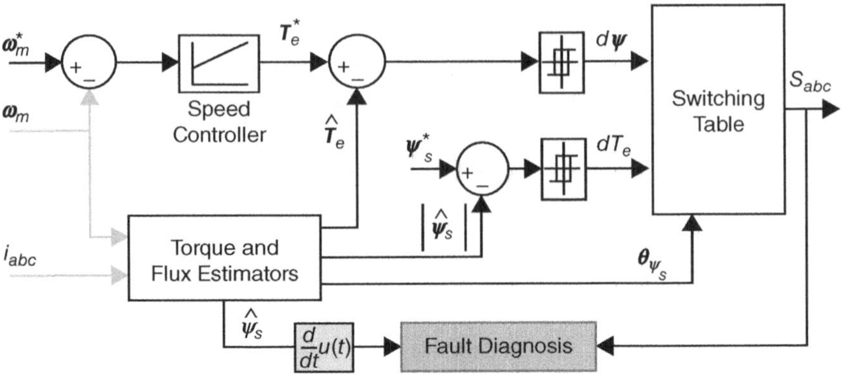

Figure 2.57 Block diagram of the DTC strategy applied to the PMSM-side converter, including the fault diagnostic technique, Freire et al. (2014).

of shorted turns of the machine at standstill. The machine is excitated through a low sinusoidal voltage in order to obtain the winding resistance and synchronous inductance by current response. In an experiment, the amplitude of the applied voltage is limited to 10V so that the short-circuit current can be kept in a safe range, which reduces the thermal impact on the estimation as well. However, in actual applications, it can be selected based on the resolution of the voltage measurement. The frequency of applied voltage is set to $200Hz$.

Several methods were developed to detect and identify turn-to-turn faults in PMSMs. A winding fault creates an unbalanced operating condition, from which measurable information such as the voltages, sequence components, impedance changes, and harmonic content, many similar to those for induction drives, and could be used for detection purposes. Detection is an important part of a fault tolerant system, the mitigation of the fault current assumes that a fault detection and estimation algorithm exists and is already implemented.

Most of these techniques analyze one or several typical frequency components at steady state. However, at transient states, the frequency is varying with time and the spectrum is distorted. Conventional frequency analysis methods such as fast Fourier transform (FFT), short-time Fourier trans-form (STFT) become invalid or limited, and more advanced, but more complex methods are used, such as wavelet transform (WT), Hilbert-Huang transform (HHT). However, their demand for large memory and computation time usually makes it difficult to be implemented in real time.

Since most PM machines are driven by inverter with pulse-width-modulated voltages, both the voltages and currents will contain HF signals (harmonics), which can also be used to detect turn fault. The advantage of employing HF signals for fault detection is that the effect of back EMF can be eliminated since it only contains low-frequency components, namely, the fundamental and a few its integer multiples. In addition, the effect of speed and load variations on the high-frequency current ripples is less pronounced and can be eliminated by employing suitable fault indicators.

Hu et al. (2017) focus on the turn fault detection technique based on PWM HF currents. They are extracted through a bandpass filter, and a root-mean-square (RMS) detector is utilized to measure the RMS value. The three RMS currents from all the phases in a 3-phase permanent magnet machine (SPM) are compared, and two fault indicators are proposed and studied, with the consideration of phase imbalance. Analytical models are established, and simulation are conducted to evaluate the feasibility of the method.

For predicting HF currents, the back EMFs can be ignored, because they are at much lower frequency. Based on these considerations, the voltages equations at high frequency can be written as Equation 2.73. The subscript 'HF' in voltages and currents denotes their HF components.

$$[u_{abcN\_HF}] = [R_{abcN}][i_{abcf\_HF}] + [L_{abcN}]\frac{d}{dt}[i_{abcf\_HF}], \tag{2.73}$$

where

$$[u_{abcf\_HF}] = [u_{a\_HF}u_{b\_HF}u_{c\_HF}u_{f\_HF}]^T \tag{2.74}$$

$$[i_{abcf\_HF}] = [i_{a\_HF}i_{b\_HF}i_{c\_HF}i_{f\_HF}]^T. \tag{2.75}$$

The inputs to the state space equation are the HF components of u_a and u_b. Based on this state space equation, the output HF currents can be predicted and the influence of a turn fault on the HF currents can be assessed. Although the RMS current i_{a-rms} changes when turn fault occurs, there is no simple way to define a threshold, since $i_{a-rms-h}$ and $i_{a-rms-f}$ in healthy and fault conditions are dependent on modulation index M. Thus, the threshold must be determined as a function of modulation index, rather than a constant, which, will increase the complexity and noise susceptibility. Using the ratios of the currents in phases a and b, b and c and c and a, abnormal changes in all three phase HF currents due to a fault are captured in the fault indicator, thus enhancing the detection effectiveness.

Breuneval et al. (2017) used a hybrid model-data method for the diagnosis of a shorts in PMAC machines, and a Support Vector Machine (SVM) combined with Fuzzy Membership Functions (MBF) for fault severity classification. The technique first diagnoses the fault, and if the observation is classified as faulty. Then, a membership degree in the observation in a severity class is determined. The idea, here, is to compute a degree of membership of an observation to a class. Therefore, an observation can belong to several classes, with different membership degrees. To implement this idea, the fuzzy sets theory was used.

The membership functions (MBF) are computed from the distance to the hyperplane which are, themselves, computed by classical SVM. Depending on the membership degree in a class a threshold is used to classify this observation. The method was shown to be accurate and computationally efficient.

Measurements often used for the detection of a short and the estimation of its severity are related to the send harmonic of the reactive power, derived from the third harmonics of currents and voltages. This has shown to be a good indicator, with the drawback that as the drive is based on a voltage source inverter with current controllers, the bandwidth of the controller affects the results differently at different power frequencies. Huang et al. (2020) proposed a new ITSC indicator based on the Rayleigh quotient function, combining the second harmonics of currents and voltages, see figures 2.58, 2.59.

For a given complex Hermitian matrix **X** and nonzero vector w, the Rayleigh quotient function $R(X; w)$ is defined as:

$$R(\mathbf{X}, \mathbf{w}) = \frac{\mathbf{w}^*\mathbf{X}\mathbf{w}}{\mathbf{w}^*\mathbf{w}}. \qquad (2.76)$$

The second harmonics in dq-axis currents and voltages are not weighted together directly. Instead, the ratios of the second harmonics in i_{dq} and v_{dq} after ITSC to those before the fault, here called current and voltage ratios, are normalized. The ratios of i_d, i_q, v_d and v_q are denoted as a_{id}, a_{iq}, a_{vd}, and a_{vq}, respectively. Then the ratio matrix, **X**, in the Rayleigh quotient function is:

$$\mathbf{X} = \mathbf{A}'\mathbf{A} \qquad \mathbf{A} = [\mathbf{a_{id}}\ \mathbf{a_{iq}}\ \mathbf{a_{vd}}\ \mathbf{a_{vq}}]. \qquad (2.77)$$

The authors performed co-simulation with Finite Element analysis and experiments to evaluate the analysis for a broad range of fault severities.

The detection time depends primarily on the speed of the machine. As the speed increases, the periods of currents and voltages decreases, and hence the time available to extracting the 2nd harmonics decreases, Overall, the proposed indicator can detect one-turn faults under different ω_c within 2.5 electrical cycles, and the detection of fifteen-turn faults is much faster due to the higher indicator values.

Intermittent faults in PMAC drives manifest themselves in ways similar to those of induction machines. In their early stage they last too short to affect the performance or the main variables to be detected. Usually they can be detected through the transients they cause in the phase currents, since they are fast enough not to be compensated by the fast current controllers.

Zanardelli et al. (2007) discussed two types of electrical stator faults in Permanent Magnet AC machines. The first was a momentarily increased resistance in one phase, due to a bad connection between the motor and the controller. The second electrical fault is a turn-terminal short, simulating an insulation failure

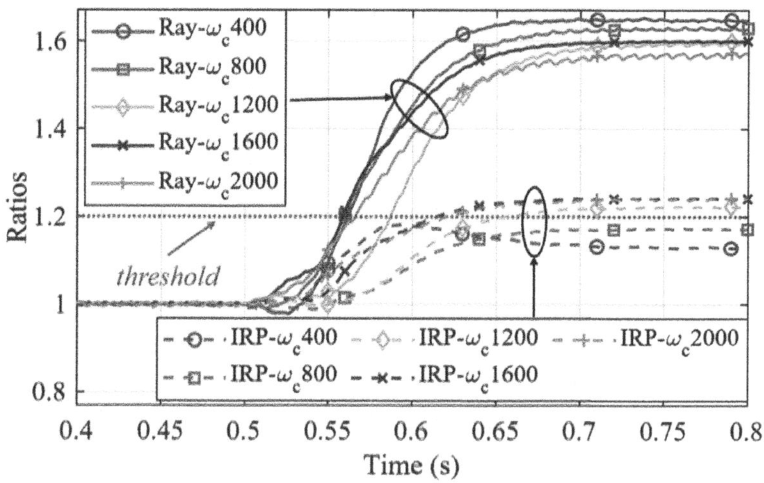

Figure 2.58 Ray ratios and IRP ratios for online ITSCs detection under different ω_c, at 10A and 300 rpm in experiment. Severity: 1 turn directly shorted, Huang et al. (2020).

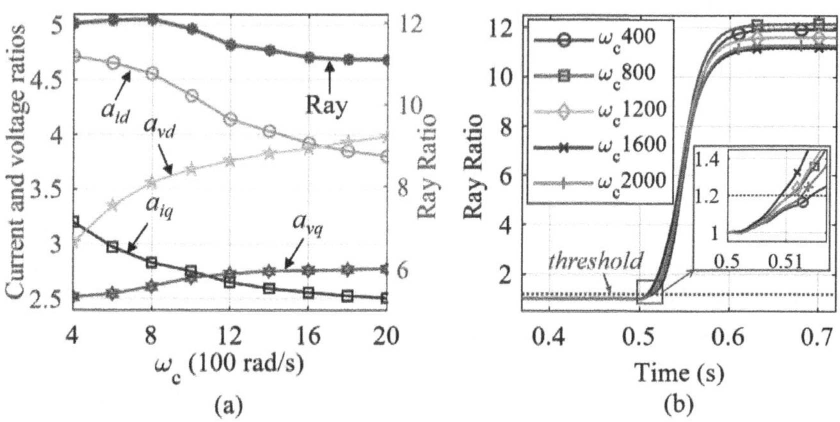

Figure 2.59 Current and voltage ratios and Ray ratios (a) and online ITSCs detection under different ω_c (b), at 10A with 600 rpm in experiment. Severity: 15 turn shorted by $rf = 0 : 5$, Huang et al. (2020).

in the stator windings of the motor. The intermittent nature of the faults and the high frequency of the oscillations they result, led the authors to continuous monitoring with time-frequency tools, Short Time Fourier Transform and shift invariant Discrete Wavelet Transform. They used the direct and quadrature currents for categorization, which are of high enough frequency not to be masked by the controller action. They used linear discriminant classifier and they allocated the faults in four categories: inception and clearing of a short and an open circuit.

Haje Obeid et al. (2017) studied the intermittent fault in a permanent Magnet AC machines, as a means to predict a steady fault. They used distortions in the stator currents as well as in the reference voltages.

They identified the intermittent fault using a pattern detection method. The pattern to identify is the fault signature (FS) and the identification should be feasible independently of its scale. The fault signatures are the deformation generated at the end of the intermittent short circuit. If the FS scale was known, a convolution product between a measured signal (for example $V_{q,ref}$) and the FS pattern in this signal would be enough. The fault identification will be seen as a correlation between the FS and the pattern. But as the FS scale is related to the fault to detect, it is an unknown parameter. Because of that they opted for an adapted wavelet transform. They noticed that for an intermittent fault in phase a, u_{bc} is more influenced by the other line-line voltages. The fault localization method was based on the amplitude comparison of the phase to phase voltages distortions. Its purpose is to identify the faulty phase through a difference in the peak amplitudes in these voltages.

Theoretical and experimental tests have proven the effectiveness of this method not only in identifying and localizing the intermittent fault but also in improving the signal to noise ratio. The authors recognize two major limits leading to the absence of detection of the intermittent fault: a very small number of turns involved in the fault (μ), and a very high short-circuit resistance ($R_{f_s c_O N}$).

Yan et al. (2019) proposed a technique based on the average current Park's vector method of the phase currents, and a fuzzy-based fault diagnosis for the VSI in the three-phase permanent-magnet synchronous motor drive. The fuzzy logic approach was applied to process the fault symptom variables and obtain the fault information of power switches. Compared with other fuzzy logic methods, here the fuzzy logic design, fuzzy inputs, and fuzzy rules are different. The proposed fault diagnosis method can detect and locate not only the single or multiple open-circuit faults, but also the intermittent faults in power switches.

When the PMSM is operated in the normal condition, the three-phase currents are sinusoidal waves. Then, a constant value ξ is introduced, equal to the value of normalized phase current i_{kN} with $k = a, b, c$.

$$i_{kN} = \frac{i_k}{I_m} \tag{2.78}$$

with I_m the maximum amplitude of the phase currents.

Under healthy condition, $\xi = 2/\pi = 0.6366$. To realize the fault diagnosis, three diagnostic variables e_k are introduced, which are equal to the errors of the normalized currents average absolute values and are expressed as:

$$e_k = \xi - \langle |i_{kN}| \rangle \tag{2.79}$$

Under normal operating condition, the three diagnostic variables e_k are all near zero. When the open-circuit faults in the PWM-VSI occur, at least one of these three values will become a distinct positive value, which can be used to identify the faults.

To implement the fault diagnosis, fault symptom variables are formulated as follows:

$$E_k = \begin{cases} N, & e_k < -\delta_0 \\ Z, & -\delta_0 \leq e_k < \delta_0 \\ PS, & \delta_0 \leq e_k < \delta_1 \\ PL, & e_k \geq \delta_1 \end{cases} \tag{2.80}$$

$$Z_k = \begin{cases} N, & \langle i_{kN} \rangle < -\mu_0 \\ Z, & -\mu_0 \leq \langle i_{kN} \rangle \leq \mu_0 \\ P, & \langle i_{kN} \rangle > \mu_0 \end{cases}, \tag{2.81}$$

where δ_0, δ_1 μ_0 are the threshold values. N, Z, P, PS and PL are the values of the fuzzy domains, where N indicates negative, Z indicates zero, P indicates positive, PS indicates positive small, and PL indicates positive large. The fault symptom variables E_L and Z_k can be used to detect and locate the open-circuit faults in the PWM-VSI, but the intermittent faults in the power switches, cannot be detected with the average current Park's vector method. To deal with this problem, these fault diagnosis variables are processed with a fuzzy control algorithm.

The diagnosis procedure is based on the analytical and heuristic knowledge symptoms of the inverter. Heuristic knowledge in the form of qualitative process models can be expressed as if–then rules. In this paper, the analytical and heuristic symptoms are the fault symptom variables E_L and Z_k, which are the input of the fuzzy-based fault diagnosis (FFD) block. Based on the principle of fuzzy control theory and the six fault symptom variables E_k and Z_k should be fuzzified as:

$$\begin{cases} E_k \in \{N, Z, PS, PL\} \\ Z_k \in \{N, Z, P\} \end{cases}, \tag{2.82}$$

The method can detect and locate different kinds of open-circuit faults within $90ms$ after the fault occurrence, and sudden load change will not cause a misdiagnosis.

2.5.3 Sensor Faults

Diao et al. (2013) described a fault detection and diagnosis structure of mechanical and current sensors faults of a Permanent Magnet Synchronous Machine (PMSM) drive.

In PMSM drives the position sensor is used to detect the rotor magnetic field position or at least, to measure the speed. The objective is to provide an efficient back-up solution when the mechanical sensor fails.

The method is based on two interconnected observers: an Extended Kalman Filter (EKF) and a Model Reference Adaptive System (MRAS) observer.

The EKF, thanks to its optimality, is designed to estimate the position in case of mechanical sensor fault and despite sensor current fault. Starting from the classic model of a field-oriented PMAC machine, the Kalman Filter-based estimator uses the equivalent two-phase currents measurements (i_α, i_β) in the fixed frame and the calculated reference voltages (V_α, V_β) to estimate the rotor position and speed, respectively $\hat{\theta}$ and $\hat{\omega}$. The estimated position and speed given by EKF are used with reference voltages to estimate the three-phase currents ($\hat{i}_a$, $\hat{i}_b$, $\hat{i}_c$) using the theory of MRAS.

The MRAS estimates the phase currents using the actual position and speed. The computation and sort of the residuals (difference between measured and estimated values) allows the fault isolation.

Yu et al. (2018) proposed a scheme that integrates three independent observers, taking a-phase current, b-phase current, and c-phase current as inputs, respectively. The observers are capable of online monitoring the state of current sensors under normal operation. Furthermore, after one or even two current sensors getting faulty, the observers can detect and localize the faults and switch the system to tolerant vector control mode even with only one healthy phase current sensor available.

In this scheme, the estimated signal replaces the faulty current signal without changing the original system structure.

When all the current sensors are fault-free, the three-rotor flux amplitudes observed by three state observers are equal, and the estimated currents ($\hat{i}_a$, $\hat{i}_b$, $\hat{i}_c$) of the three state observers track three actual phase currents, respectively. When any one of the three current sensors is faulty, a phase current, for example, i_a becomes null instantly and $\hat{i}_\beta$ only changes little in the period of estimation. By making full use of this phenomenon, one current sensor fault diagnosis is realized. A judgment principle for the first sensor fault diagnosis is established, based on the observer outputs to determine the health of all three sensors.

After the first current sensor fault occurrence, the fault can be detected and localized, and the system will continue work in the fault-tolerant control mode with the other two healthy current sensors. Unfortunately, there is still some possibility that the two remaining healthy current sensors get faulty. In this part, the fault diagnosis method for the second current sensors will be used.

Chakraborty and Verma (2015) present a technique for fault detection and isolation to make the traditional vector-controlled induction motor drive fault tolerant against current and speed sensor failure. The proposed current estimation uses d- and q-axes currents and is independent of the switching states of the three-leg inverter. While the technique introduces a new concept of vector rotation to generate potential estimates of the currents, speed is estimated by one of the available model reference adaptive system (MRAS) based formulations. A logic-based decision mechanism selects the right estimate and reconfigures the system by rejecting the signal from the faulty sensors.

The technique is based on first aligning the α stationary axis with that of the phase. If the a-phase sensor is faulty, both α and β measurements will be wrong, but if the b-phase sensor is defective the current in α-phase will be correct and the current in phase β will be wrong.

If the α - β system is rotated by 120°, for a fault in a-phase, the measurement in the new α' phase will be correct, while phase β' will provide wrong data.

Using the actual measurement of a- and b- phase currents and the corresponding reference-magnitude of the same in the dq rotating reference frame, eight estimates of currents in the α - β reference frames are thus possible. Hence, the system can be made fault tolerant if a mechanism can be developed to switch over from the faulty measurement to the corresponding correct estimate.

For the speed sensor fault detection, an X -MRAS (X for $\vec{V} \times \vec{I}^*$ - See Chakraborty et al. (2010)) based method is proposed. The actual values of d- and q-axes currents are obtained from current sensors, whereas the estimated values are obtained from the current estimation algorithm (i.e. $\hat{i}_\alpha$ and $\hat{i}_\beta$). The estimated current signals (in α - and β -reference frame) and currents in the dq axes rotating reference frame are obtained with the help of the vector rotator, which, in turn, depends on speed. The error, $X_r - X_s$ is fed to the adaptation mechanism to generate the speed signal.

$$X_r = v_{sq}^* i_{sd}^* + v_{sd}^* i_{sq}^* \tag{2.83}$$

$$X_s = v_{sq}^* i_{sd} + v_{sd}^* i_{sq}. \tag{2.84}$$

2.5.4 Fault Mitigation and Management

There have been several published methods regarding turn-to-turn fault mitigation, primarily in Permanent Magnet machines. The simplest solution is to short the full damaged phase, but this may cause more problems than it resolves. In the majority of the techniques, the stator current is controlled to produce a magnetic flux linkage that is equal in magnitude but opposes the flux linkage produced by the magnets. This reduces the induced electromotive force (EMF) voltage in the faulted turns and suppresses the fault circulating current. Depending on the speed and load, the machine design and inductances, the time to detect, decide, and respond (a few *ms* usually is allowed before thermal damage), the injected current, and the resulting torque reduction can be estimated in the control algorithm.

Welchko et al. (2006) proposed a control method to null the magnet flux in an interior permanent-magnet (IPM) motor following short-circuit-type faults in either the inverter drive or motor stator windings. Phase-based control is employed to implement the flux-nulling control method so that it is possible to take advantage of a zero-sequence current in order to minimize the current in the shorted phase. One requirement of fault-tolerant PM machine drives is the necessity of having individual control of the phase currents. In effect, this requires that each phase be driven by an H-bridge inverter or, alternatively, employing a six-leg inverter. The basic concept is to set the d-axis current command equal to the motor characteristic currents so that it will null the magnetic field:

$$i_d e^* - \frac{\lambda_m}{L_d}. \tag{2.85}$$

Another possibility is to set the zero-sequence current to null the current in the faulted phase:

$$i_d 0^* - i_d^{s^*}. \tag{2.86}$$

Mitcham et al. (2004) proposed a current injection scheme for open slot, bar wound, large PM machines that cancels the flux associated with the faulty term and forcing an adequate predetermined current wave shape in the remaining healthy turns. The analysis was under the assumption of minimal mutual coupling between phases.

Arumugam et al. (2015), separate from current injection as discussed by Welchko et al. (2006) discussed terminal shorting and mechanical and electrical shuns that require additional electrical or mechanical components and considerations

Cintron-Rivera et al. (2015) analyzed the fault current in a three-phase permanent magnet machine using the electrical model of three-phase PMAC machine.

The fault current is:

$$\vec{i}_f = \overbrace{\vec{i}_a \cdot \frac{\{r_{af} + j\omega \cdot (L_{af} + M_{a_f a_h})\}}{Z_{a_f}}}^{\vec{i}_a\text{contribution}} + \underbrace{\frac{\Delta\lambda_{pm}\omega}{Z_{a_f}}}_{\text{Back EMF contribution}} . \tag{2.87}$$

The proposed mitigation method consists of the following steps following the detection of a fault:

1. Limit the operating speed; if conditions allow, drop the speed below the base value.
2. Restrict the amount of torque the machine can develop.
3. Start field-weakening to reduce the amount of voltage induced in the fault. This limit helps to keep the fault current within acceptable values.

Jiang et al. (2018) demonstrated the feasibility of fractional slot concentrated winding permanent magnet SG (PMSG) in dealing with turn-to-turn SC fault. The law of FSCW PMSG SC fault is analysed and verified by finite element analysis (FEA), including the influence on turn-to-turn SC current by the number of the shorted turns and the coil position in the slot. The three phase current injection control is employed to mitigate the short-circuit fault in PMSG when a turn-to-turn one occurs.

When an open fault occurs, whether in a winding or the inverter, the control algorithm has to be modified to allow the continued operation of the drive. For both Y- and Δ-connected motors, the concept is that if the faulty phase is isolated, then the inverter can produce line currents phase shifted with respect to each other by 60°, resulting in a rotating MMF with only a forward component. For example, if phase-C is open circuited due to an internal or external fault, the stator rotating MMF will be:

$$\mathcal{F}_s = N_s \cos(\theta) i_a + N_s \cos(\theta - 2\pi/3) i_b, \tag{2.88}$$

where θ is the electrical angle of the rotating magnetic field, N_s is the effective number of windings per phase, and i_a and i_b are controlled by the inverter such that they are phase shifted by 60°, e.g.

$$i_a = I_m \cos(\omega t - \pi/6) \text{ and } i_b = I_m \cos(\omega t - \pi/2) \tag{2.89}$$

Substituting i_a and i_b:

$$\mathcal{F}_s = (\sqrt{3}/2) N_s I_m \cos(\theta - \omega t). \tag{2.90}$$

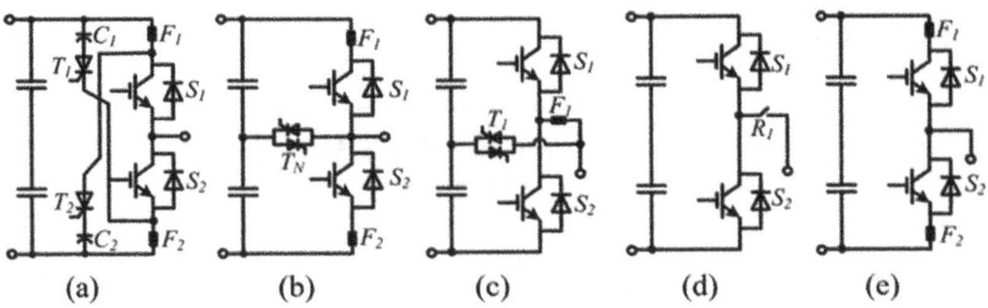

Figure 2.60 Five typical fault-isolation schemes, Zhang et al. (2014).

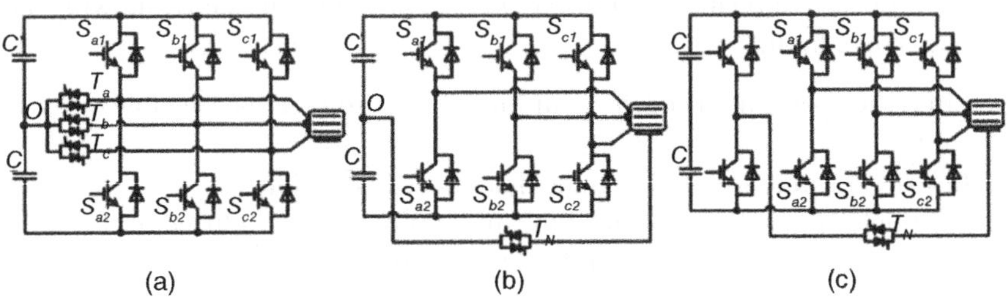

Figure 2.61 Three-phase open circuit tolerant drives Zhang et al. (2014).

If the rated power is needed, the amplitude of remaining phase currents must be increased by a factor of $1/\sqrt{3}$, meaning a 73.2% increase in current.

If one phase fails, to allow the remaining two phases to maintain continuous operations, three typical fault-tolerant topologies with additional switches employed in motor applications are presented by Zhang et al. (2014), in Figures 2.60 and 2.61.

The first fault-tolerant topology shown in 2.61 (a) forces the faulty phase to connect to the midpoint of the dc-link via the additional TRIACs. After faults, the reconfigured system is similar to the structure where only four switches are used to drive a three-phase machine. Note that since the inverter is still capable of providing the full rated current, the torque production will be preserved.

The second method connects the neutral point of the motor to the dc-bus midpoint via the incorporated TRIAC TN, as shown in Figure 2.61(b). Note that only one TRIAC is added for the fault tolerance. In case of loss of one phase, the magnitude of the other currents in the other two phases has to increase by $\sqrt{3}$, and the phase shift between these two is regulated to 60° in post-fault operations. Note that the modified control strategy leads to a zero-sequence current. Therefore, TN needs to be turned on after faults to flow the neutral current.

The third method connects the neutral point of a motor to an additional fourth leg. As shown in Figure 2.61(c), an extra leg and a TRIAC are incorporated. The fourth pole can be permanently connected to the converter even in normal operations without TRIAC. Compared to the previous two solutions, this topology yields the following good features:

- the system is free of the dc-bus midpoint balancing problems and minimum capacitance sizing,
- the line-to-line voltage can be increased to 1 p.u. compared to previous methods.

A similar principle can be applied in five-phase motors as well, with a variety of strategies, and an increase of the phase current and/or zero sequence component. Similar approach can be applied in other multi-phase motors, like four phases, six phases, etc.

A generalized optimal fault-tolerant control (FTC) technique was proposed by Mohammadpour et al. (2014) for open-circuit faults in five-phase permanent-magnet (PM) machine drives. The proposed FTC technique considers pentagon and pentacle connections of stator windings in addition to conventional star connection. Open-circuit faults in stator windings of the machine as well as switches of the inverter are analyzed. The optimization objective is to produce ripple-free electromagnetic torque with minimum ohmic losses under open-circuit fault conditions. This is achieved by a simple closed-form equation to calculate reference optimal currents in all cases where a lookup table is utilized to represent the effect of fault type. This technique is easily extended to PM machines with higher number of phases.

Mendes and Marques Cardoso (2006) realized a comparison of mitigation techniques, stator neutral point connection (SNPC) and stator phase connection (SPC) hardware reconfiguration through triac, which can be applied to both Rotor Field Oriented and Direct Torque control. For SPC and SNPC topologies, the available electromagnetic torque is reduced to 50% of its rated value. After an inverter fault detection, the drive can keep running, although at a lower speed, until an opportunity comes to solve the problem.

Under an SNPC drive reconfiguration mode, the control strategy imposes motor line currents $\sqrt{3}$ times higher than the ones obtained under the normal operating mode. For this reason, the semiconductors rated current has to be also increased $\sqrt{3}$.

An alternative method to mitigate faults and increase reliability is the use of multiphase machines to and continue operation under the loss of one, Mohammadpour and Parsa (2013), Levi (2008).

Mohammadpour and Parsa (2013) proposed a generalized optimal fault-tolerant control (FTC) technique for open-circuit faults in five-phase permanent-magnet (PM) machine drives. The proposed FTC technique considers pentagon and pentacle connections of stator windings 2.62 in addition to conventional star connection.

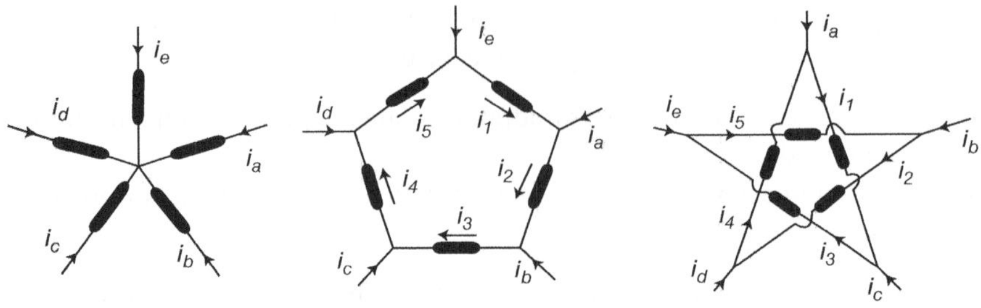

Figure 2.62 Three possible configurations for stator windings of a five-phase machine: (a) star, (b) pentagon, and (c) pentacle. Mohammadpour et al. (2014).

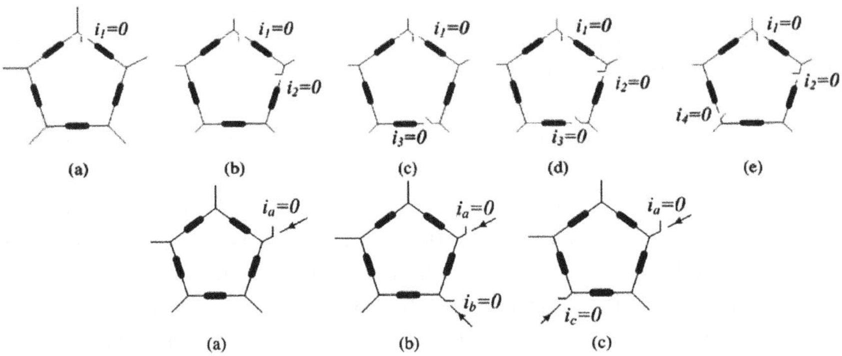

Figure 2.63 Open-circuit faults in phase windings of a five-phase motor with pentagon connection Mohammadpour et al. (2014).

In general, all possible fault conditions in Figure 2.63 can be categorized in the following eight groups:

1. single phase (1Ph);
2. adjacent double phase (A2Ph);
3. nonadjacent double phase (NA2Ph);
4. adjacent triple phase (A3Ph);
5. nonadjacent triple phase (NA3Ph);
6. single line (1L);
7. adjacent double line (A2L);
8. nonadjacent double line (NA2L).

From this list, the first five cases are faults in stator windings of motor, while the last three ones are related to faults in inverter switches.

The optimization objective is to produce ripple-free electromagnetic torque with minimum ohmic losses under open-circuit fault conditions.

This is achieved by a simple closed-form equation to calculate reference optimal currents in all cases where a lookup table is utilized to represent the effect of fault type. This FTC technique is easily extended to PM machines with higher number of phases.

For induction multi-phase machines under Rotor Field Oriented control, the operation of a drive with one open phase can be modified to achieve minimum stator peak current, minimum losses or minimum ripple Tani et al. (2012).

2.6 Power Electronics and Systems

Power electronics converters become an essential part of electrical systems. They take an increasing place in various applications such as electric drives, power conversion, industrial systems, etc. However, their increasing complexity and performance are leading to reliability issues. They are submitted to numerous stresses such as thermal cycling, electrical disturbances, moisture, vibration. For instance, Figure 2.64 shows the distribution of stresses in the case of automotive and electronic applications Winter (2008).

These stresses induce multiple faults. In the same applications, the different failures are distributed as follows (see Figure 2.65) [Choi et al. (2015)]:

- the failure of capacitors takes part for 30%;
- the failure of PCB takes part for an amount of 26%;
- the failure of semiconductor takes part for an amount of 21%;

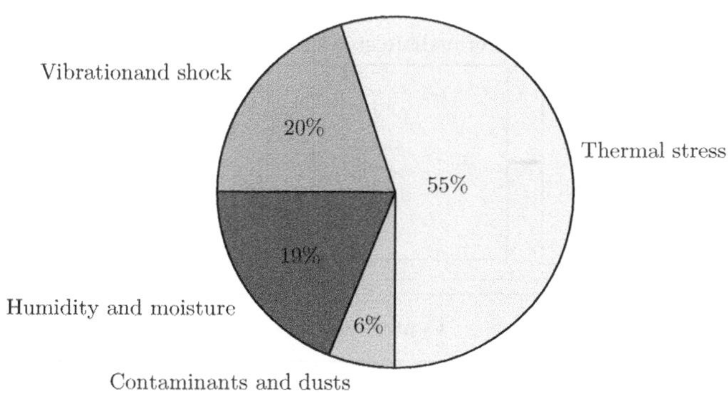

Figure 2.64 Power electronic source of tresses.

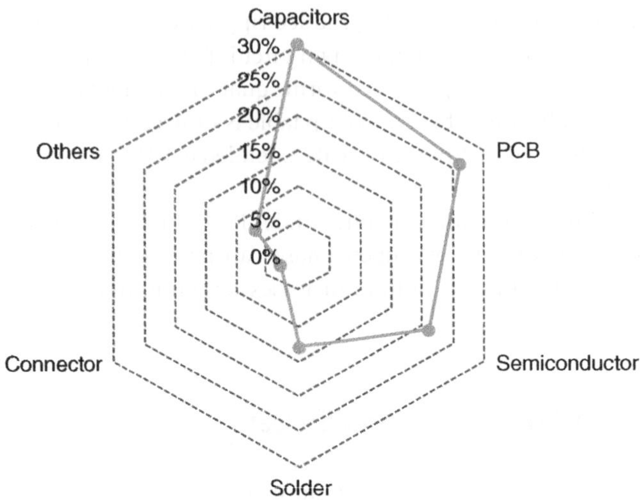

Figure 2.65 Power converter failure distribution.

- the failure of connectors takes part for an amount of 3%;
- the rest is 7%.

2.6.1 A Brief Description of Power Electronics in AC drives

The power stage of electric drives are often composed of four parts (see Figure 2.66):

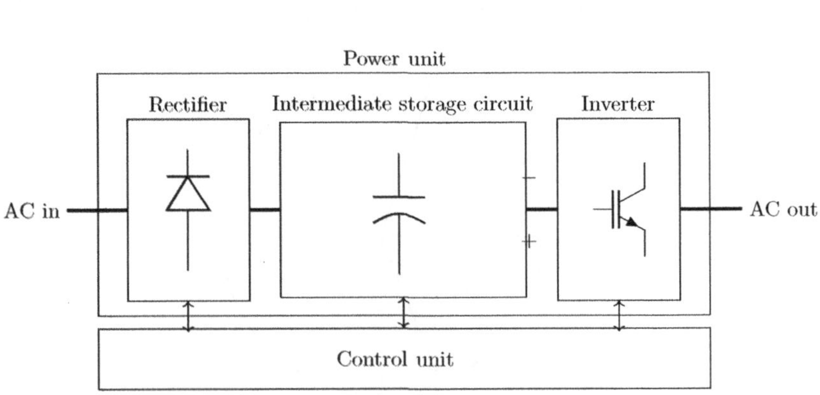

Figure 2.66 Motor Drive.

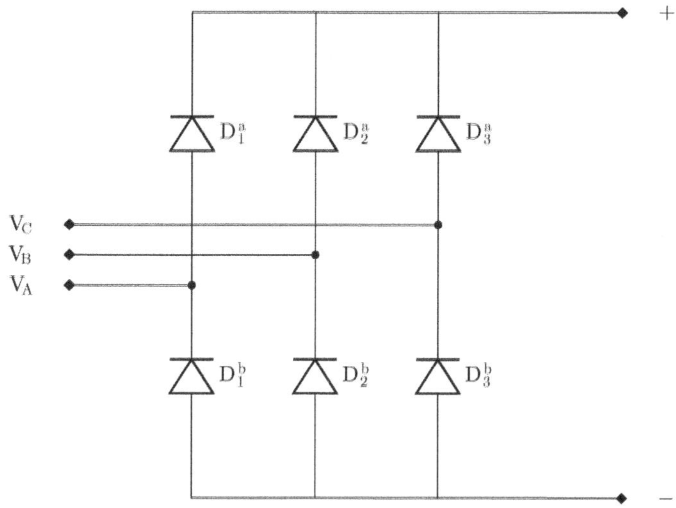

Figure 2.67 Three-Phase Diode Rectifier.

- the front-end rectifier, which converts the three-phase AC voltages (or current) to a DC current (or voltage);
- the intermediate storage stage (inductive or capacitive one according to the nature of the rectifier);
- the back end inverter: it converts the DC voltage (or current) into three phase AC voltages (or currents);
- the control unit which drives the two converters.

The front end rectifier can be a three-phase diode bridge (see Figure 2.67). Diodes are turned ON when the cathode-anode voltage becomes positive and are turned OFF when the current falls under a certain level.

The following Figure 2.68 shows the three-phase voltages and currents and the DC voltage for this uncontrolled rectifier.

In case of balanced three-phase voltage, (eq. 2.91) gives the AC supplied direct voltages,

$$\begin{cases} V_A = \hat{V}\sin(\omega t) \\ V_B = \hat{V}\sin(\omega t - \frac{2\pi}{3}) \\ V_C = \hat{V}\sin(\omega t - \frac{4\pi}{3}) \end{cases} \tag{2.91}$$

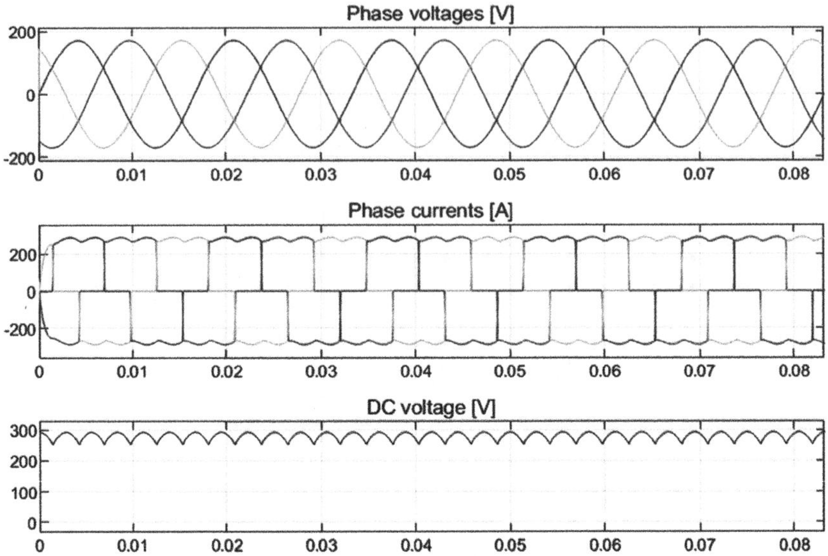

Figure 2.68 Three Phase Diode Rectifier.

The DC output voltage is given by eq. 2.92:

$$< V_{DC} >= \frac{3\sqrt{3}\widehat{V}}{\pi} \tag{2.92}$$

The front-end rectifier can consist of a three-phase controlled bridge with six thyristors (See Figure 2.69). These devices are triggered when the cathode-anode voltage becomes positive and a current gate pulse is supplied. It is turned OFF when the current falls down below a holding level and a negative cathode-anode voltage is applied during a minimum turn OFF time t_{off}.

Figure 2.70 shows the phase voltages and currents and the DC link voltage for this controlled rectifier.

In the case of balanced three-phase voltages (eq. 2.91), the DC output voltage is given by the equation 2.93:

$$< V_{DC} >= \frac{3\sqrt{3}\widehat{V}}{\pi} \cos\alpha \tag{2.93}$$

with α the firing-delay angle.

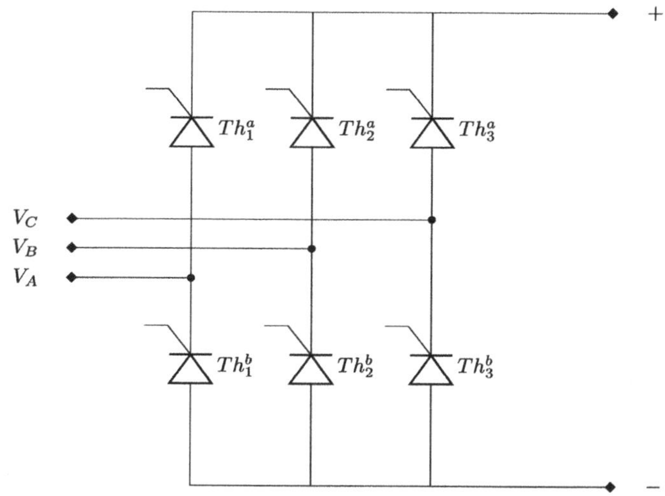

Figure 2.69 Three Phase Thyristor Rectifier.

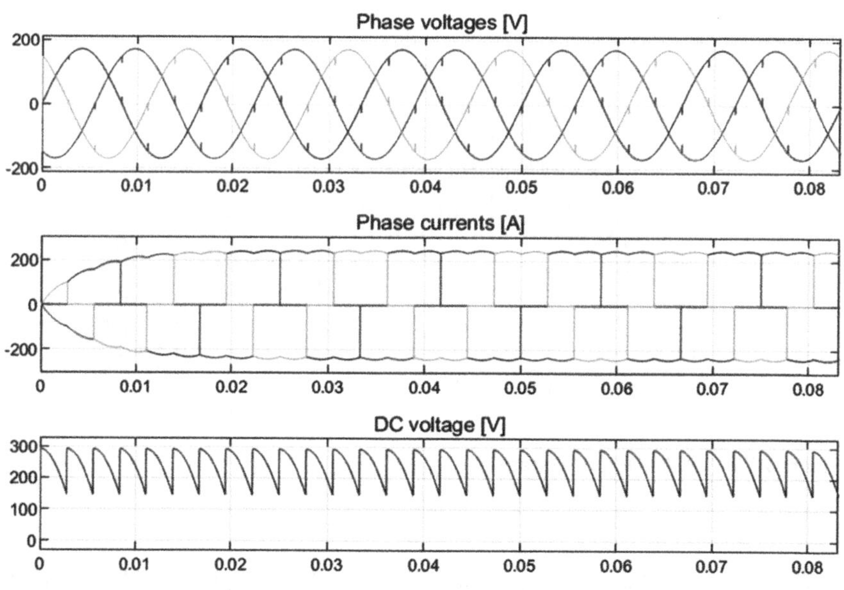

Figure 2.70 Three phase thyristor rectifier waves.

The back-end converter is an inverter (see Figure 2.71). It generally consists of six insulated-gate bipolar transistors (IGBT) or six metal–oxide–semiconductor field-effect transistor (MOSFET) with anti-parallel diodes.

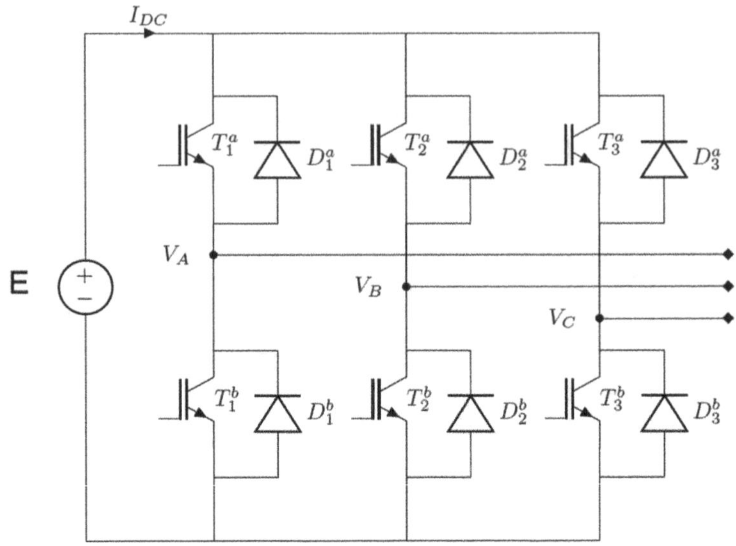

Figure 2.71 Three Phase Inverter.

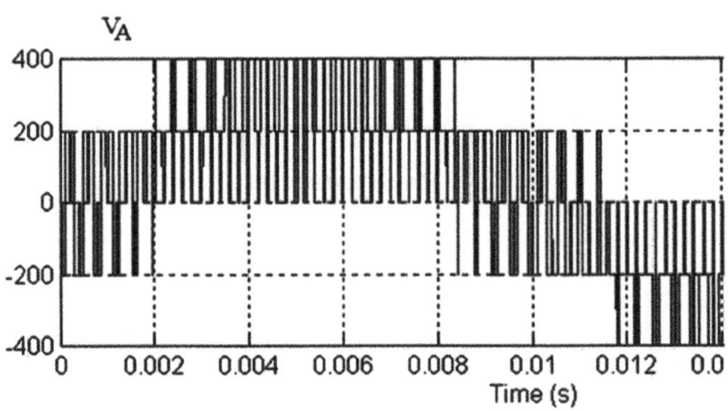

Figure 2.72 PWM Output Voltage.

The control unit generally provides the Space Vector Power Width Modulation (PWM) gate signals of the inverter (see Figure 2.72). Noise is rejected above the sampling frequency.

This Symmetrical Space Vector PWM modulation follows the following principles:

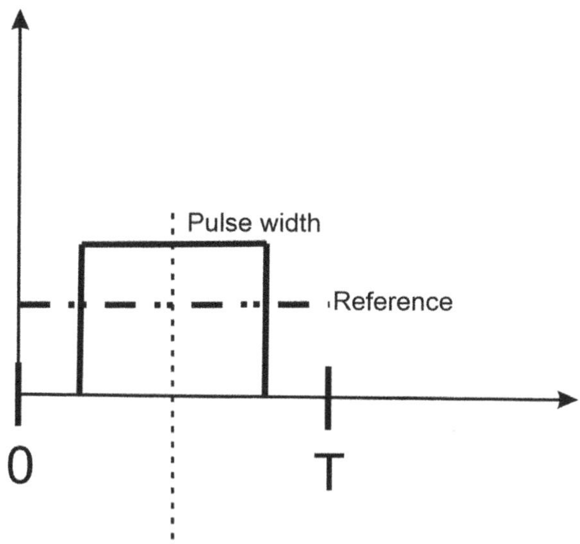

Figure 2.73 PWM principle.

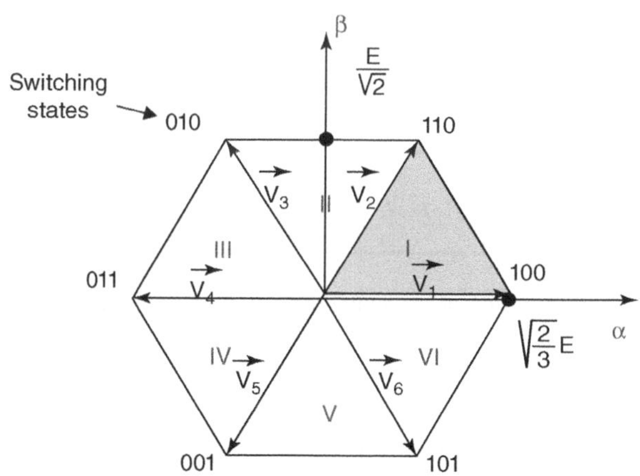

Figure 2.74 α-β representation of the vector limits.

- the reference signal is sampled at regular intervals T_e (referred to as regular PWM);

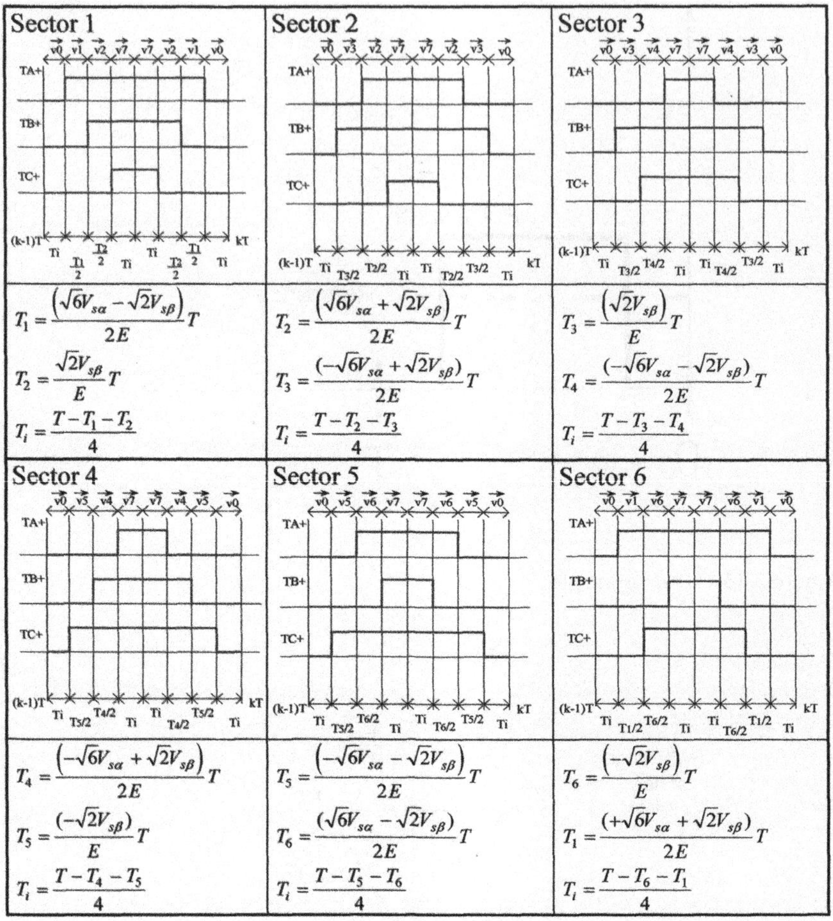

Figure 2.75 Representation of the switching times. (c) Courtesy of Eyrolles. *Source:* Grellet and Clerc (1997).

- for each phase, a pulse of width T_i, centered on the period (symmetrical PWM) is provided in order to have a mean value which is equal to reference voltage at the beginning of the sampling period;
- this modulation is carried out synchronously on the three phases.

Let us consider the different switching states in a three-phase inverter (see Table 2.5).

Table 2.5 Voltage vectors generated by a two levels inverter.

k	$\vec{v_k}$	$\Re(\vec{v_k})$ α axis	$\Im(\vec{v_k})$ β axis	S_A	S_B	S_C
1	$\vec{v_1}$	$E\sqrt{\frac{2}{3}}$	0	1	0	0
2	$\vec{v_2}$	$E\sqrt{\frac{1}{6}}$	$E\sqrt{\frac{1}{2}}$	1	1	0
3	$\vec{v_3}$	$-E\sqrt{\frac{1}{6}}$	$E\sqrt{\frac{1}{2}}$	0	1	0
4	$\vec{v_4}$	$-E\sqrt{\frac{2}{3}}$	0	0	1	1
5	$\vec{v_5}$	$-E\sqrt{\frac{1}{6}}$	$-E\sqrt{\frac{1}{2}}$	0	0	1
6	$\vec{v_6}$	$E\sqrt{\frac{1}{6}}$	$-E\sqrt{\frac{1}{2}}$	1	0	1
0	$\vec{v_0}$	0	0	0	0	0
7	$\vec{v_7}$	0	0	1	1	1

Considering the different possible states of the static switches, we can define eight vectors (two of them are null) which limit six sectors (see Figure 2.74).

with E the DC voltage.

For each reference voltage vector, after selecting the right sector, the switching times are computed in order to respect the previous properties (see Figure 2.75).

2.6.2 A Brief Description of Static Switches

The Metal–oxide–semiconductor field-effect transistor (MOSFET) and the insulated-gate bipolar transistor (IGBT) are the main controlled static devices used in electric drives. They mainly use Silicon or Silicon Carbide semiconductors.

2.6.2.1 MOSFET

Source: Correvon (2020b), Fairchild (2000), Pathak (2001).

The MOSFET is an unidirectional voltage and bidirectional current switch (see Figure 2.76).

It is an unipolar structure (see Figure 2.77). In order to analyze the behavior of this device, let us consider an horizontal N-MOSFET with a Silicon technology. Two heavily doped N type regions are diffused in a P type Silicon substrate. A thin insulating layer (generally in SiO_2) isolates the gate connection from the body. The current flowing across the MOSFET is regulated by the charge carrier concentration in the region below the gate (the channel). The carrier concentration and the electrical strength depend on the gate voltage V_{GS} and the Drain source voltage V_{DS}.

If $V_{GS} < 0$, the internal NP and PN back to back body diodes hold the drain source voltage (see Figure 2.78(a)). No current flows.

If $0 < V_{GS} < V_T$, due to an impoverishment of carriers in this depletion state, the channel remains highly resistive and no current flows (see Figure 2.78(b)).

If $V_{GS} > V_T$, by electrostatic influence, the P carriers under grid are pushed out from the channel and the electrons are attracted to this region. This area is called the inversion layer. The n-induced channel becomes conductive. If a positive Drain Source voltage is applied, a current flows. The device enters into a linear behavior. As V_{DS} grows, electron concentration decreases near the drain area (see Figure 2.78(c)). If V_{DS} continues to grow, the inversion layer is no longer connected to the drain region and the device enters into a saturation mode (see Figure 2.78(d)).

Thus, N-MOSFET exhibits the following characteristics I_{DS} versus V_{DS} (see Figure 2.79).

In power applications, in order to decrease the on-state losses, the MOSFET works in the saturation mode.

However, horizontal structure is not suitable for high-voltage applications. Vertical topology performs a higher voltage withstand. Several ones have been designed. Figure 2.80 details some of them.

The N-MOSFET brings several parasitic elements:

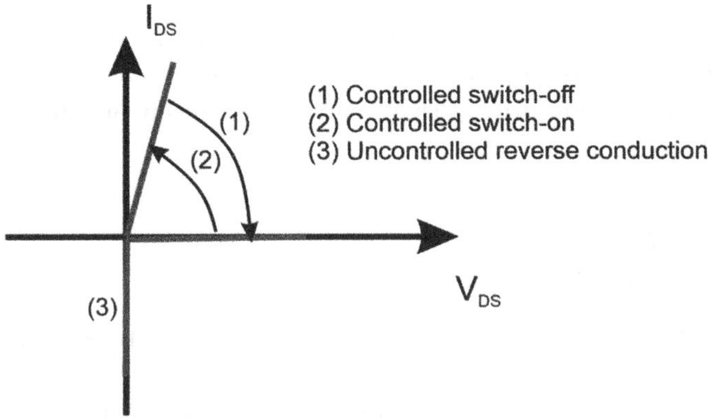

Figure 2.76 MOSFET Switching characteristic.

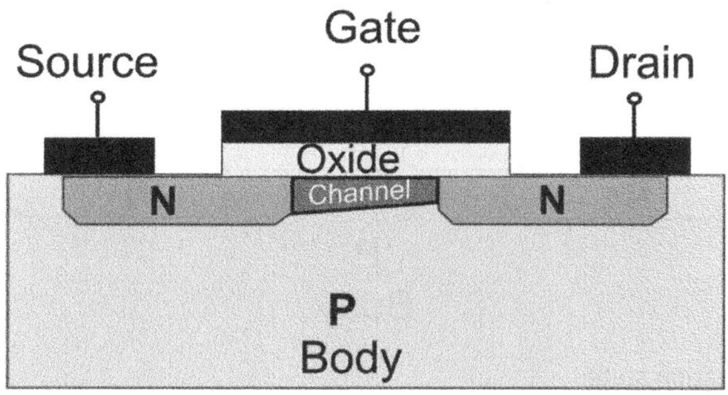

Figure 2.77 N-MOSFET Horizontal Structure.

- A NPN transistor;
- A PN-diode: Due this parasitic component, the MOSFET cannot sustain an inverse polarization;
- An on-state resistance;
- Some parasitic capacitances.

The on-state resistance is due to (Figure 2.81):

- R_{N+}: N+ zone resistance;
- R_{ch}: channel resistance;
- R_A: the resistance of this accumulation layer under the gate;
- R_J: the resistance of the N- epitaxial layer;
- R_D: the resistance from the P zone to the substrate;
- R_S: substrate resistance.

The structure also brings different parasitic capacitances (Figure 2.82):

- C_{DS}: drain-source capacitance = C_O+C_{N+}+C_P;
- C_{GS}: gate-source capacitance;
- C_{GD}: drain-source capacitance.

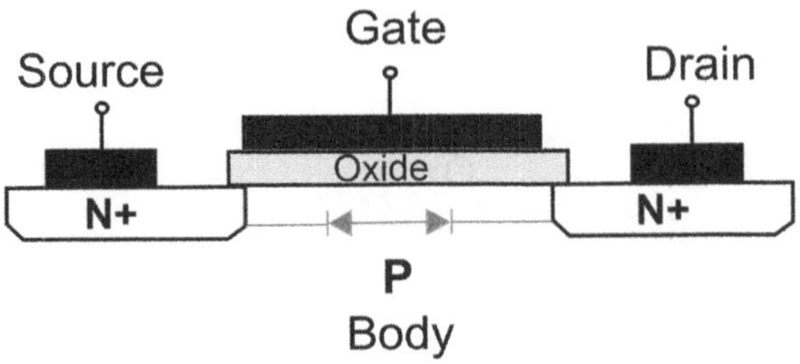

(a) Accumulation state.

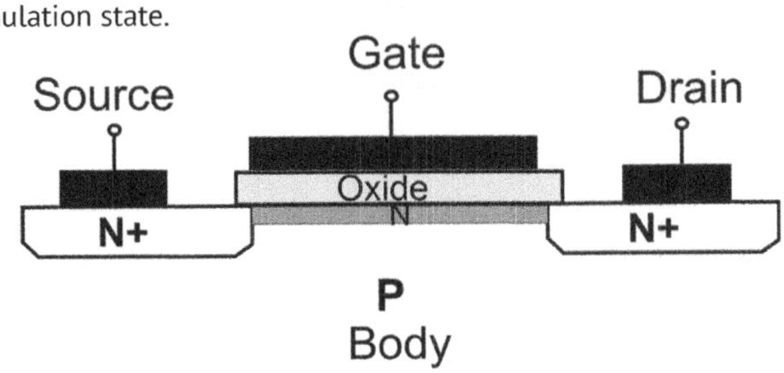

(b) Depletion state.

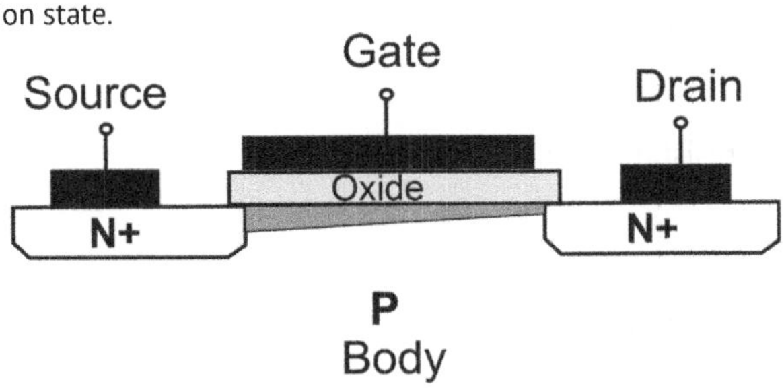

(c) Linear state.

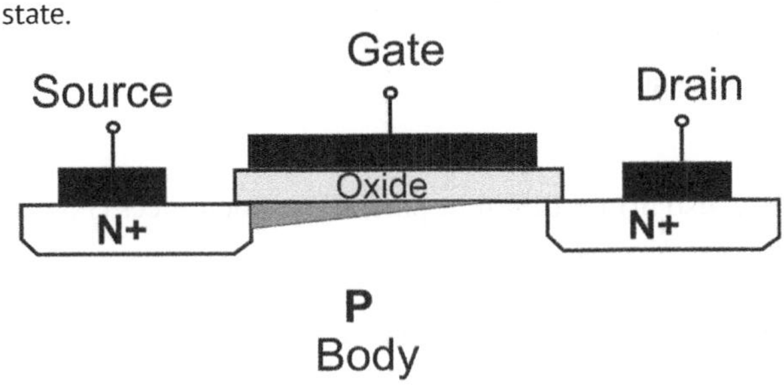

(d) Saturation mode.

Figure 2.78 N-MOSFET behaviour.

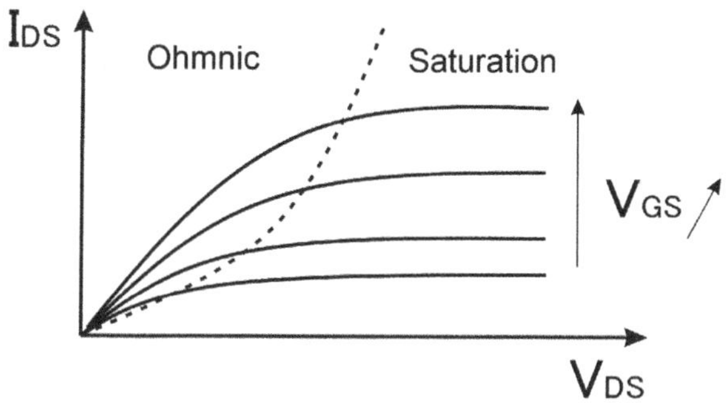

Figure 2.79 N-MOSFET Characteristic.

(a) V N-MOSFET Structure.

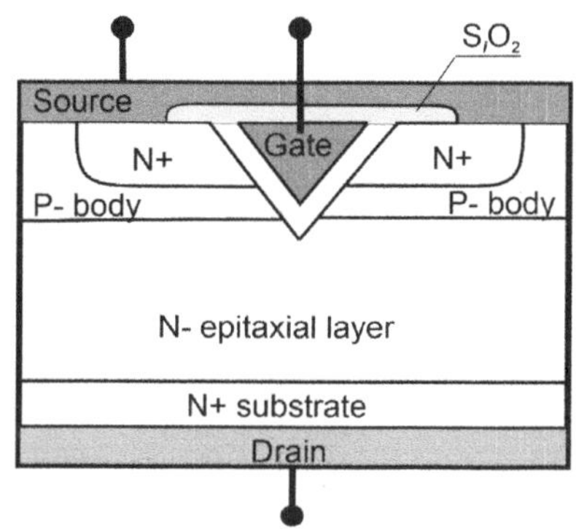

(b) D N-MOSFET Structure.

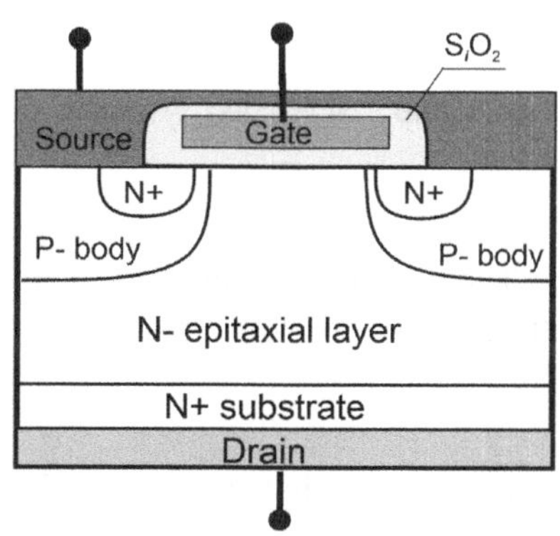

In addition, the various connecting wires and bondings also carry parasitic inductances. Thus, the equivalent N-MOSFET scheme can be represented as in the Figure 2.83.

(c) U N-MOSFET Structure.

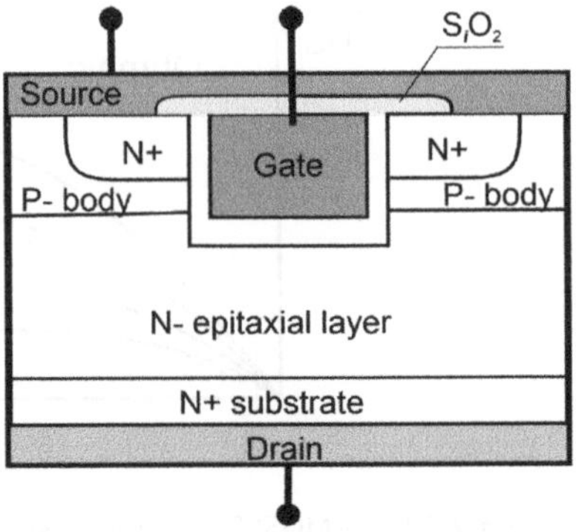

(d) Super Junction N-MOSFET Structure.

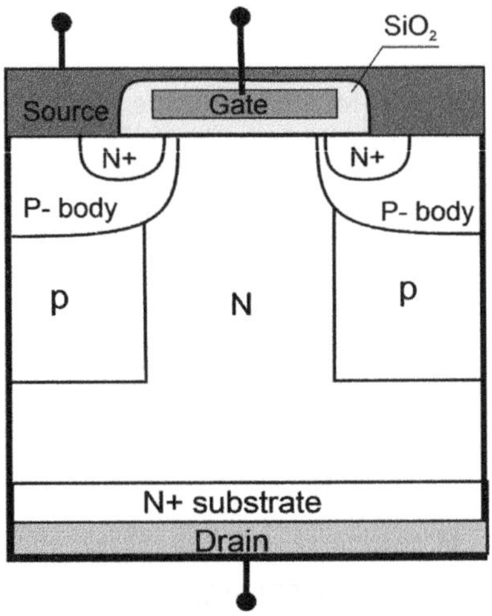

Figure 2.80 N-MOSFET vertical structures. *Source:* Fairchild (2000) Zhang et al. (2018).

These various parasitic elements affect the switching characteristics of the MOSFET at turn-on (see Figure 2.84(a)) and turn-off (see Figure 2.84(b)).

Turn-on transient with an inductive load:

Let us consider t_0=0, the beginning of the switch-on transient of the N-MOSFET. The different turn-on transient times are defined as follows:

- From t_0 to t_1: V_{GS} is below the threshold voltage V_{TH}. Thus, the channel is not still established. I_{DS} remains almost equal to 0; or.

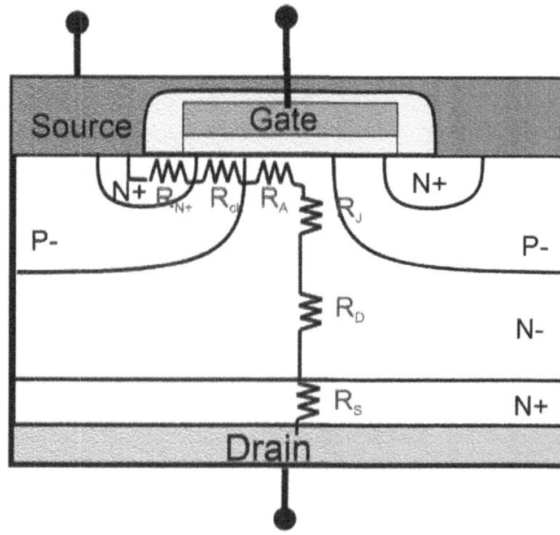

Figure 2.81 Parasitic resistances within a MOSFET.

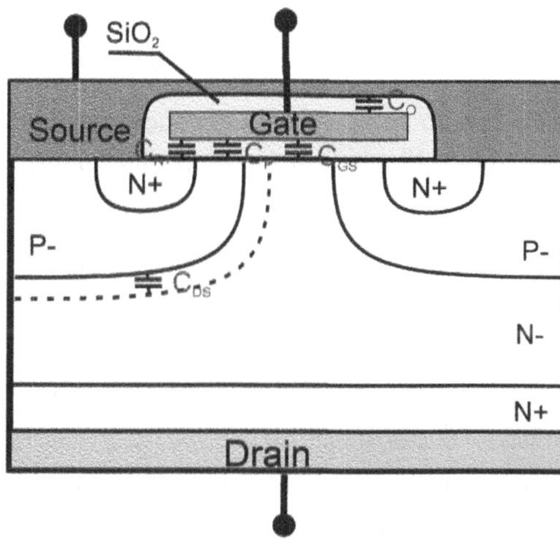

Figure 2.82 Parasitic capacitances within a MOSFET.

- From t_1 to t_2: I_{DS} increases until the on-state current. The body diode is still conducting. or;
- From t_2 to t_3: V_{GS} remains almost constant. It is called the "Miller Plateau". The gate circuit must provide the load of the capacitance C_{GD}.
- From t_3 to t_4: V_{DS} decreases down to the on-state value R_{DSon}I. V_{GS} decreases down to 0.

Turn-off transient with an inductive load:

Let us consider t_0=0, the beginning of the switch-off transient of the N-MOSFET. The different turn-off transient times are defined as follows:

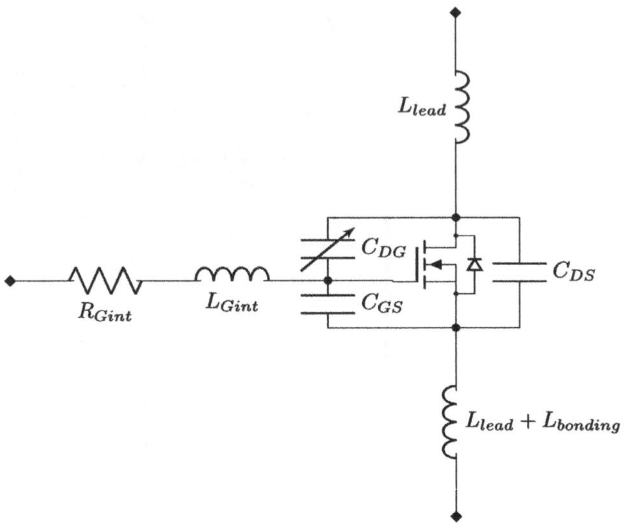

Figure 2.83 Equivalent circuit of a N-MOSFET.

- From t_0 to t_1: V_{GS} decays exponentially to the Miller Plateau level. The gate current is flowing through C_{GS} and C_{GD}. I_{DS} remains constant. or;
- From t_1 to t_2: I_{DS} doesn't change. V_{DS} is slightly increasing up to the clamping voltage of free wheeling diode necessary in an inductive circuit. or;
- From t_2 to t_3: I_{DS} doesn't vary. V_{GS} remains almost constant. The gate circuit must discharge C_{GD}. This is the "Miller Plateau".
- From t_3 to t_4: I_{DS} decreases towards to 0. V_{GS} falls down to V_{TH}. C_{GS} is discharged through any external impedance between Gate and Source terminals. The MOSFET is in the linear region.

Some of these parasitic elements or switching characteristics, as we shall see below, provide a useful information about the junction temperature or the aging of the components.

2.6.2.2 IGBT

Source: Correvon (2020a), Pathak (2001)

The IGBT can be an unidirectional voltage and a bidirectional current switch (see Figure 2.85(a)) or bidirectional voltage and unidirectional current switch according to their structure (see Figure 2.85(b)).

The insulated-gate bipolar transistor (IGBT) consists of four layers such as represented in Figure 2.86.

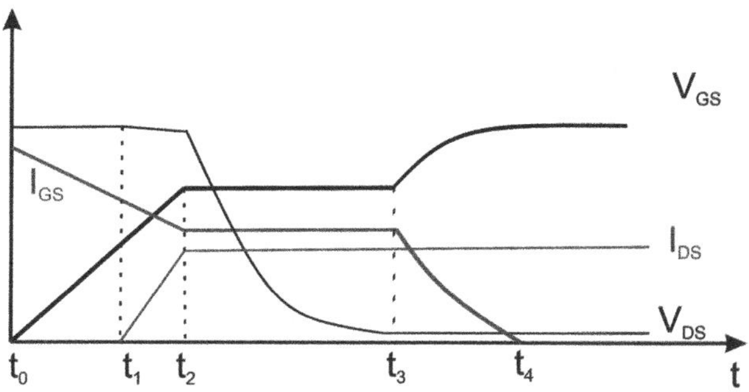

(a) Switch-on Waveforms.

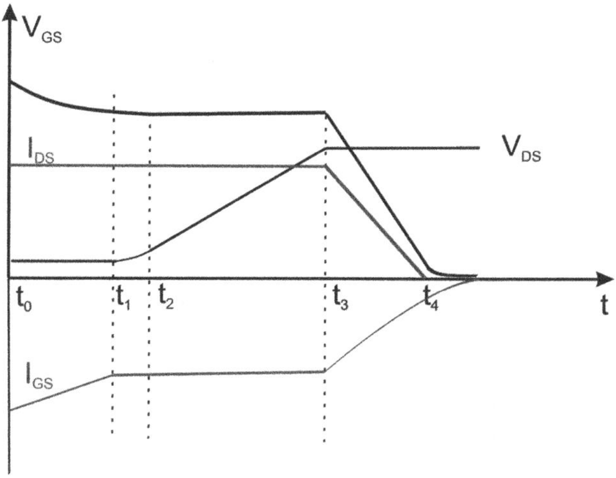

(b) Switch-off Waveforms.

Figure 2.84 Switching characteristics of a N-MOSFET.

The structure and behavior of an IGBT are similar to a power MOSFET. But the N^+ substrate layer is replaced by a P^+ one. This allows a minority carries injection (holes) into the N^- epitaxial layer. This lowers the on-state voltage V_{CE}.

If the current is too high, the parasitic NPN and PNP transistors can trigger NPNP thyristor structure and thus create a latch-up. This can be met also in certain case in a dynamic behavior.

Two main structures exist: the non punch through structure (Figure 2.87(a)) and the punch through structure (Figure 2.87(b) or 2.87(c)). The last one has an additional N+ buffer layer.

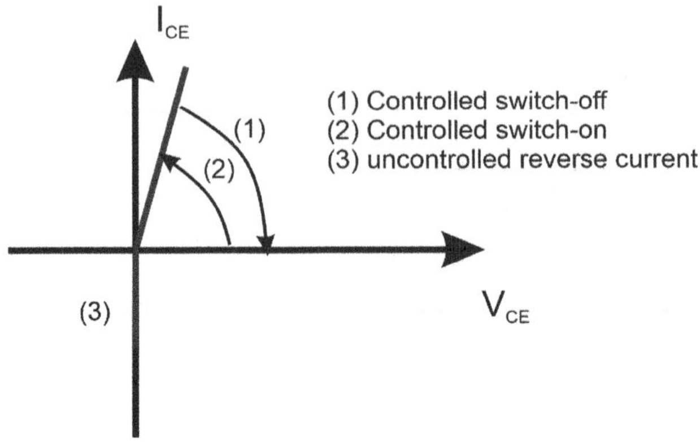

(a) Asymmetric voltage blocking capabilities structure.

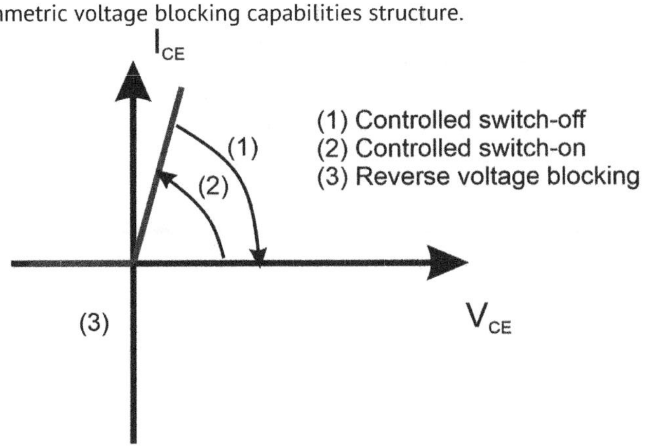

(b) Symmetric voltage blocking capabilities structure.

2.85 IGBT Switching characteristic.

Let us compare punch through and non-punch through IGBT structures (Yilmaz et al. (1986)).

Non-Punch Through IGBT:

- it has symmetric blocking capabilities and, thus, it can block a reverse voltage;
- switching loss are almost unaffected by temperature.

Punch Through IGBT:

- it cannot block a reverse voltage;
- it has faster turn-off times;
- it has a lower V_{CEsat};

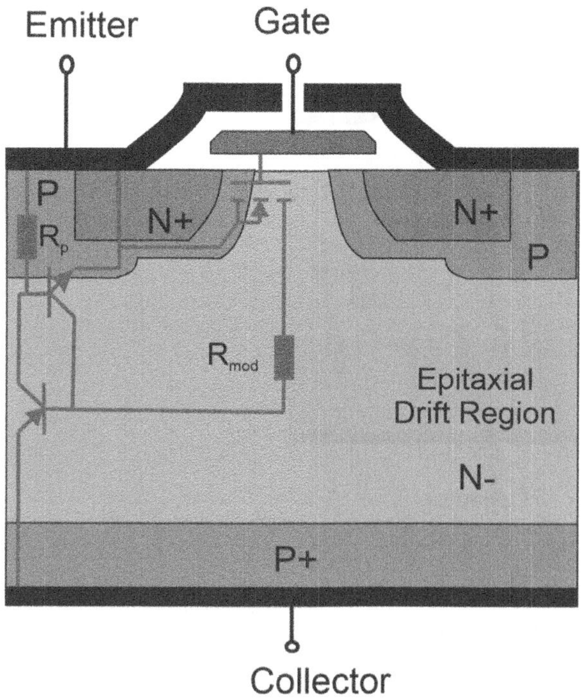

Figure 2.86 IGBT structure.

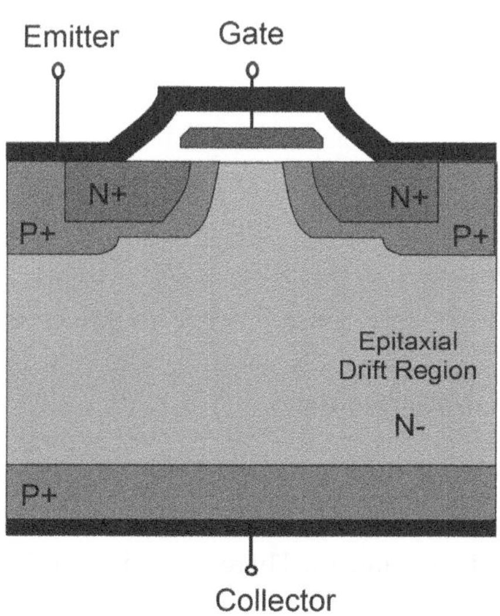

(a) Non Punch Through IGBT Structure.

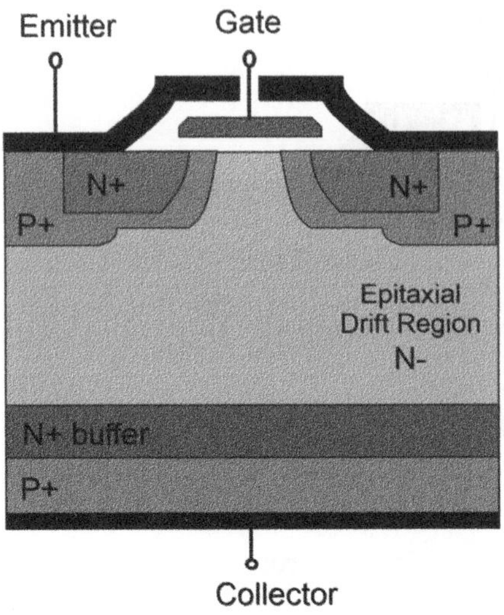

(b) Punch Through IGBT Structure.

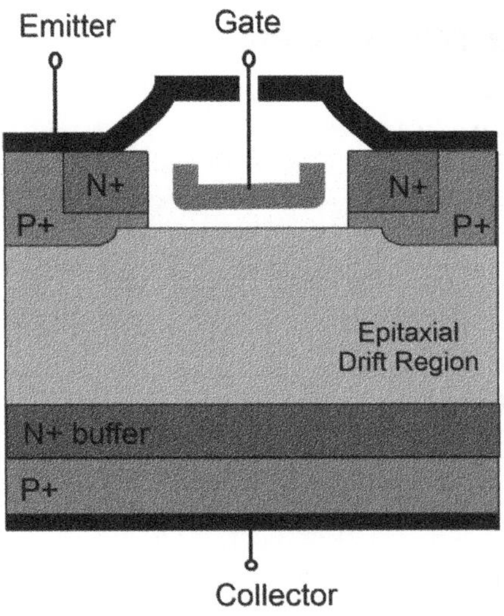

(c) Trench Punch Through IGBT Structure .

2.87 IGBT Punch through and non punch through Structures.

- turn-off switching loss depends on temperature.

The Figure 2.88 illustrates the parasitic resistances. There are quite similar to the parasitic ones in the MOSFET structure.

The Figure 2.89 shows the parasitic capacitances as in the MOSFET structure. The Gate-Emitter capacitance is: $C_{GE}= C_O+C_{N+}+C_P$.

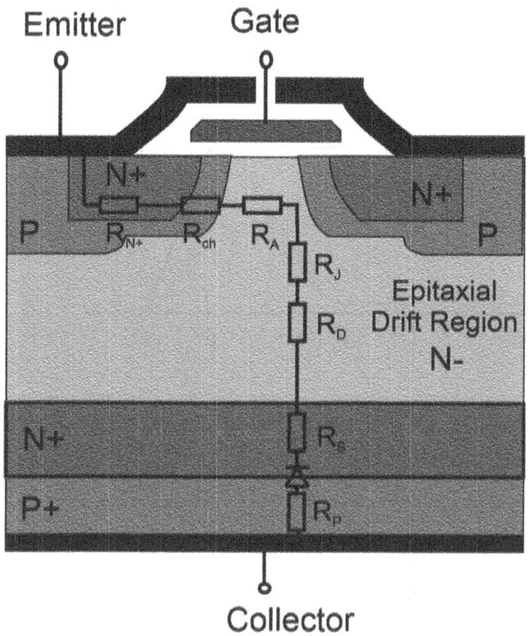

Figure 2.88 Parasitic resistances with IGBT.

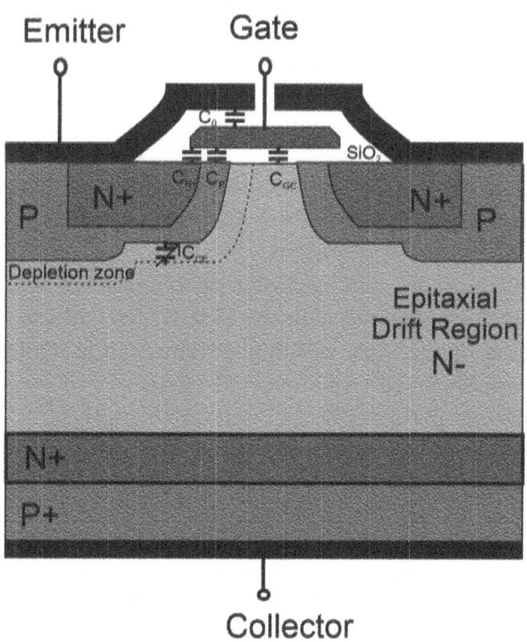

Figure 2.89 Parasitic capacitances with IGBT.

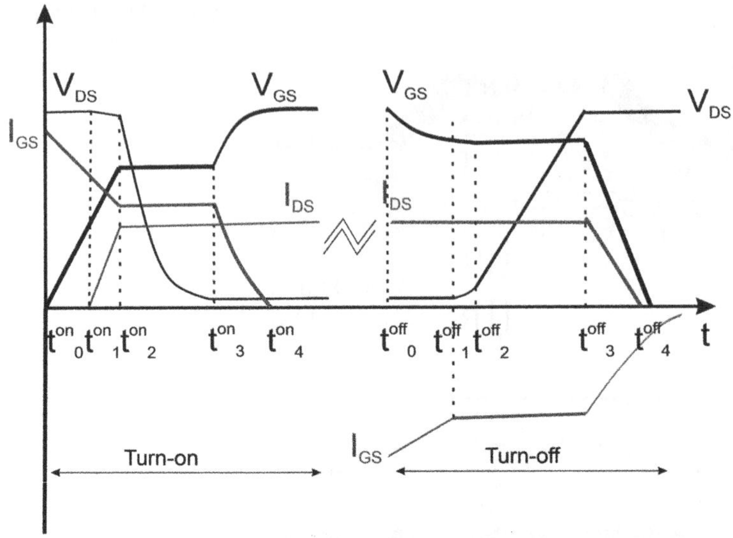

Figure 2.90 IGBT turn-on and turn-off transient waveforms.

All parasitic elements interfere with the switching process.

The waveform shapes, during switching, are quite similar to those of the MOSFET (2.90). We can also notice a "Miller Plateau".

2.6.2.3 Si and SiC Technology

MOSFET and IGBT power devices can be made from a silicon or silicon carbide substrate or in some new applications in GaN (Gallium Arsenide). The latter is used in high voltage or high temperature applications.

Thus, for the same assigned values, SiC IGB has faster switching speed and lower switching and on-state losses compared with the Si IGBT. In addition, Si IGBT switching losses increase significantly with higher temperature while they remain almost constant with SiC IGBT (Albanna et al. (2016)).

Figure 2.91 compares the main properties of Si, SiC, and GaN materials.

Table 2.6 represents the main requirements according to the type of application.

2.6.2.4 Thermal Behavior

If we consider a body subject to internal losses, the power losses P_a cause, on the one hand, the heating of the body and, on the other hand, a heat exchange

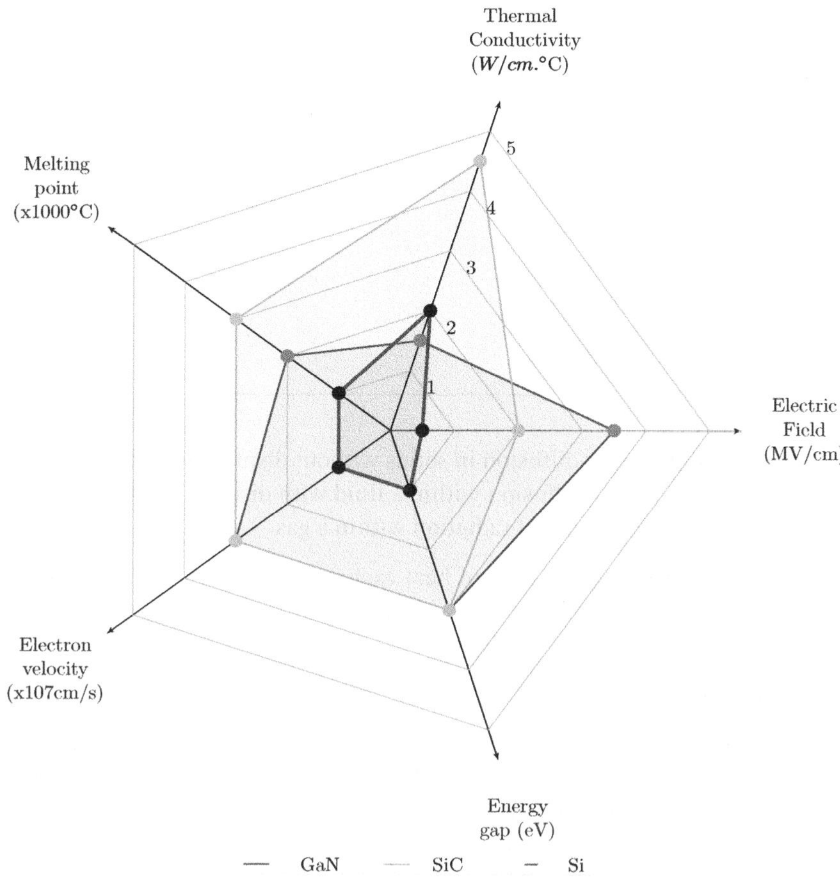

Figure 2.91 Summary of Si, SiC, and GaN material properties*Source:* Millán et al. (2014).

with the outside. Thus, the thermal transfer during a time dt is represented by eq. 2.94

$$P_a \cdot dt = Q_1 + Q_2 \tag{2.94}$$

Q_1 represents the amount of heat which is used to heat up the body:

$$Q_1 = M.c \cdot d\theta \tag{2.95}$$

It is usually written: $M.C = C_T$. This is the thermal capacity of the body expressed in $J.K^{-1}$.

Q_2 is the amount of heat exchanged with the exterior in three very different processes:

Table 2.6 Semi-conductor requirements according to the kind of applications.

Application	Electric field withstand	Thermal conductivity	Melting point	Electron velocity	Energy gap
High-temperature operation		X	X		
High-frequency switching				X	
High-voltage operation	X				X

- the conduction, i.e. by diffusion in solids without displacement of matter;
- the convection, i.e. by diffusion within a fluid with displacement of matter;
- the radiation, by emission of radiation within a gas.

The laws of these three modes of heat exchange are different. But for temperatures below 250°C, these three laws, as a first approximation, can take the form:

$$Q_i = S \cdot \alpha_i \cdot \theta \cdot dt \tag{2.96}$$

where S is the heat exchange surface normal to the heat flow and heating. The coefficient α_i is characteristic of the mode of transmission. The quantity of heat can be obtained by summing the three transmission coefficients and writing as follows:

$$Q_2 = \ S \cdot \alpha_t \, \theta \, dt \tag{2.97}$$

where $\alpha_t = \sum \alpha_i$. α_t can be expressed according to the thermal conductivity λ and the geometrical transfer length ℓ in the direction of the heat flow.

$$\alpha_t = \frac{\lambda}{\ell} \tag{2.98}$$

Finally, we define the thermal resistance with eq. 2.99:

$$R_t = \frac{1}{S\,\alpha_t} = \frac{1}{\lambda}\frac{\ell}{S} = \rho_t \frac{\ell}{S} \tag{2.99}$$

Thus, for a single body, the thermal transfer could be represented by the equation eq. 2.100:

$$P_a = C_t \frac{d\theta}{dt} + \frac{\theta}{R_t} \tag{2.100}$$

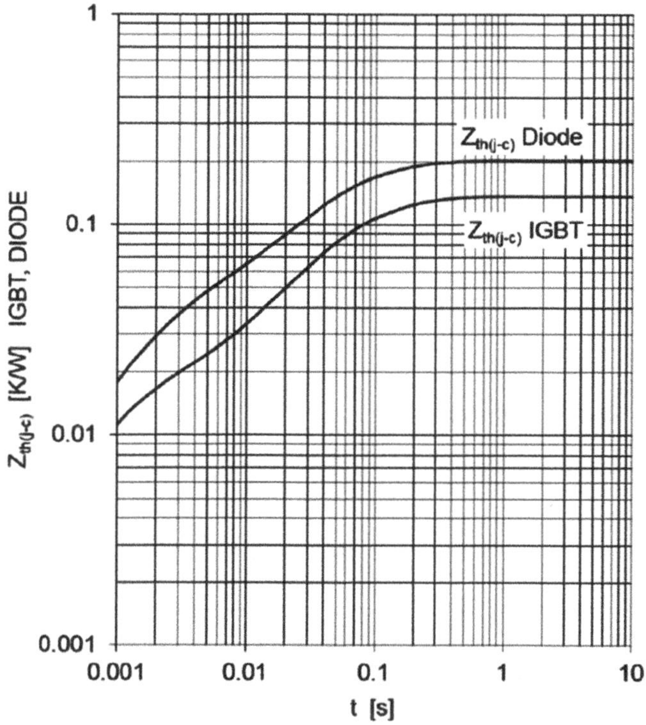

Figure 2.92 Example of thermal impedance from 5SNG 0150Q170300 ABB module . Courtesy of ABB Power Grids Switzerland Ltd., Semiconductors *Source:* ABB (2016).

The thermal impedance is the transfer function between the power loss and the temperature variation (eq. 2.101):

$$Z_{th} = \frac{\Delta\theta}{\Delta P} \tag{2.101}$$

In the case of a single body, it can be written (eq. 2.102):

$$Z_{th}(s) = \frac{\Delta\theta}{P_a} = \frac{R_T}{1 + R_T C_T s} \tag{2.102}$$

The transient thermal impedance can be represented,in the data-sheet, as in Figure 2.92.

The module can be considered as a stack of bodies and the thermal behavior can be represented by different networks (Figure 2.93).

There are two kinds of networks that are used here: Cauer and Foster ones.

Cauer model is based on the physical behavior (Figure 2.94).

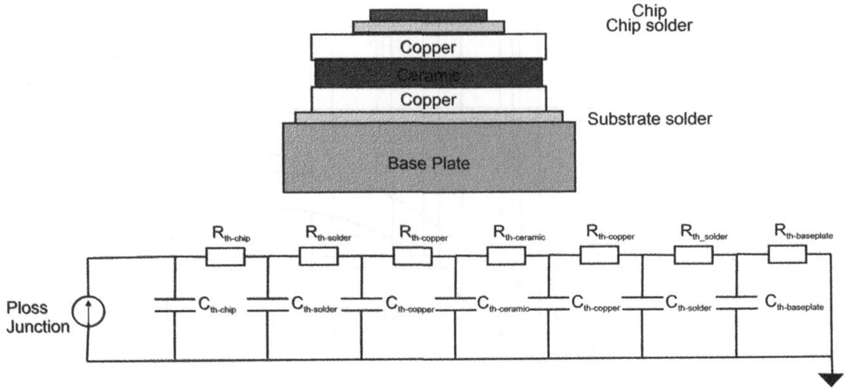

Figure (a) Multi-stack thermal structure.

Figure (b) Equivalent thermal network *Source:* Alavi et al. (2017).

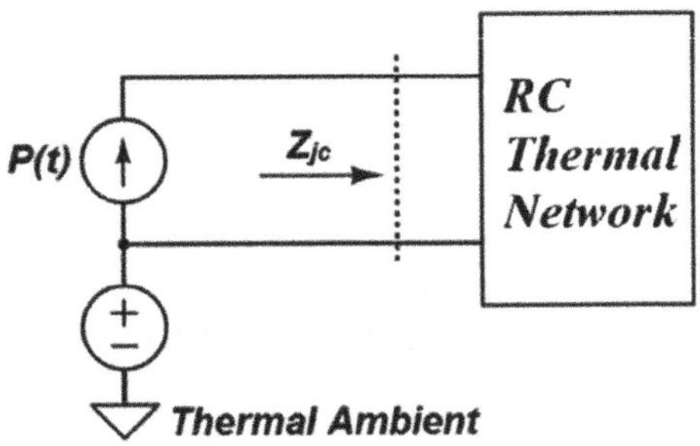

Figure 2.93 Multi-stack body.

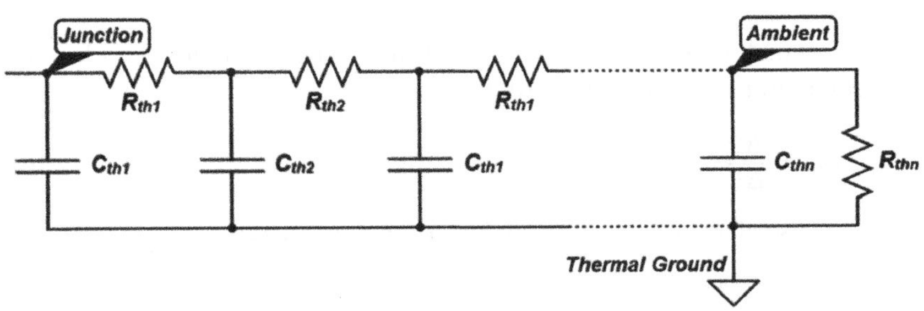

Figure 2.94 The Cauer thermal model of the IGBT*Source:* Alavi et al. (2017).

Thus, the transfer function can be defined by eq. 2.103:

$$Z_{th}(s) = \frac{1}{sC_{th_1} + \frac{1}{R_{th_1} + \frac{1}{sC_{th_2} + \ldots + \frac{1}{R_{th_n}}}}} \tag{2.103}$$

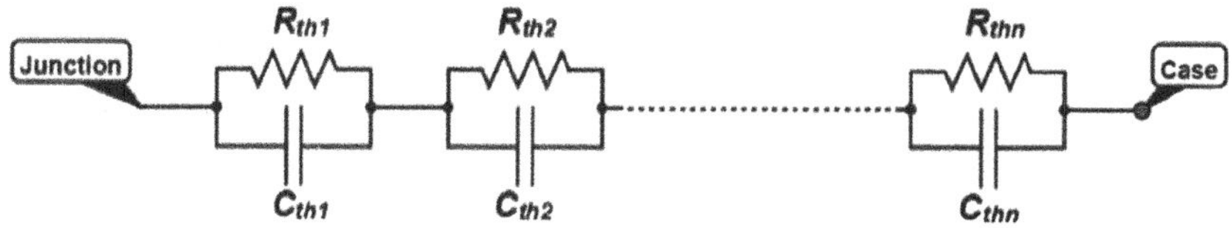

Figure 2.95 The Foster thermal model from junction to case*Source:* Alavi et al. (2017).

Foster Model is represented as in Figure 2.95.

The thermal impedance is represented by the equation 2.104:

$$Z_{th(j-c)}(t) = \sum_{i=1}^{n} R_i(1 - e^{-t/\tau_i}) \tag{2.104}$$

with $\tau_i = R_i.C_i$

We can transform one model to the other (Murthy and Bedford (1978)). From the data-sheets, the different networks parameters can be identified. They also can be measured.

2.6.3 A Brief Description of Capacitors

2.6.3.1 General Description

The main function of a capacitor is the storage of electrical charge between two conductive plates (electrodes) separated by an insulating called a dielectric (Figure 2.96). When a potential difference is applied between the conductive plates, an electrical charge proportional to the applied voltage and the capacity is accumulated in the capacitor.

The energy E stored for an ideal capacitor of capacity (C_{AK}) between the anode and the cathode under a voltage U is given by:

Figure 2.96 General constitution of a flat capacitor.

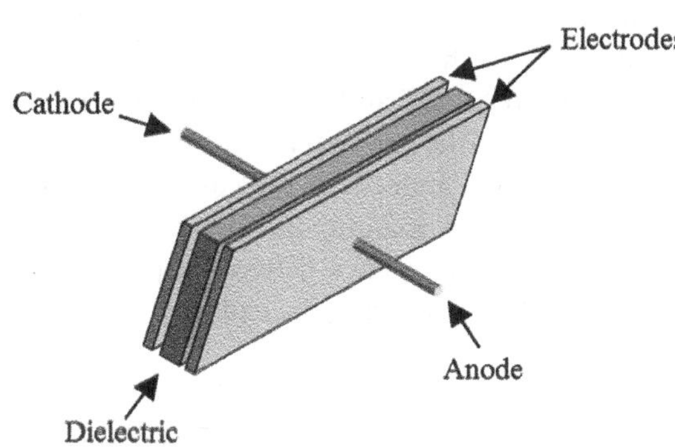

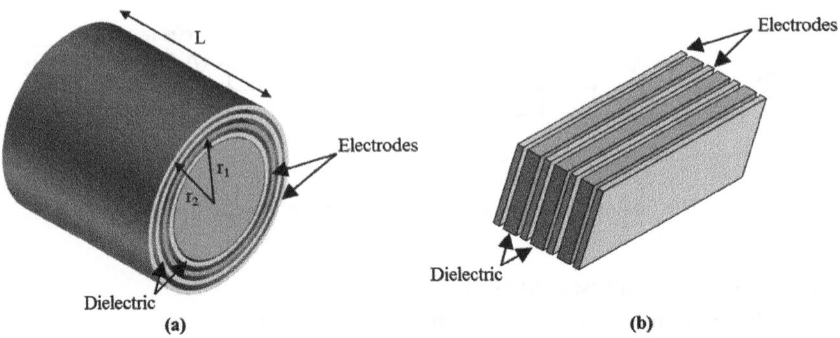

Figure 2.97 Cylindrical geometry (a) and stack geometry (b) of real capacitors.

$$E = \frac{1}{2}\, C_{AK}\, U^2 \tag{2.105}$$

As shown in equation (2.105). For a planar capacitor, the value of the capacity (C_{AK}) is directly proportional to the surface of the facing electrodes and inversely proportional to the thickness of the dielectric:

$$C_{AK} = \varepsilon_0\, \varepsilon_r \frac{S}{e} \tag{2.106}$$

with:

- ε_0 = absolute permittivity of the vacuum ($\varepsilon_0 = 8.854 \times 10^{-12}$ F/m);
- ε_r = relative permittivity of the dielectric (dimensionless);
- S = surface facing anode-cathode (m^2);
- e = thickness of the dielectric in (m).

From the two previous equations, it is noticed that the energy stored at the terminals of the capacitor depends on the surface of the electrodes S, on the applied voltage U as well as on the physical and geometric characteristics of the dielectric used (ε_r, e). From the previous equation, it is possible to increase the capacity of the capacitors by increasing the area of the armatures. The geometries in "stack" and "coil" are often used (see Figure 2.97).

Regardless of the geometry used, the resulting capacity will be similar to a planar capacitor. Indeed, in the case of cylindrical capacitors with one turn, and by taking into accounts the fineness of the thickness of the dielectric ($r_2 - r_1 \ll r_1$), the equivalence between a plane and cylindrical geometry will be established by considering the equation below (Perisse (2003)):

$$C_{cyl} = \varepsilon_0 \varepsilon_r \frac{2\pi L}{\ln\left(\frac{r_2}{r_1}\right)} = \varepsilon_0 \varepsilon_r \frac{2\pi L}{\ln\left(1 + \frac{r_2 - r_1}{r_1}\right)} \approx \varepsilon_0 \varepsilon_r \frac{2\pi r_1 L}{r_2 - r_1} \approx \varepsilon_0 \varepsilon_r \frac{S}{e} \tag{2.107}$$

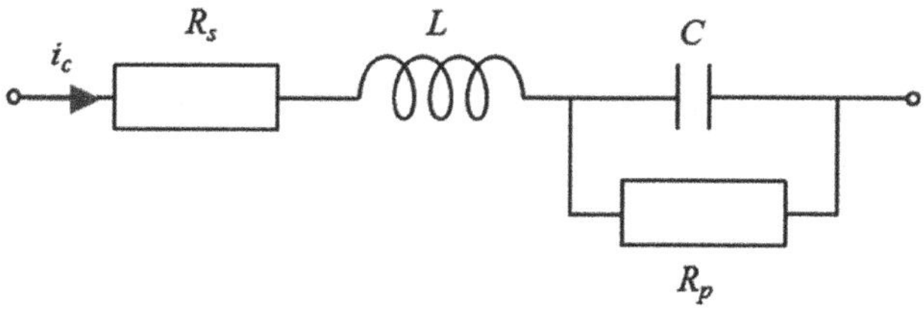

Figure 2.98 Equivalent electrical diagram of a capacitor.

In the case of stacked capacitors, the total capacity of the capacitor is equivalent to the sum of the different elementary plane capacitors placed in parallel:

$$C_{stack} = n.\,\varepsilon_0\,\varepsilon_r\frac{S}{e} \tag{2.108}$$

with n the number of layers placed in parallel.

The construction of the capacitor is completed by adding the connections. It is usually wrapped in a plastic case. These manufacturing processes, as well as the defects linked to the various materials constituting it, imply that the equivalent diagram of a capacitor cannot be represented by a simple capacity. Losses related to connections, electrodes, and dielectric must be taken into account. Most often, a capacitor is modeled by the equivalent diagram in Figure 2.98: where, R_s represents the resistance of the electrodes and connections, L the equivalent series inductance due to the connections and the geometry of the windings, C the nominal capacitance of the capacitor and R_p the resistance taking into account the dielectric losses and the leaks between the electrodes.

When a complex impedance capacitor $\overline{Z}$ is powered by a sinusoidal voltage $u(t)$ with a frequency lower than the resonance frequency of the component, the phase shift φ (i_c, u) between the current i_c passing through it and the voltage at its limits are slightly greater than $-\frac{\pi}{2}$ (see Figure 2.99). This is due to losses in the capacitor, related to the angle of losses. The latter is defined as the complement of the phase shift φ (i_c, u) to $-\frac{\pi}{2}$ (Venet et al. (2002)).

The total losses P in the capacitor are equivalent to the dielectric losses P_d and to the losses by Joule effect due to the connections and electrodes (Nishino (1996)):

$$P = P_d + R_s i_c^2 \tag{2.109}$$

The capacitor loss factor is defined by expressing the tangent of the loss angle which is equivalent to expressing the ratio between the real and imaginary parts of the capacitor's impedance $\overline{Z}$:

$$\tan\delta = \frac{\text{Real (Z)}}{\text{Imaginary (Z)}} = \tan\delta_d + R_s C\omega\left(1 + \frac{1}{R_p^2 C^2 \omega^2}\right) \tag{2.110}$$

with

$$\tan\delta_d = \frac{1}{R_p C\omega} \tag{2.111}$$

where $\tan\delta_d$ is the loss factor of the dielectric at the pulsation ω.

By neglecting the losses due to connections and electrodes, the loss factor of the capacitor $\tan\delta$ is exclusively due to the dielectric $\tan\delta_d$. From the previously listed equations, it is clear that the performance of a capacitor is largely based on the properties of its dielectric.

Capacitors are passive components representing approximately 40% of components used on electronic circuits (Cho and Paik (1999)). They are essential elements in most electronic and power electronics equipment. Therefore, choosing a suitable capacitor for a given application means choosing the characteristics and parameters of all of the materials making it up. Several types of capacitor exist in the literature making it possible to meet the diverse electrical, thermal and mechanical requirements of the user. Figure 2.100 shows the world consumption of capacitors for the most used technologies. We can distinguish three main families, namely:

- Aluminum or tantalum electrolytic capacitors;
- Ceramic, mica or glass capacitors;
- Plastic or paper film capacitors.

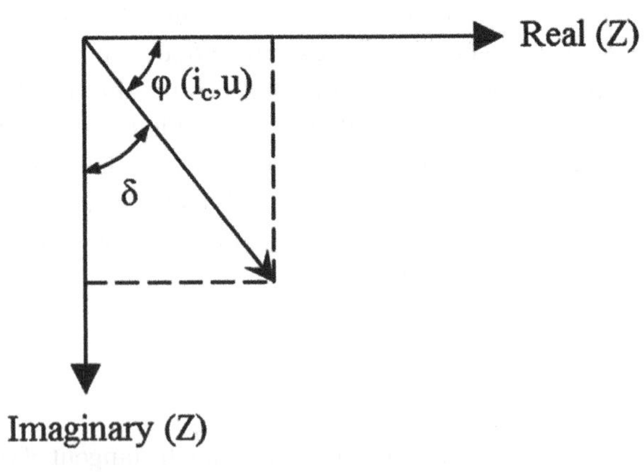

Figure 2.99 Phase difference between i_c and u of a real capacitor.

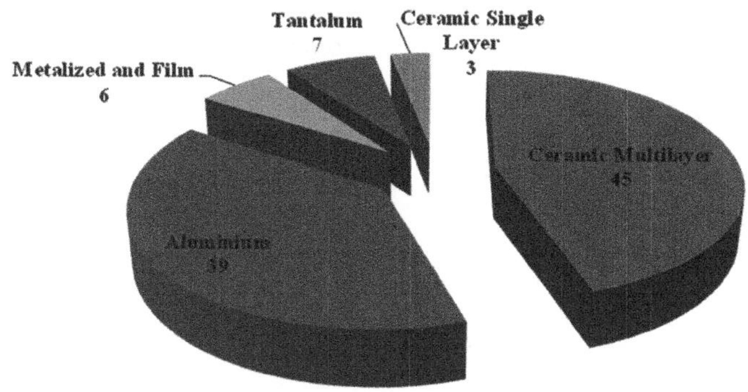

Figure 2.100 World consumption by type of capacitor (2012).

2.6.3.2 Different Kinds of Capacitors

2.6.3.2.a Non-polarized Capacitors

(1) Air

Synonymous of design simplicity, air capacitors use the air as insulation. This configuration leads naturally to the apparition of failures like short circuits when the capacity value is very low (of the order of one hundred pF). To avoid failures, the distance between the two electrodes must be relatively large. In addition, mechanical rigidity must be ensured for the same reason. These types of capacitors are especially used for applications requiring a variable capacity, for example, the compensation of an oscilloscope probe. Indeed, since there is no contact between the two electrodes, it is easy to vary without friction the facing surfaces as shown in Figure 2.101 and, thus, vary the capacity.

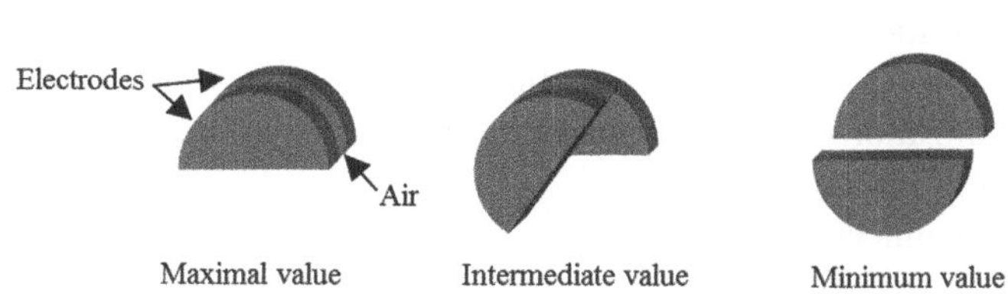

Figure 2.101 Variable air capacitor–different capacity values.

(2) Ceramic

This type of capacitor (Figure 2.102) uses a ceramic whose composition can vary as an insulator. This latter makes it possible to separate a stack of electrodes allowing, thus, obtaining a large surface and therefore a high value of capacity. They are inexpensive but have very poor temperature stability. This constraint limits their use to applications whose temperature resistance is not a predominant criterion. There are two distinct types:

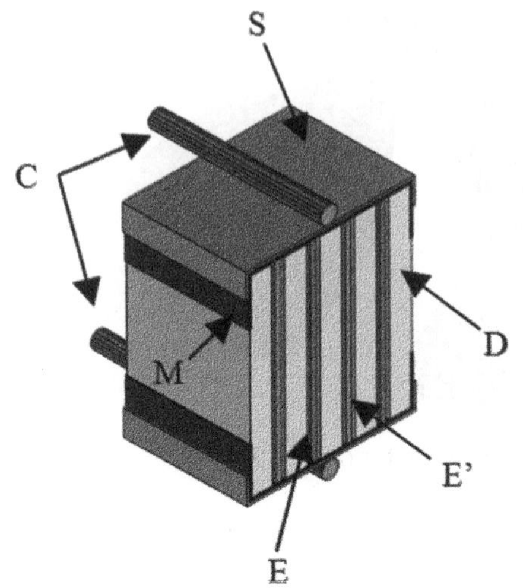

Figure 2.102 Composition of a ceramic capacitor.

- type 1 (or class 1): with determined temperature coefficient;
- type 2 (or class 2): with an unknown temperature coefficient.

The high relative permittivity of ceramics makes it possible to reach capacitance values of a few nF up to ten mF. They also have the advantage to obtain a very low series inductance and are therefore particularly suitable for high-frequency applications. However, since the structure is made of ceramic, they are very fragile. In addition, temperature has a very strong influence on their capacity for high value models (called class 2 such as for example: X5R, X7R, ...). with

- D: ceramic dielectric;
- E and E': electrodes, armatures of the capacitors;
- M: metallization connecting the electrodes to each other;
- S: solder of connections;
- C: radial connections.

The following Table 2.7 shows the influence of the type of ceramic on the characteristics of the capacitor. The types of ceramic correspond to different materials both in terms of their chemical composition and their manufacturing process.

(3) Plastic

Plastic capacitors use several types of plastics as insulation (these are often metallized). They are composed of a stack of thin layers of metallized plastic (the electrodes). The successive metallizations of this stacking are connected alternately by metal deposition on two of the sheets of the component, and then

Table 2.7 Characteristics of a ceramic capacitor.

Dielectric	metal oxides and titanates	Titanates and Zirconates	Titanates and Niobates
Capacity	low, less than a few nF	up to $1\mu F$	High
Insulation resistance (MW)	>20000	>10000	
Loss factor tan δ	10	250 to 300	500 to 600
Temperature coefficient (ppm/°C)	−1500 to +100, NP0/C0G	irregular	irregular
Temperature range (°C)	−55 to +125	−55 to +125	
Tolerance	5 to 10%	> 10%	
Permittivity (ε_r)	> 200	200 to 10000	> 2000

Figure 2.103 Photos of polyester capacitors.

the pins are powered. This type of capacitor is used for decoupling the signals because it is very stable compared to the temperature. As said previously, there exist several types of plastics as insulation. For example, polyesters (MKT or MKS, see Figures 2.103 and 2.104), polystyrenes (MKJ), polypropylenes (MKP, see Figure 2.105), or even polycarbonates (MKC). Polyesters (or polyethylenes) are appreciated because it is possible to obtain very thin layers of insulation and therefore a fairly high capacity. Polystyrenes do not have this advantage but are the most stable among all plastic film capacitors. Polypropylenes are the most used in switching power supplies because they have a very low series resistance (which therefore allows high currents), however, they can be quite bulky due to the impossibility of obtaining very thin layers.

(3)1 Polyester capacitors (MKT) Well known under the name of Mylar, the polyester makes it possible to produce capacitors of general purposes at a

Figure 2.104 Photo of a polycarbonate capacitor.

Figure 2.105 Photo of a polypropylene capacitor.

reduced price and whose performances are fairly average. The stability of the characteristics (capacity, losses, and insulation resistance) as a function of the temperature and the frequency is rather poor; this is why they are used especially in applications of decoupling, filtering, interference suppression ... where the exact value of capacity hardly matters.

Some characteristics

- Nominal value: from $1nF$ to $250\mu F$,
- Tolerance: from 1%to 20%,
- Operating voltage: from 40V to 10000V,
- Temperature coefficient: positive. $+300 \times 10^{-6}/°C$,
- Relative permittivity: 3.2,
- Dielectric rigidity: $> 275kV/mm$,

- Insulation resistance: $> 30G\Omega$ – it decreases with frequency and when the temperature increases,
- Loss factor: $< 70 \times 10^{-4}$ going up at low and high temperature, increasing with frequency,
- Temperature range: -55 to $+125°C$,
- Use: Low frequency circuits, decoupling, filtering and applications where the performance/price and capacity/space ratios must be the greatest.

(3)2 Polycarbonate capacitors (MKC) This type of dielectric makes it possible to produce capacitors whose capacity remains very stable at ordinary temperatures. Metallic film capacitors are found for low power applications and aluminum frame capacitors for high-voltage applications used from -55 to +125 °C. The dielectric is then a mixed paper + polycarbonate.

Characteristics

- Nominal value: from $1nF$ to $250\mu F$,
- Tolerance: from 1% to 2%,
- Operating voltage: from 40V to 5000V,
- Temperature coefficient: positive. + or - $75\times10^{-6}/°C$ around 20°C,
- Relative permittivity: 2.8,
- Dielectric rigidity: $> 180kV/mm$,
- Insulation resistance: $> 30G\Omega$,
- Loss factor: low losses. $\tan \delta_d < 20\times10^{-4}$ increases with frequency and for low temperatures,
- Temperature range: -55 to $+125°C$,
- Use: LF circuits, decoupling, filtering.

(3)3 Polypropylene capacitors (MKP) The dielectric is either paper-polypropylene or a metallized polypropylene film (metallization thickness 0.000,020 mm). The insulation resistance is very high since it reaches 10^5 MW.μF. Dielectric constant: 2.2.

Polypropylene capacitors are mainly used at low frequency (decoupling, filtering, switching power supply) because their properties (loss factor, admissible effective voltage) deteriorate very quickly as soon as the frequency exceeds a few kilohertz and for applications requiring capacities exceeding a few tens of μF under a few hundred volts.

Characteristics

- Nominal value: from $0.1nF$ to $250\mu F$,
- Tolerance: from 10 % to 20 %,

Figure 2.106 Photo of a polystyrene capacitor.

- Operating voltage: from 160V to 3500V,
- Very low series resistance,
- Use for circuits in pulse regimes, switching power supply,
- Temperature coefficient: positive above 20°C, negative below,
- Loss factor: low losses in LF tan δ_d <10×10^{-4} increases with frequency and temperature,
- Temperature range: -55 to 100°C,
- Use: Power supplies, LF.

(3)4 Polystyrene capacitors (MKS) Polystyrene capacitors (eg styroflex) are well suited for radiofrequency applications, with very low losses, very high insulation resistance (>10^{+6} MW.μF), a negative temperature coefficient. We find this kind of capacitors for voltages below 1000 volts and capacities ranging from a few picofarads to a few microfarads with tight tolerances.

Characteristics

- Temperature coefficient: negative. -120.10-6/°C,
- Relative permittivity: 2.5,
- Dielectric rigidity: > $200kV/mm$,
- Insulation resistance: > $100G\Omega$,
- Loss factor: low losses. tan δ_d <5×10^{-4} changing little with frequency and temperature,
- Temperature range: -55 to +70°C the polystyrene loses its capacities above 85°C,
- Use: Power supplies, LF, timer, filters.

(3)5 Other Many other materials are (or have been) used as insulation, such as paper, glass or mica. However, although having the advantage of not being polarized, all these capacitors using a relatively thick film, do not allow very high capacitance values to be reached (always while maintaining reasonable dimensions). This observation led to the search new techniques and ultimately led to polarized capacitors.

2.6.3.2.b Polarized Capacitors

The technique for achieving the most important capacitance values (4700μF, 10000μF or 22000μF) is the creation of an insulator directly on an electrode by a chemical oxidation reaction. For this, two metal electrodes immersed in an electrolyte are put under continuous voltage. The ions, which are present in the electrolyte, will then react with one of the two electrodes (depending on the polarity chosen) and oxidize the latter. The dielectric layer then formed is very thin (100nm for an aluminum electrolytic capacitor of 63V DC), which makes it possible to reach capacitances up to several mF. The problem is that, once the dielectric layer is formed, it becomes impossible to apply a reverse voltage. Indeed, this would cause the opposite chemical reaction (reduction of the oxide to create the metal), destroying the dielectric at the level of the anode, which would create on the one hand a short circuit and on the other hand would cause the creation of a layer of dielectric at the level of the cathode releasing dihydrogen and strongly increasing the internal pressure leading to an explosion. Thus, this type of capacitor is said to be polarized.

(4) Tantalum capacitors

Capacitors using tantalum can have a liquid or solid electrolyte. The latter (which are historically called "drop capacitors" because of the shape of the first versions) are the most used because they have better durability and a lower series resistance than those with a liquid electrolyte. Tantalum capacitors have the highest capacity per unit volume, due to their design and the higher relative permittivity of tantalum oxide compared to aluminum oxide. Indeed, unlike the other elements which put two plates opposite, here the tantalum is in the form of powder (the particles having an order of magnitude of 10 mm) agglomerated by "intering". These particles (which will act as an anode) are oxidized and then connected to the cathode either by manganese oxide, or directly by the electrolyte when the latter is liquid. The collector for connecting this electrolyte (which therefore becomes the cathode) to the outside is made of silver. This is justified by the very good electrical conductivity of this metal but also, in the case of liquid electrolytes, by its resistance to corrosion. In fact, the latter use acids such as sulfuric acid as a solvent. Thus, the effective contact surface between the anode and the cathode is immense (thanks to sintering) and the relative permittivity of the tantalum oxide (Ta_2O_5) is very large (27 against less than 10 for oxide of aluminum). They also have a relatively low series resistance (greater for liquids) and a very low equivalent series inductance thanks to the absence of winding.

However, despite these qualities, these capacitors are subject to defects. The first being cost. In fact, tantalum is a rare metal, therefore expensive, to which is added the cost of silver. Then there is the very high risk of ignition during a malfunction which, in addition to the obvious risk posed by the flames, also leads to the release of toxic vapors. This is why their use in an on-board assembly (electric vehicle for example) is highly discouraged. In addition, an important voltage safety coefficient must be respected with a rating which must be chosen equal to twice the specified useful voltage (while this is practically nonexistent for aluminum electrolytic technology) (O'Connor (2014)).

There are two Tantalum capacitor technologies:

(4)1 The tantalum capacitors with gelled electrolyte The anode is a pellet of tantalum powder pressed and sintered. The cathode consists of the silver case. The electrolyte is based on sulfuric acid.

Characteristics

- Nominal value: from $1\mu F$ to $1000\mu F$,
- Tolerance: from 10% to 20%,
- Operating voltage: 6V to 150V,
- Leakage current: from 1 micro Ampere,
- Frequency limit: less than $10kHz$,
- Use: Filtering, decoupling in Low frequency,
- Disadvantage: Used only for low operating voltage value.

(4)2 The tantalum capacitors with solid electrolyte The dielectric is composed of alumina, immersed in boric acid. The electrode and cathode are made of aluminum.

Characteristics

- Nominal value: from $10nF$ to $500\mu F$,
- Tolerance: from 10% to 20%,
- Operating voltage: 2V to 125V,
- Leakage current: from 1 micro Ampere,
- Use: Filtering, decoupling in Low frequency,
- Disadvantage: High series resistance,
- Advantage: Long service life. It occupies a smaller volume than aluminum.

(4)3 The aluminum capacitor with liquid electrolyte This type of capacitor is composed of an aluminum sheet, paper, and an electrolyte (conductive liquid) and is characterized by a high volume capacity and a compact design. They have many advantages over other non-polarized capacitors, starting with a very high power density and therefore a high capacity even if at a given volume, this remains lower than tantalums. However, they support much higher voltages (above 400V for some). The other advantage compared to tantalum capacitors is that despite the risk of explosion in the event of failure, the appearance of flame is impossible and the elements released are only slightly or not toxic (at least nothing comparable with tantalum capacitors). Consequently, the latter will always be preferred in embedded applications despite a larger volume at given capacity. In addition, they are cheaper to manufacture, given the metal used (aluminum) and the much simpler manufacturing process.

The electrolytic capacitors are produced by winding on the same principle as those with paper or dielectric formed from a plastic film. Two aluminum sheets, where one has been treated to become the anode (etching, oxidation) and the other the cathode, are separated by two absorbent papers. The role of these absorbents is to avoid the electrical contact between the two electrodes and being impregnated with the electrolyte. This paper is not the dielectric since it is the alumina layer which takes its place and its thickness (3 to 7 hundredths of mm) will determine the final volume of the capacitor. The alumina layer has a thickness which is of the order of a hundred nanometers.

Characteristics

- Nominal value: from $1\mu F$ to $150{,}000\mu F$,
- Operating voltage: from 5 to 500 volts max,
- Temperature range: -25 to + 85 °C (current model) -55 to + 125°C (high temperature model),
- Tolerance: usually $\pm$ 20% or -10 to + 50%,
- Alumina relative permittivity: 7 to 10 (24 for tantalum oxide),
- Electrical thickness: 1.5 nm/V (0.15 microns per 100 volts),
- Lifespan: 50,000 to 100,000 hours under normal operating conditions.

It follows, from the previously listed characteristics, different electrical constraints which will guide the choice of the designer towards a technology. Capacitor technologies are numerous and the following Figure 2.107 and Table 2.8 show the range of use of the different families of capacitors as well as the use of different technologies.

Figure 2.108 shows the percentage failure distribution of the different components in a static converter. In 60% of cases, the electrolytic capacitors are

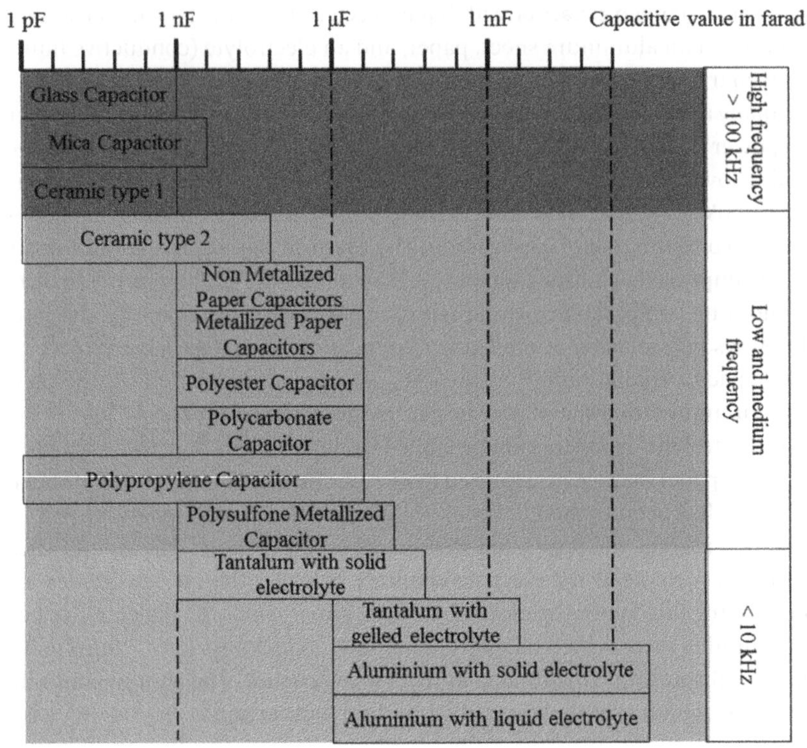

Figure 2.107 Range of use of the different families of capacitors.

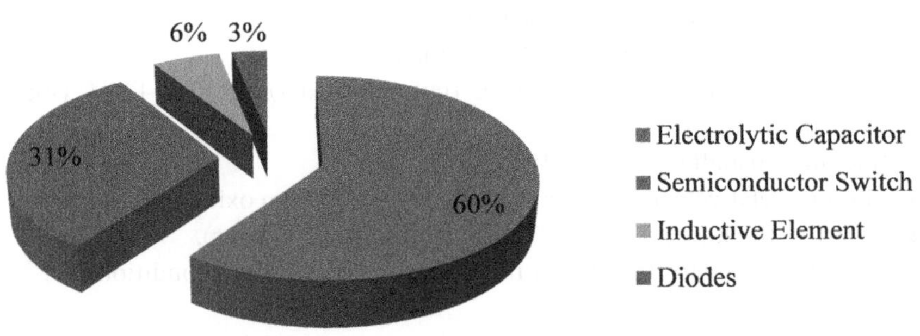

Figure 2.108 Faults repartition for different power components (Abdennadher (2010)).

responsible for the failure of the static converter. This shows that by monitoring the state of the electrolytic capacitor, one can considerably improve the reliability of the static converter.

Table 2.8 Capacitor Families.

	Use	Family of capacitors	Characteristics
High frequency > 100 kHz	Radar, TV ...	Mica (1pF to 200nF) Glass (1pF to 10nF)	Accurate, good performance temperature, replaced more and more by Micas.
	Tuning circuit	Ceramic type 1 (1pF to 2nF)	Accurate and stable.
	Decoupling circuit	Ceramic type 2 (100pF to 470nF)	Imprecise and unstable.
	For high voltages	Non-metallized paper (1nF to 100μF)	Used up to 10000 Volts.
Low and medium frequency	Tuning circuit	Metallized paper (10nF to 200μF) Polyester (1nF to 250μF)	Increasingly replaced by plastic films, also used for interference suppression circuits.
	Tuning circuit oscillator, integrator	Polycarbonate (1nF to 250μF)	Very stable, very reliable precision capacitor.
	Impulse regime switch-mode power supply	Polypropylene (100pF to 250μF)	Low series resistance supports high effective currents.
	For high operating temperatures	Metallized Polysulfone (1nF to 250μF)	Operates at high temperatures, great stability.
< 10 kHz	Filtering, decoupling	Aluminum with liquid electrolyte (1μF to 150000μF)	Leakage current of a few micro amps, operating voltage until 550 volts.
	Filtering, decoupling	tantalum with gelled electrolyte (1μF to 1000μF)	Low volume compared to aluminum, output voltage limited to150 volts.
	RC circuit, oscillator	Tantalum with solid electrolyte (10nF à 500μF)	Limited operating voltage to 125 volts, low volume.

2.6.4 Device Faults and Their Manifestation

Converters are one of the weak points of the drive. Many parts are subjected to different stresses such as thermal stress, vibration and shock, electrostatic discharge, electromagnetic cosmic ray, ...

All these stresses induce an aging process and several defects in the drivers, the power static devices, the capacitors and the inductors, the connectors, the fans and finally lead to an unwanted stop of the drive. We will focus on two

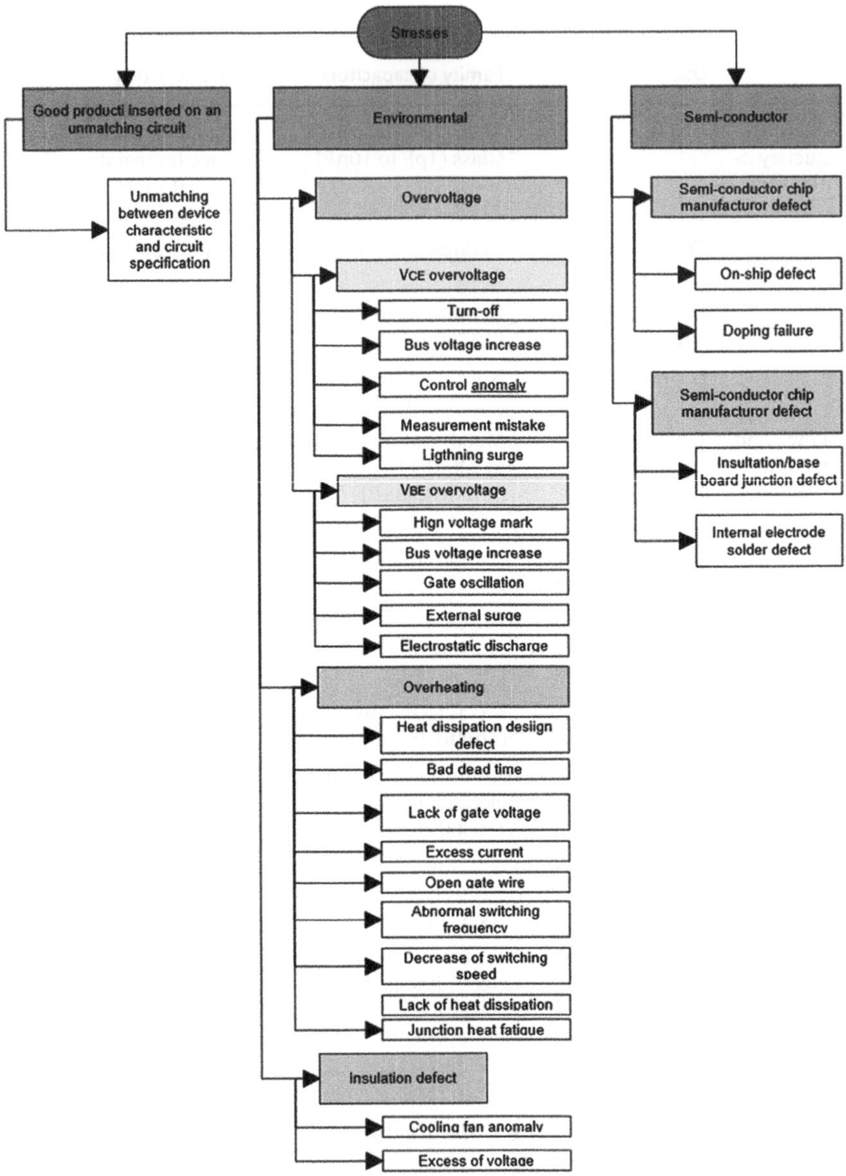

Figure 2.109 Stress upon power module. *Source:* Mitsubishi (2019).

of the main contributors of the power converter failures: the controlled static switches (IGBT and MOSFET) and the capacitors.

For instance, Figure 2.109 represents the main stress of the power modules.

2.6.4.1 Basic Notion

All the different stresses can provide abrupt fault or contribute to the aging process. Peyghami et al. (2019) give a complete analysis of the different failure modes and mechanisms on semiconductors and capacitors in a power drive (Table 2.9).

Table 2.9 Analysis of the different failure modes and mechanisms on semiconductors and capacitors. (c) Courtesy of IEEE. *Source:* Peyghami et al. (2019).

Device	Failure Type	Failure rate	Failure Mode	Failure mechanisms
Semi-conductors	Catastrophic failures	Constant	Open circuit	Device Failure in gate driver. Driver board short and open circuit.
				Bond wire lift-off. Bond wire rupture after short circuit
			Short circuit	High-voltage breakdown
				Dynamic latch-up
				Second breakdown
				Impact ionization
				High temperature due to power dissipation
	Wear-out failures	Non- constant	Parameter drift	Chip solder joint cracking
				Baseplate solder joints cracking
				Wire bonds lift off/cracking
Aluminum electrolytic capacitors	Catastrophic failures	Constant	Open circuit	Self-Healing dielectric breakdown
				Disconnection of terminals
			Short circuit	Dielectric breakdown of oxide layer
	Wear-out failures	Non- constant	Parameter drift	Electrolyte vaporization
				Electrochemical reaction
Metalized polypropylene film capacitors	Catastrophic failures	Constant	Open circuit	Self-Healing dielectric breakdown

(Continued)

Table 2.9 *(Continued)*

Device	Failure Type	Failure rate	Failure Mode	Failure mechanisms
				Connection instability by heat contraction of dielectric film
				Connection instability by heat contraction of dielectric film
				Reduction in electrode area caused by oxidation of evapored metal due to moisture absorption.
			Short circuit	Dielectric film beakdown
				Self Healing due to over current
				Moisture absorption by film
	Wear-out failures	Non con-stant	Parameter drift	Dielectric loss
Multi layer ceramic capacitors	Catastrophic failures	Constant	Short circuit	Dielectric breakdown
				Cracking. Damage to capacitor body.
	Wear-out failures	Non- con-stant	Parameter drift	Oxide vacancy migration. Dielectric puncture. Insulation degradation. Micro-crack within ceramic.

A more detailed description of the various defects is given in the following paragraphs.

2.6.4.2 On Chip Failures

The Grid oxide of power MOSFET or IGBT, especially if it is thin as in SiC MOSFET, can be punched through if it submitted to high electric fields (see Figure 2.110) (Hologne (2018), Ouaida et al. (2014), Ouaida (2014)). This degradation can be brought by:

- over-current or over-voltage in case of overload or hard switching transient;
- electromigration of hot carriers or charges;
- parasitic triggering;
- electrostatic discharges (ESD).

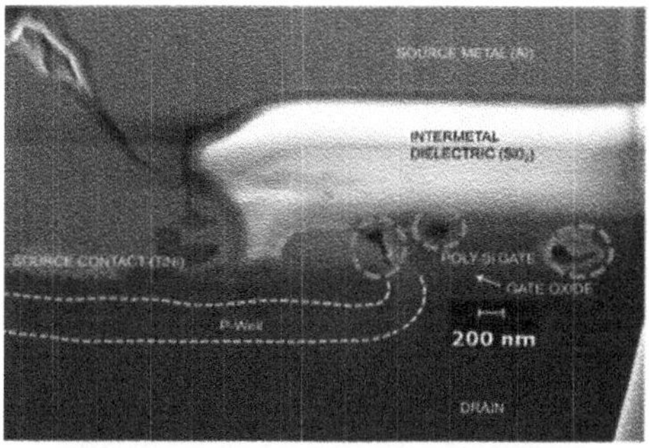

Figure 2.110 Scanning Electron Microscopy with a Focused Ion Beam Gate cut of an oxide defect (Ouaida (2014)).

Collector-Emitter breakdown of IGBT can be caused by some crystal damage, different doping defects, a defective field stop implantation profile etc... [Lee et al. (2016)] and, consequently, can produce:

- a leak at low voltage or near the avalanche region;
- an early breakdown.

2.6.4.3 Packaging and Chip Environment Failures

Let us consider the different faults on power module. It is often built up on the basis of Figure 2.111.

Many of these parts are sources of failures. Main failures are due to:

- **Bonding wires lift-off or cracking** (Figure 2.112(a)): it is due to an intermetallic formation at a weakened bond joint when the metal of wire bonds and metallization are different or thermal cycling when the different parts have unmatched coefficients of thermal expansion (CTE). (Bower et al. (2008),Zhang (2012)).
- **Electromigration in connectors** (Figure 2.112(b)): This a general movement of metallic atoms through conductive paths in the device. It can appear

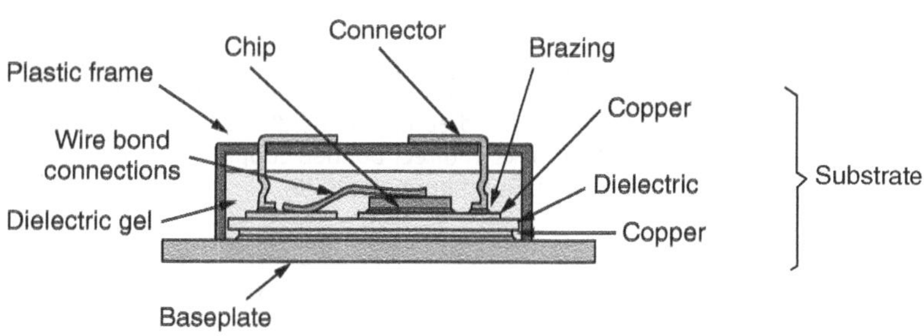

Figure 2.111 Cross-sectional view of a power module. *Source:* Zhang (2012).

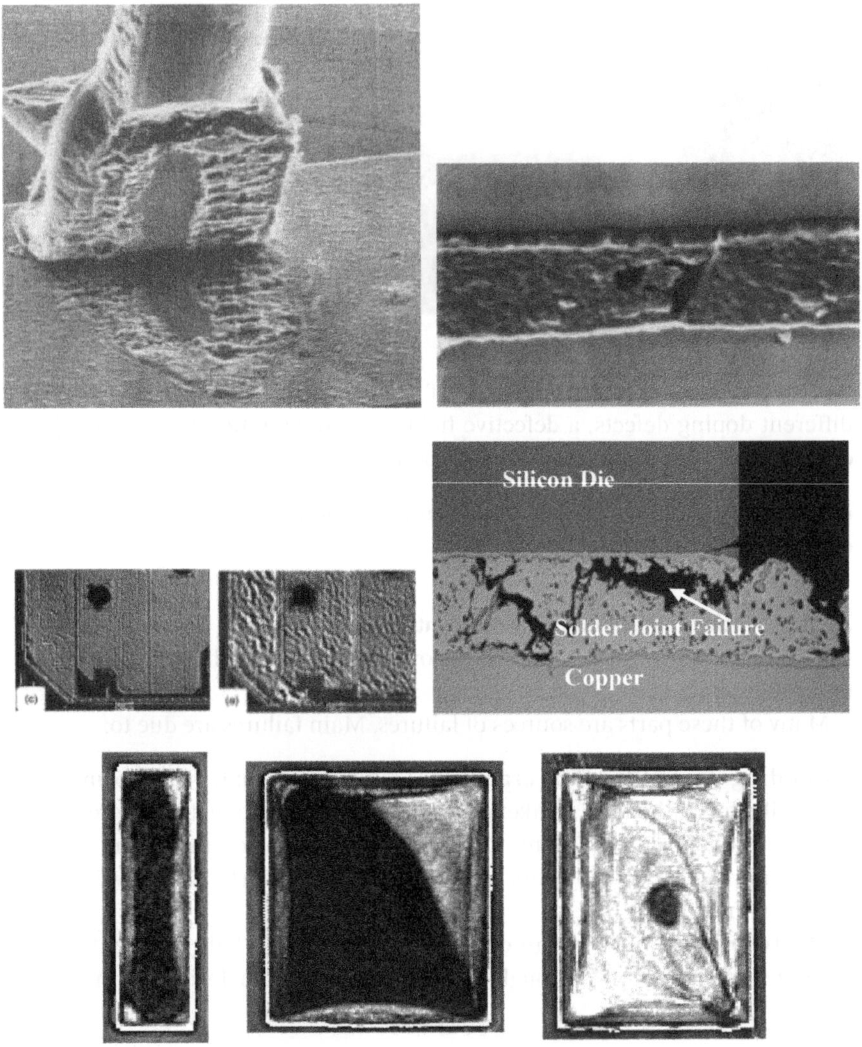

Figure 2.112 Packaging and chip failure mechanisms. (a) Bond Wire Lift-off. *Source:* Bower et al. (2008). (b) Electromigration in connectors. *Source:* Bower et al. (2008). (c) Metallization reconstruction *Source:* Durand et al. (2013). (d) Crack in solder layer. *Source:* Musallam et al. (2010). (e) Cracks in direct bonded copper observed by scanning acoustic microscopy. *Source:* Gunther et al. (2006).

under high temperature and current operating conditions. It can lead to a local increase of resistivity and, thus, overheating which finally create voids (Bower et al. (2008)).

- **Metallization reconstruction** (Figure 2.112c): mismatched CTE lead to metallization compression stresses during thermal cycling which lead to plastic deformation at the grain boundaries and finally to some cracks between the semi-conductor and the metallization. This triggers a local overheating. (Hologne (2018),Durand et al. (2013)).
- **Cracks in the solder layer** (Figure 2.112(d)): The solder layer undergoes very high thermal and mechanical cyclic stresses. It is the main failure mode for IGBT (Musallam et al. (2010)). It is quite similar to bonding failure mechanism (Bower et al. (2008).
- **Cracks in direct bonded copper (DBC)** (Figure 2.112(e)): Cracking can appear at the edge of the interface between copper and ceramic in direct copper bonding substrate (Hologne (2018), Xu et al. (2013), (Gunther et al. (2006)).

2.6.5 Capacitor Failure Modes

There are two main types of failure: degradation and catastrophic failure (see Figure 2.113). These two types are defined below:

2.6.5.1 Failure by Degradation

It is defined by a change in the characteristics of the component which no longer guarantees the functionality for which it is intended. The most common degradations are an increase in the equivalent series resistance (ESR) and a decrease in the capacity.

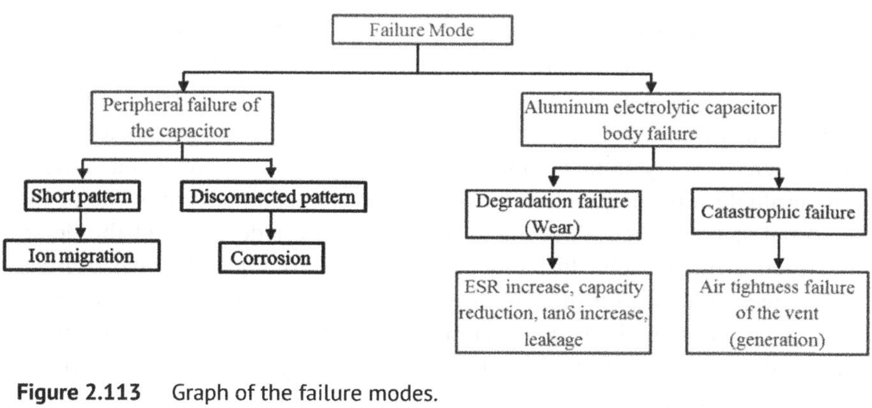

Figure 2.113 Graph of the failure modes.

2.6.5.2 Catastrophic Failure

A catastrophic failure for a capacitor is most often characterized by a short circuit or an open circuit. A visible defect such as an electrolyte leak or a material deterioration of the component will be considered here as a catastrophic failure even if under these conditions, the required function can be ensured for some time.

2.6.6 Diagnosis and Prognosis Techniques for Power Devices

2.6.6.1 Introduction

The relevance of diagnostic methods relies mainly on two key points:

- **The design of a pattern vector**. It is made of fault sensitive parameters. They allow the discrimination of the various operating states:
  - For diagnosis, there are several choices such as raw data (currents, voltages across devices, temperature), temperature-sensitive electrical parameters (TSEP), statistics values, model features.
  - For prognosis, they mainly consist of thermal sensitive parameters correlated to the "junction temperature".
- **A tool** that takes advantage of this pattern vector:
  - In case of diagnosis, the choice of the classifier. It could be based on pattern recognition, neural network, etc...
  - In case of prognosis, two approaches are mainly used: Statistical methods based on off-line accelerated tests and time series extrapolators. These tools must be able to estimate the incertitude of the prediction. But in fact, today, almost no method is capable of doing so.

2.6.6.2 Failure Modes Indicators and TSEP for Power Electronic Devices

The direct measurement of the junction temperature is delicate. An approximate value can be obtained using thermal sensors or optical probes. However, indirect measurement using temperature-sensitive electrical parameters is often preferred. Many of the failure modes indicators are directly linked to them. The evaluation of the junction temperature of the components is one of the essential step for the failure diagnosis of the power devices and the determination of their useful life. Indeed, the junction temperature has a direct or indirect influence on the parameters used to monitor aging and it is also used in counting methods such as rain flow. Therefore, we will devote a paragraph to it. First of all, we have to define what is junction temperature T_j. There are many definitions. We will use the one given by the standard IEC

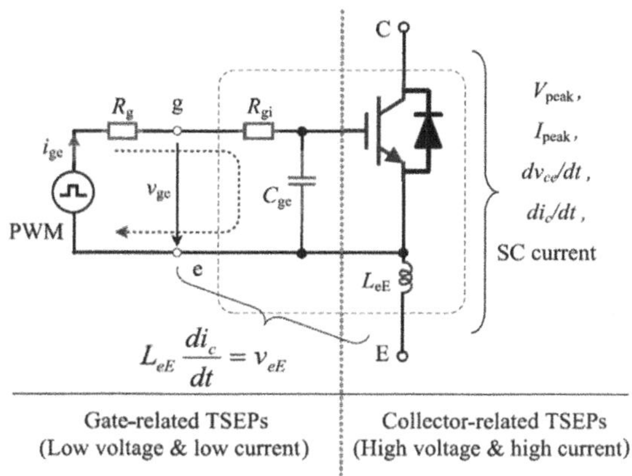

Figure 2.114 Gate-related and collector-related TSEP parameters. (Luo et al. (2018)).

60747-9: the "virtual semiconductor junction temperature." It means the temperature of a junction area of an IGBT or a diode chip, but this one is not uniform in the chip. It is introduced to define a thermal impedance junction to case.

The junction temperature could be given by several electrical quantities called TSEP. We will classify them into 6 main categories (Table 2.10).

- voltage under low current;
- voltage under high current;
- threshold voltage;
- gate emitter voltage and Miller plateau;
- saturation current;
- switching times.

Some of these parameters are related to the gate-emitter voltage, the others are correlated to the collector emitter voltage or current (Figure 2.114). If they need a high collector current I_c, they can induce an undesired self-heating and, thus, depends on the operating conditions.

The parameters related to switching times (t_{don}, t_{on}, t_{off}, t_{if}, t_{vf}) could be used in laboratories but are not suitable to online applications since these transient times are difficult to be measured. They could be of a few tens of nanoseconds for SiC components. See Figure (Figure 2.115).

As an example, Figure 2.116 shows a comparison of TSEP based on switching times and their dependence to the collector current I_c. Thus, they need to be calibrated.

Table 2.10 TSEP parameters.

Categories	Subcategories	Description
Voltage under low current	V_{ce} at on stateand low current	It is done to avoid self-heating. See Avenas et al. (2012), Luo et al. (2017).
Voltage under high current	V_{ce} at on state and high current	It is measured off line by Avenas et al. (2012). Ghimire et al. (2014) has realized a measurement system for V_{ce}(on) which gives an almost standard estimation of T_j. Eleffendi (2016) has designed an online measurement system.
	V_f	Reverse voltage on diode. See Luo et al. (2017), Khatir (2012).
Threshold voltage	V_{ge} Vth	The threshold voltage of the IGBT is defined as the minimum gate-emitter voltage required to form a conductive channel at the metal-oxide semiconductor gate, so that current can flow between collector and emitter can be established. See Eleffendi (2016), Avenas et al. (2012).
	V_{ge} quasi Vth	There is a TSEP estimation method proposed which uses the voltage across the stray emitter inductance, at the time when the collector current begins to rise from zero, to trigger a measurement of V_{ge}. See Hoeer et al. (2015).
	V_{ge} pre Vth	V_{ge} is measured at a fixed point of time during the turn-on process and the voltage level changes with temperature. This effect can be used as TSEP. See Mandeya et al. (2018).
V_{ge}	Gate emitter voltage	V_{ge} is near Vth but in this case the current is higher and creates a non-negligible self-heating. See Avenas et al. (2012).
V_{ge} Miller Plateau	Vge Miller turn-on	See Van der Broeck et al. (2018), Van der Broeck (2018), and Liu et al. (2019).
	V_{ge} Miller turn-off	See Sundaramoorthy et al. (2013).
Saturation Current	I_{sat}	For this measurement, a voltage is applied to the power electrodes. The gate-emitter voltage value V_{ge} is slightly higher than the thresholdvoltage Vth of the device. A current probe measures the resulting saturation current Isat. See Avenas et al. (2012).

(Continued)

Table 2.10 *(Continued)*

Categories	Subcategories	Description
	V_{eE}	The integration of the voltage on Emitter Stray Inductance gives a collector current estimation. See Luo et al. (2018).
Transient TSEP Switching times	t_{don}	Turn-on delay time. See Luo et al. (2018), Kuhn and Mertens (2009).
	t_{doff}	Turn-off delay time. See Luo et al. (2018).
	t_{on}	Turn-on time. See Luo et al. (2018).
	t_{off}	Turn-off time. See Luo et al. (2018), Kuhn and Mertens (2009).
	t_{if}	Current fall time. See Luo et al. (2018).
	t_{vf}	Voltage fall time. See Luo et al. (2018).

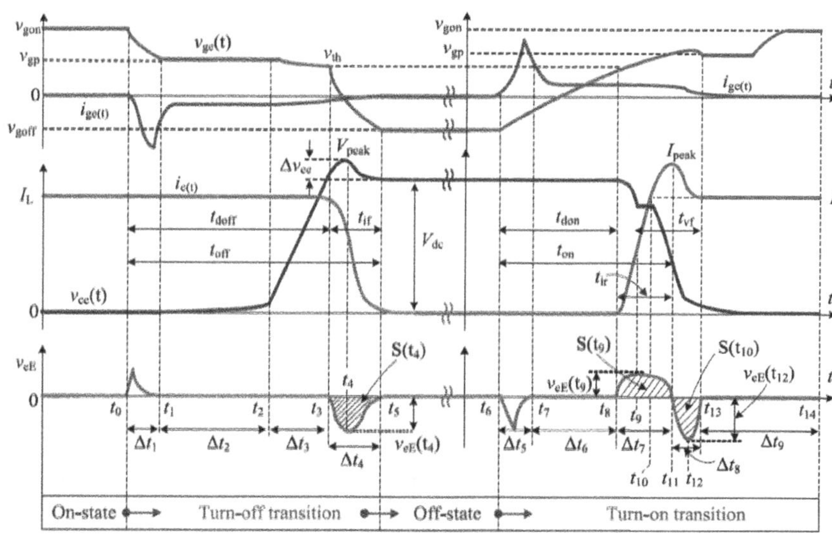

Figure 2.115 Transient waveforms of IGBT module under inductive load (Luo et al. (2018)).

The integration of V_{eE} gives an image of the collector current I_c (Figures 2.115 and 2.117). Its measurement is easily done with low-voltage sensors.

We can compare the performances of theses different TSEP (Figure 2.119) by taking into account:

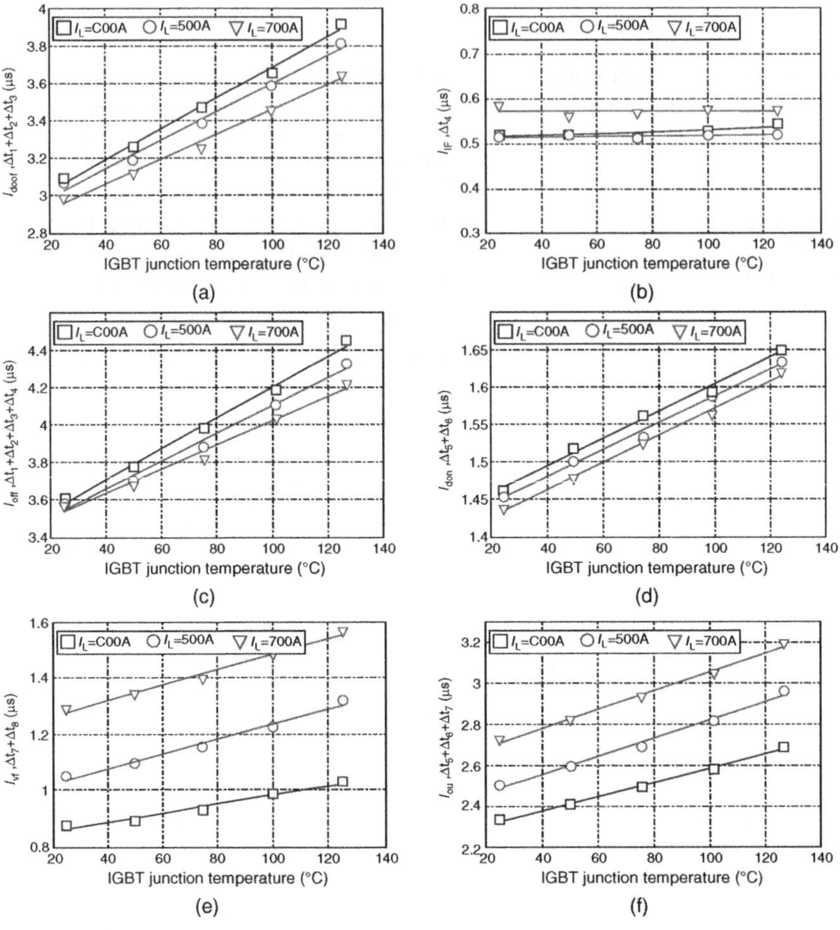

Figure 2.116 Comparisons of six Time-based TSEPs under different load currents at fixed $V_{dc} = 1800V$. (a) Turn-off delay time t_{doff}; (b) current fall time t_{if}; (c) turn-off time t_{off}; (d) turn-on delay time t_{don}; (e) voltage fall time t_{vf}; and (f) turn-on time t_{on} (Luo et al. (2018)).

- linearity;
- sensitivity

$$\text{sensitivity} = \frac{|s|}{s_{\max}} \text{ with s in } [V/°C/V];$$

$$\text{sensitivity ratio} = \frac{s_{\max} - s_{\min}}{T_{\max} - T_{\min}} \text{ in } [V/°C];$$

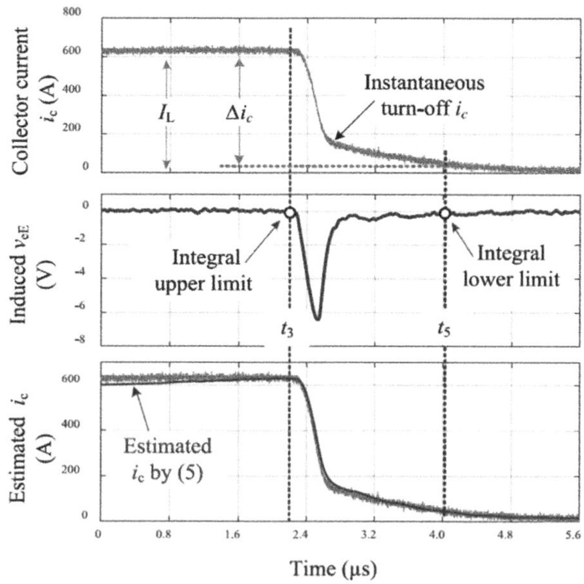

Figure 2.117 Collector current estimation using integration of V_{eE} ($V_{dc} = 1800V$, $I_L = 600A$, and $T_J = 25°C$) (Luo et al. (2018)).

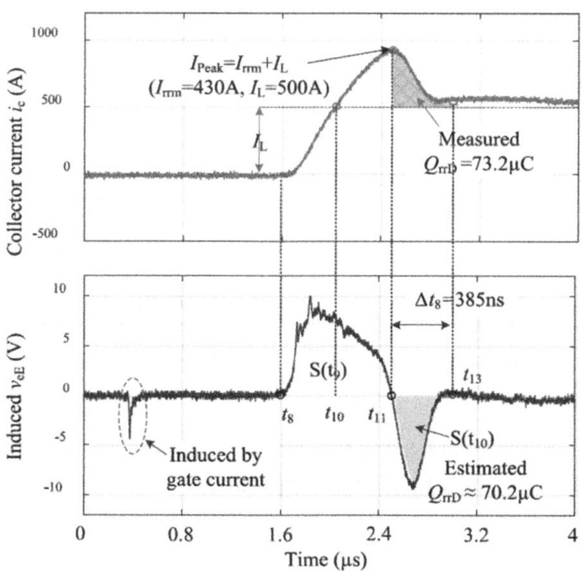

Figure 2.118 Waveforms of turn-on i_c and related induced V_{eE} (Luo et al. (2018)).

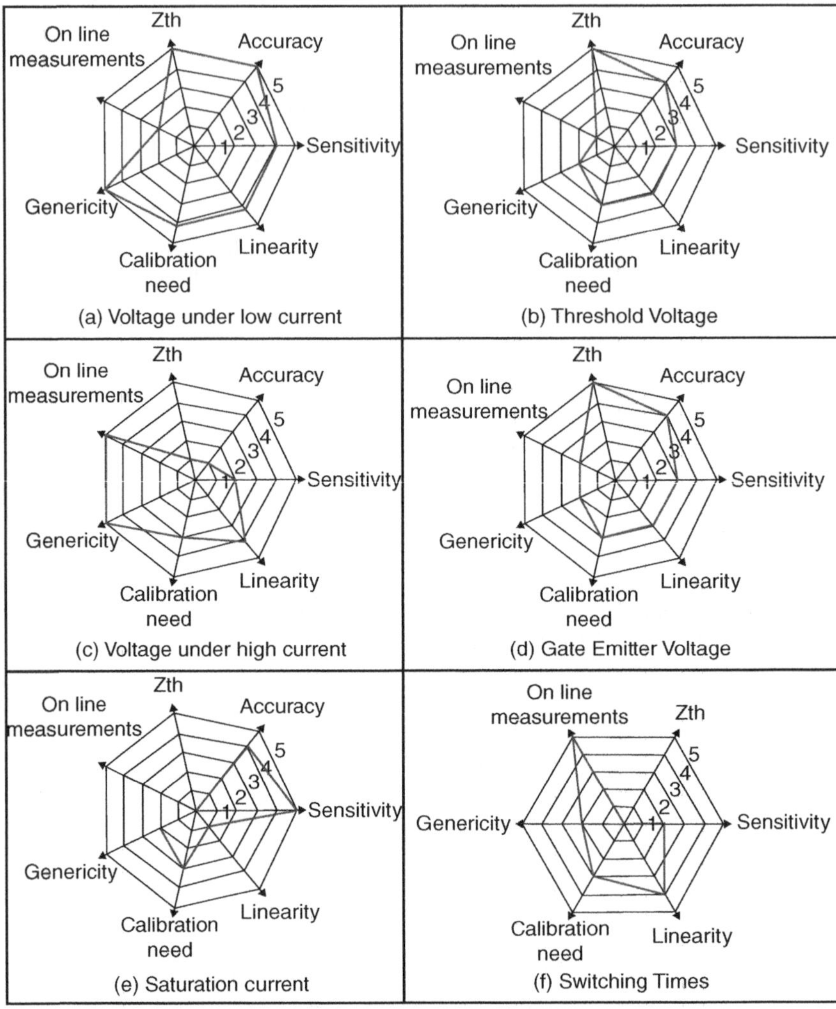

Figure 2.119 Comparison of the different TSEP. *Source:* Avenas et al. (2012).

- accuracy;
- genericity (ability to be used with every semi-conductor chip);
- Z_{th} (possibility to give the thermal impedance);
- online measurements;
- calibration need.

Some TSEPs are not really suitable for SiC devices. For instance, wide temperature dynamics make calibration of these components more difficult and the very short Miller Plateau is not suitable for on line measurement.

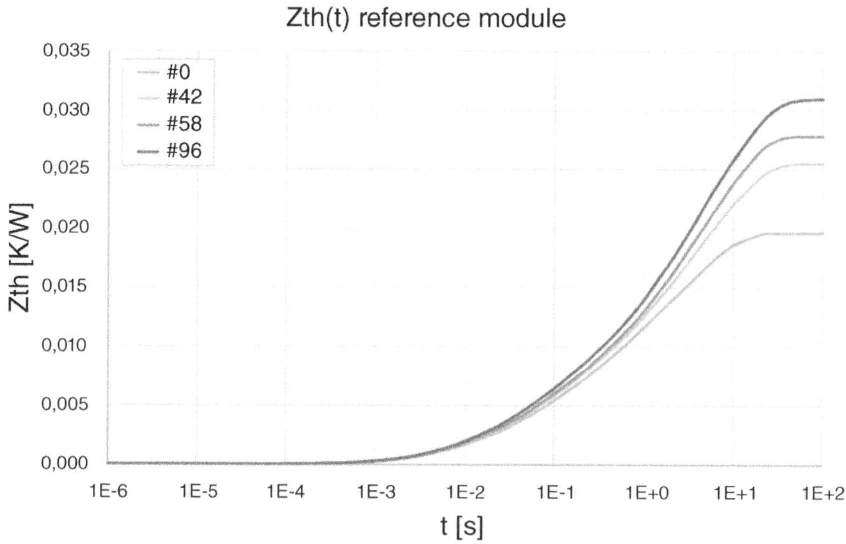

Figure 2.120 Zth curves for different cycle steps (v. Essen et al. (2016)).

The thermal impedance Z_{th} or one of their component R_{th} the thermal resistance and C_{th} the thermal capacitance are also often used as a sensitive parameter with aging (Ji et al. (2015), van der Broeck et al. (2018), Hiller et al. (2015), Sathik et al. (2015), v. Essen et al. (2016)).

Many of these TSEP depend on each other. For instance, the sensitivity factor of the Collector-Emitter voltage depends on the collector current (Figure 2.121).

$$k_{TS}(I_c) = \frac{V_{CEsat}(T, I_c) - V_{CEsat}(T_0, I_c)}{T - T_0} \tag{2.112}$$

Thus, a calibration process is often necessary (Figure 2.122). It can, moreover, evolve with aging.

2.6.6.3 Diagnosis of Failure Modes

The diagnosis of power modules generally consists of one of the following methods:

- Diagnosis based on the direct analysis of the current: these methods used the 3-phase current or their projection into the dq Park frame, their mean values or instantaneous ones or their slopes.
- Diagnosis based on the direct analysis of the TSEP: the saturation voltage V_{CE} or an other TSEP is used in order to detect the occurrence of a failure.

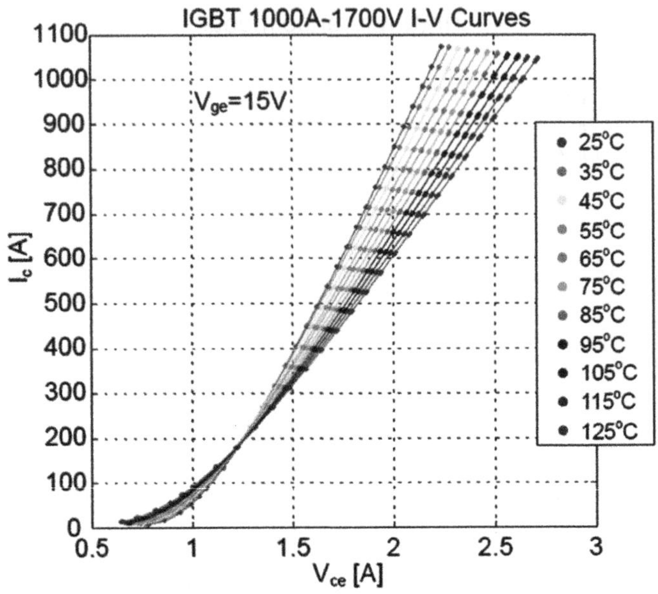

(a) I-V characteristics curves.

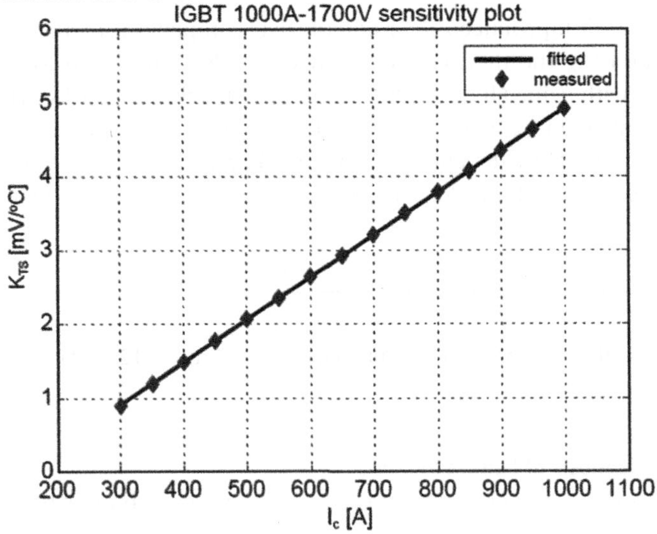

(b) sensitivity factor k_{TS}.

Figure 2.121 IGBT Saturation Voltage dependencies. *Source:* Amoiridis et al. (2015).

- Diagnosis based on signal processing: The individual raw signals are shaped by a signal processing method such as Fast Fourier Transform, wavelet, etc.
- Diagnosis clustering: Base on a relevant pattern vector, a clustering technique is applied such as pattern recognition, support vector machine, etc.
- Diagnosis based on neural network: neural networks alone or in conjunction with a signal processing technique can be used.

2.6.6.3.a Diagnosis based on the Direct Analysis of the Current

(1) Park's transformation:

As we have previously seen, it is possible to represent the three-phase inverter voltage as a 2D vector that can take eight values at the most (2.5). Two of these are equal to zero. So, a mathematical expression of the voltage vector is given as follows:

$$\vec{v_k} = \begin{cases} E\sqrt{\frac{2}{3}}\exp j(k-1)\pi/3 & with\, k = 1\, to\, 6 \\ 0 & for\, k = 0\, and\, 7 \end{cases} \tag{2.113}$$

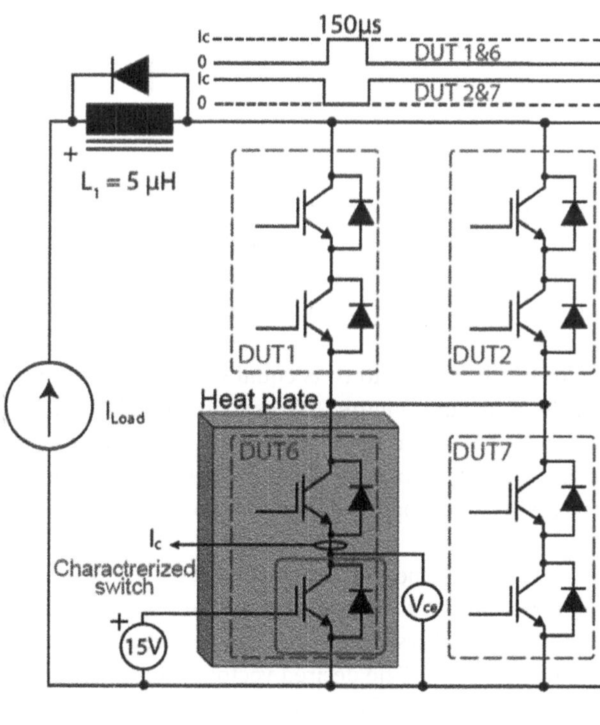

(a) PC test setup.

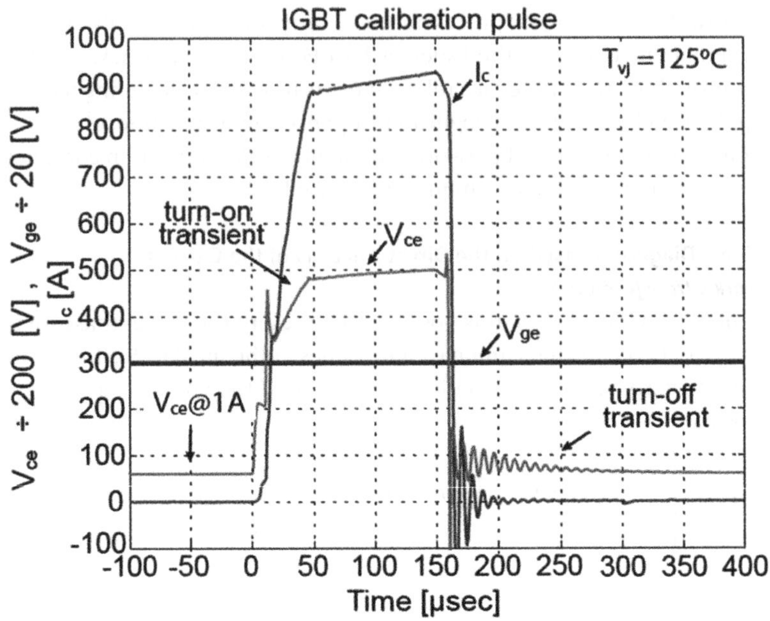

(b) Typical calibration pulse at T_j=125°C sense current of 1A and load current at around 900A.

Figure 2.122 V_{CEsat} versus T_j calibration. *Source:* Amoiridis et al. (2015).

So the modulus of this vector is $E\sqrt{\frac{2}{3}}$ with E is the value of the DC bus voltage. Therefore, the voltage vectors naturally depend on the state of the components of the voltage inverter, i.e.: ON or OFF state. For this purpose, the Table (2.5) presents the vectors as a function of the states of the components (1=ON, 0=OFF).

The modulus of the vectors $\vec{v_1}$ to $\vec{v_7}$ is equal to $E/\sqrt{2}$ and that's regardless of the vector in the α/β plan. Figure 2.123 shows these. We distinguish and enumerate six sectors that will allow to apply a combination of two vectors (in addition to the null vector) to place a desired voltage vector on the electrical system. This naturally calls upon the well-known principle of Space Vector Modulation (SVM).

In case of defect of one component in the voltage inverter, a problem occurs because one can not applied the desired vector $\vec{v^*}$, and instead a disturbed vector is applied to the electrical system. As an example, let us consider a problem of firing of the component S_A. The desired vector is composed of the application of the three vectors $\vec{v_1}$, $\vec{v_2}$ and $\vec{v_0}$ as an example. But, in our case, the real

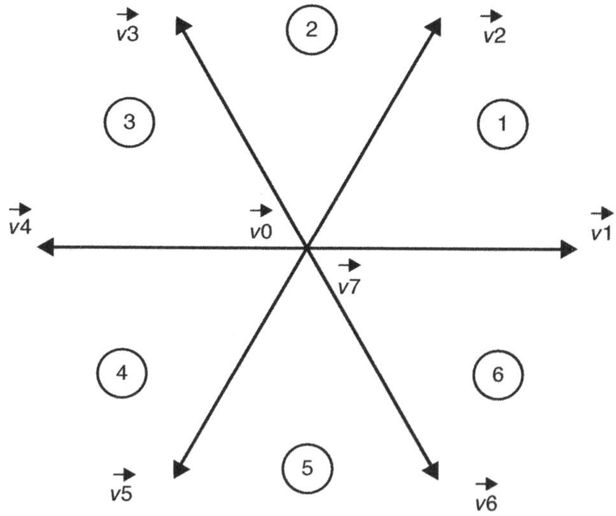

Figure 2.123 Representation of voltage vectors and the 6 sectors.

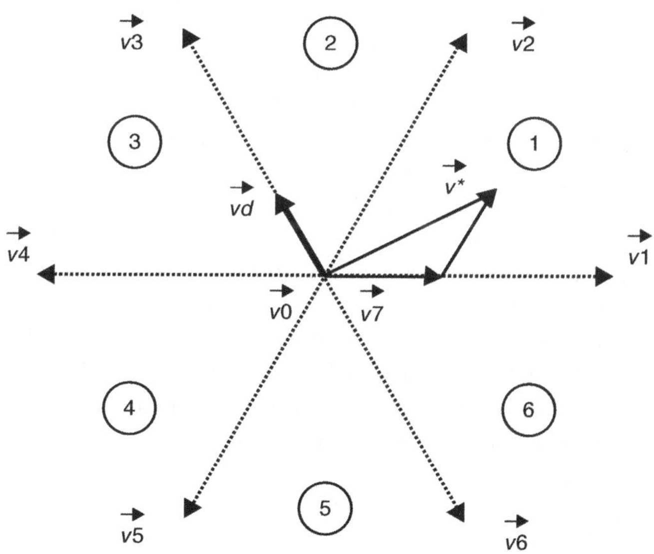

Figure 2.124 Representation of one voltage vector in case of defect.

vector is now composed of the application of the three vectors $\vec{v_3}$, $\vec{v_0}$ and $\vec{v_0}$. So, the defective vector applied, $\vec{v_d}$, is represented in the Figure 2.124. It can be seen there that it is far form the desired vector $\vec{v^*}$.

Inverter Fault

Several types of faults can occur; see de Araujo Ribeiro et al. (2003a) Razik et al. (2011). Among these, one can classify as:

- a problem of short-circuit from the dc link to the ground;
- a problem of short-circuit in the dc link capacitor;
- a problem of damage (open or short circuit) of switch;
- a problem connection of line of the induction motor to the ground;
- etc.

Naturally, the consequence of fault can result in:

- a total breakdown of the process;
- an emergency intervention with the altered process;
- an intervention to repair the process as soon as possible.

The gate misfiring fault can result in a breakdown of the inverter. One of the six components may stay open due to a problem coming from the drive circuit. Due to high switching frequency in the inverter, the lifetime of the components may decrease drastically. The monitoring of this kind of defect, the misfiring, is more complicated because it is an intermittent process. So, the monitoring as the detection of this type of fault is extremely difficult. Among all possibilities as toolboxes, the best way to monitor the process is based on spectral analysis, or better, on using the time-frequency approach. For this there are two options. The first one is the continuous wavelet transform, and the second one is the discrete wavelet transform. What's the best approach to make a diagnostic?

Another possibility is considering the stator in $d-q$ frame. In fact, theoretically, the trajectory will be close to a circle when the operating condition is normal. But, in case of abnormal condition, as the stator current is impacted by the defect, the trajectory is modified. In this approach, we have to compute the current trajectory using the Park's vector transformation.

To do that, we calculate the two stator currents:

$$I_d = I_a\sqrt{\frac{2}{3}} - (I_b + I_c)/\sqrt{6} \tag{2.114}$$

$$I_q = (I_b - I_c)/\sqrt{2} \tag{2.115}$$

Figure 2.125 shows the stator current using the Park's transformation. One can see that a circle is formed, as well as the effect of the inverter on the current. In the same way, for the same condition, we present in Figure 2.126 the vector trajectory. One can easily see the effect of the misfiring gate. The vector trajectory is modified.

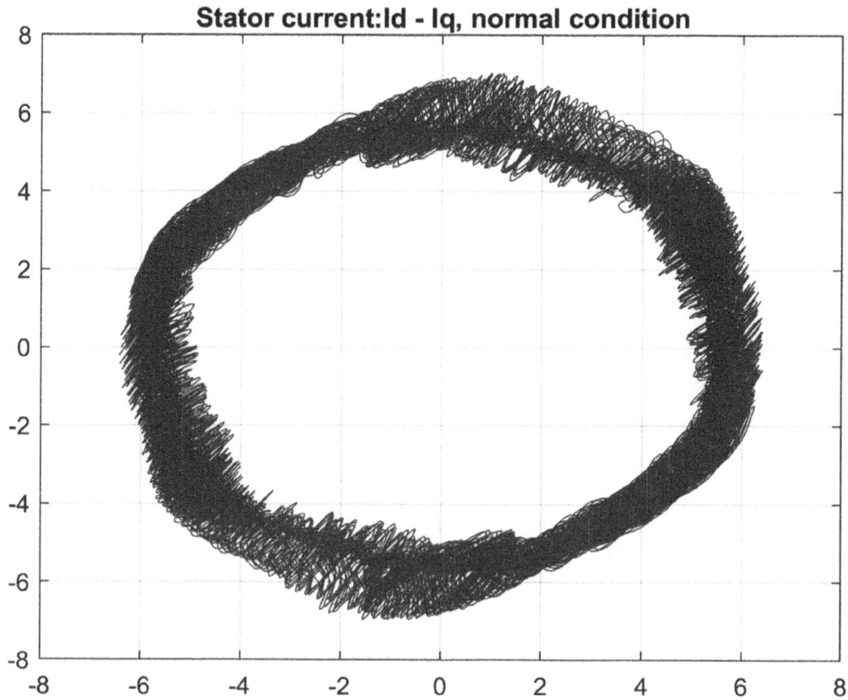

Figure 2.125 Experimental results with a misfiring during $300\mu s$ for q_1. Park's vector trajectory under normal condition.

(2) Slope of current:

Peuget et al. (1998) applied the Concordia transform to the phase currents in order to compute the currents i_α and i_β. The slope Ψ of the current trajectory, in a α, β frame allows to locate the open-leg faults in an inverter.

$$\Psi = \frac{i_{\alpha_k} - i_{\alpha_{k-1}}}{i_{\beta_k} - i_{\beta_{k-1}}} \tag{2.116}$$

Thus, the value Ψ gives the faulty opened leg. The faulty transistor is isolated by determining which alternation of the current is missing in the faulty phase.

- Faulty phase A: $\Psi = \infty$;
- Faulty phase B: $\Psi = \sqrt{3}$;
- Faulty phase C: $\Psi = -\sqrt{3}$.

(3) Average current:

Mendes and Marques Cardoso (1999) deal with a new method based on monitoring an average motor supply current using Park's Vector. It makes the link

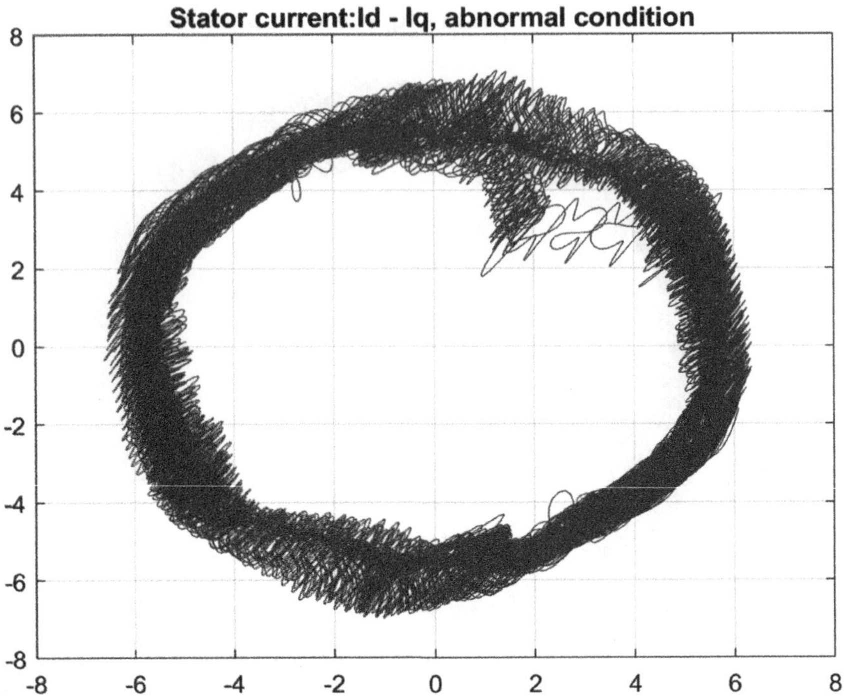

Figure 2.126 Experimental results with a misfiring during 300μs for q_1. Park's vector trajectory under abnormal condition.

between the fault type (open-circuit or short circuit) and the average motor supply current Park's Vector modulus and the faulty device with the corresponding phase value.

$$I_{S_{av}} = I_{D_{av}} + jI_{Q_{av}} = \left|I_{S_{av}}\right| \angle\theta_{S_{av}} \tag{2.117}$$

(4) Fuzzy technique:

Another way to make a diagnostic is the use of fuzzy logic (de Araujo Ribeiro et al. (2003b), Estima and Marques Cardoso (2011)). This way, one could be able to theoretically deduce the operating condition of the inverter. However, the vector trajectory is impacted by a lot of defects like broken bars fault in induction motor. This kind of fault is obviously easier to monitor as to establish a diagnostic.

In order to illustrate this, we show in Figure 2.127 the problem due to a misfiring on the transistor T1 (of the switch S1 made of a transistor and an antiparallele diode). As one can see in this example, we would like to apply

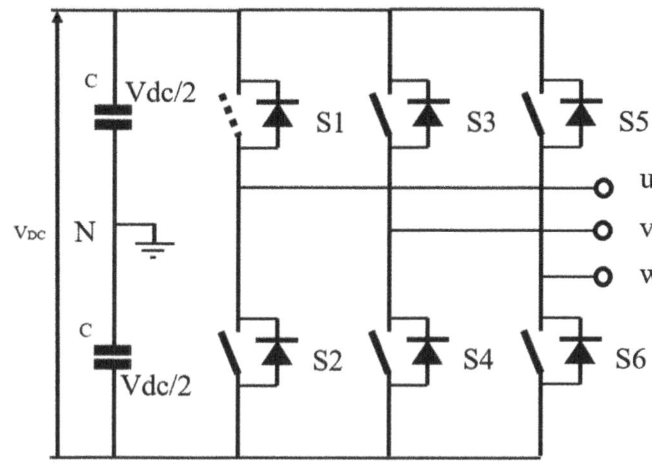

Figure 2.127 Inverter under an intermittent misfiring gate on Transistor T1.

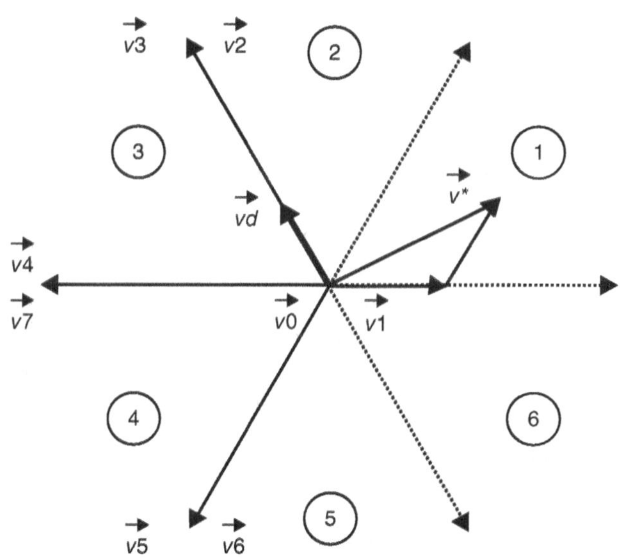

Figure 2.128 Representation of voltage vectors and the 4 sectors failed.

a desired vector $\vec{\upsilon^*}$ which is composed of a piece of the vector $\vec{\upsilon_1}$, $\vec{\upsilon_2}$ and $\vec{\upsilon_0}$. Unfortunately, the gate of T1 doesn't operate and it is not possible to apply the vectors $\vec{\upsilon_1}$ and $\vec{\upsilon_2}$ (Figure 2.128). In order to facilitate understanding, we present in Table 2.11 the original vectors and those generated by a defect in T1. The vector $\vec{\upsilon_1}$ becomes $\vec{\upsilon_0}$, and $\vec{\upsilon_2}$ becomes $\vec{\upsilon_3}$.

Table 2.11 Voltage vectors generated by a two levels inverter under misfiring on Gate 1.

k	$\vec{v_k}$	$\Re(\vec{v_k})\alpha$ axis	$\Im(\vec{v_k})\beta$ axis	S_A	S_B	S_C		S_A	S_B	S_C	$\vec{v_k}$
1	$\vec{v_1}$	$E\sqrt{\frac{2}{3}}$	0	1	0	0	$\Rightarrow$	0	0	0	$\vec{v_0}$
2	$\vec{v_2}$	$E\sqrt{\frac{1}{6}}$	$E\sqrt{\frac{1}{2}}$	1	1	0	$\Rightarrow$	0	1	0	$\vec{v_3}$
3	$\vec{v_3}$	$-E\sqrt{\frac{1}{6}}$	$E\sqrt{\frac{1}{2}}$	0	1	0	$\Leftrightarrow$	0	1	0	$\vec{v_3}$
4	$\vec{v_4}$	$-E\sqrt{\frac{2}{3}}$	0	0	1	1	$\Leftrightarrow$	0	1	1	$\vec{v_4}$
5	$\vec{v_5}$	$-E\sqrt{\frac{1}{6}}$	$-E\sqrt{\frac{1}{2}}$	0	0	1	$\Leftrightarrow$	0	0	1	$\vec{v_5}$
6	$\vec{v_6}$	$E\sqrt{\frac{1}{6}}$	$-E\sqrt{\frac{1}{2}}$	1	0	1	$\Rightarrow$	0	0	1	$\vec{v_5}$
0	$\vec{v_0}$	0	0	0	0	0	$\Leftrightarrow$	0	0	0	$\vec{v_0}$
7	$\vec{v_7}$	0	0	1	1	1	$\Rightarrow$	0	1	1	$\vec{v_4}$

Thus, the vector applied to the induction machine is completely different in practice. Moreover, this kind of defect can be unhealthy for electric motors because it generates a DC component. For example, if T1 cannot switch, then a positive i_u current cannot be possible. This DC component in the phase u will generate the same fault in the other phases because the sum of the three-phase currents is equal to zero. This provides a possibility to establish a diagnosis. Consequently, to monitor the state of health of an inverter, one can analyze the DC current in a sliding windows and determine the presence of a component defect. To be able to do this, it must be possible to measure all three currents.

In order to make a diagnostic, one can use the fuzzy logic approach. To do that, we need to define three membership functions to quantify the sign of the DC component and its existence. To do this, we use the following functions:

- The DC component is negative: DC_N;
- The DC component is zero: DC_Z;
- The DC component is positive: DC_P.

To make a diagnosis, we must establish a number of rules. The first concerns the state of health of all components. It is written as follows:
If IA_{DC} is DC_Z **And** IB_{DC} is DC_Z **And** IC_{DC} is DC_Z **Then** "Healthy Components" where IA_{DC} is the DC component of the current flowing in the electric motor in the phase A. For the detection of a mistake on a component, we must describe six other rules as follows:
If IA_{DC} is DC_N **And** IB_{DC} is DC_P **And** IC_{DC} is DC_P **Then** "T1?" **Or**
If IA_{DC} is DC_N **And** IB_{DC} is DC_N **And** IC_{DC} is DC_P **Then** "T2?" **Or**
If IA_{DC} is DC_P **And** IB_{DC} is DC_N **And** IC_{DC} is DC_P **Then** "T3?" **Or**

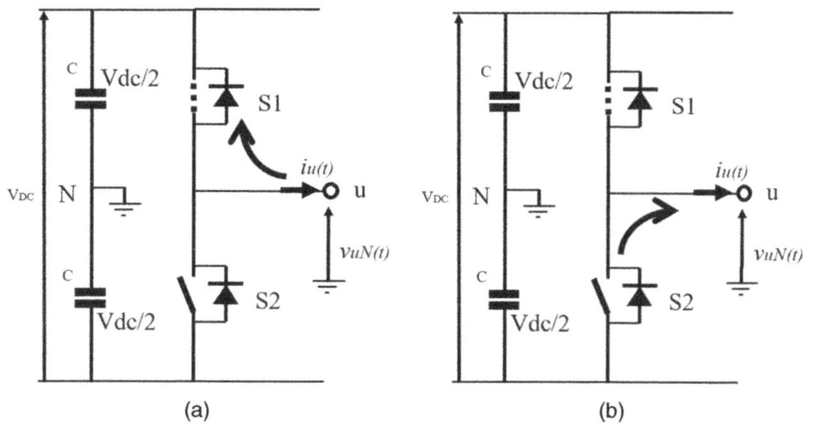

Figure 2.129 Scheme of the inverter under an intermittent misfiring in the ledge 1. T1 doesn't want to conduct. (a) G1 is under fault: the current is negative. (b) G1 is under fault: the current is positive.

If IA_{DC} is DC_P **And** IB_{DC} is DC_N **And** IC_{DC} is DC_N **Then** "T4?" **Or**
If IA_{DC} is DC_P **And** IB_{DC} is DC_P **And** IC_{DC} is DC_N **Then** "T5?" **Or**
If IA_{DC} is DC_N **And** IB_{DC} is DC_P **And** IC_{DC} is DC_N **Then** "T6?".

Figure 2.130 illustrates all these possibilities.

Even if this approach is simple to implement, this one is applicable for the detection of a permanent fault not for a random intermittent gate misfiring.

2.6.6.3.b Diagnosis based on the Direct or Indirect Analysis of Junction Temperature

The junction temperature T_j is a relevant parameter to evaluate the power module state of health. T_j can be used for diagnosis or prognosis purpose, since it is very sensitive to bond wire aging and solder fatigue (Wei et al. (2019)).

(5) Diagnosis based on TSEPs (see 2.6.6.2)

However, T_j cannot be easily measured. Thus, it is often indirectly evaluated from TSEPs.

For instance, Rds_{on} could be used for a MOSFET (see Figure 2.131).

Anderson et al. (2011) propose to measure Rds_{on} from the current ripple generated by the phase voltages of a space-vector modulation:

$$Rds_{on} = \frac{\Delta v_u}{\Delta i_u} \tag{2.118}$$

where Δi_u is the ripple current during the measurement interval and Δv_u is the corresponding voltage ripple.

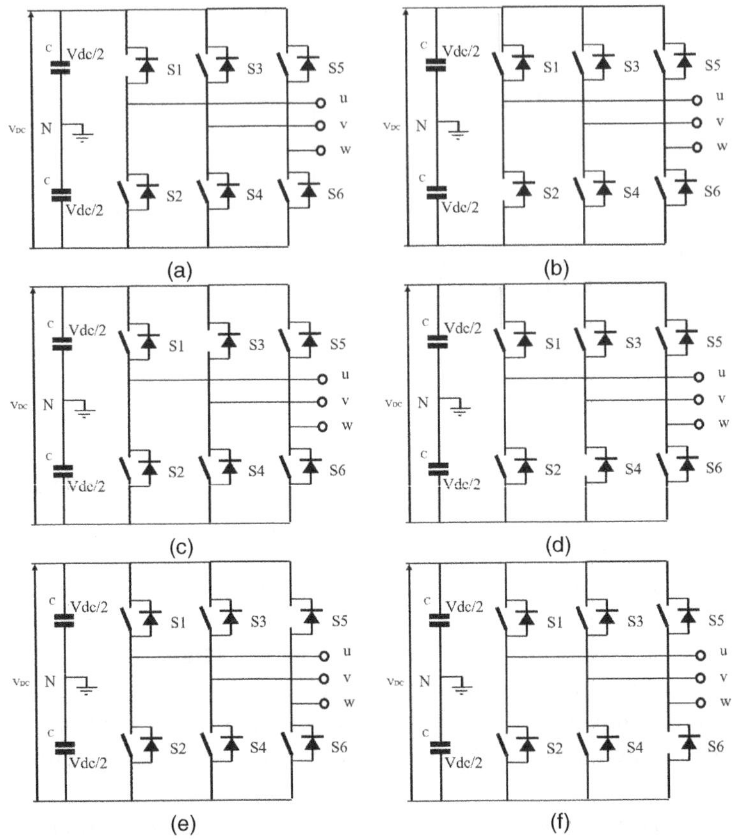

Figure 2.130 Scheme of the inverter under one component fault. (a) G1 is under fault. (b) G2 is under fault. (c) G3 is under fault. (d) G4 is under fault. (e) G5 is under fault. (f) G6 is under fault.

In case of an IGBT, the on-state emitter collector voltage Vce_{on} and the reverse diode drop voltage V_f are often used.

For instance, Jinsong Han et al. (2016) give an application for a Three-Level Neutral point-clamped Inverter in order to diagnose IGBT bond wires and diode bond wires failures. It should be noted that these TSEP are also sensitive to the load current. Thus, calibration must be proceeded. In their paper, a failure is detected when Vce_{on} and V_f exceed a faulty baseline (see Figure 2.132).

But this approach needs a complex diagnosis circuit (see Figure 2.133) and requires a modification of the PWM.

In a similar way, Wang et al. (2020) realize the diagnosis of bond wires failures in a Three-Level Neutral point-clamped inverter. It is done by finding the idle on-state IGBTs under the specific load current direction and switching

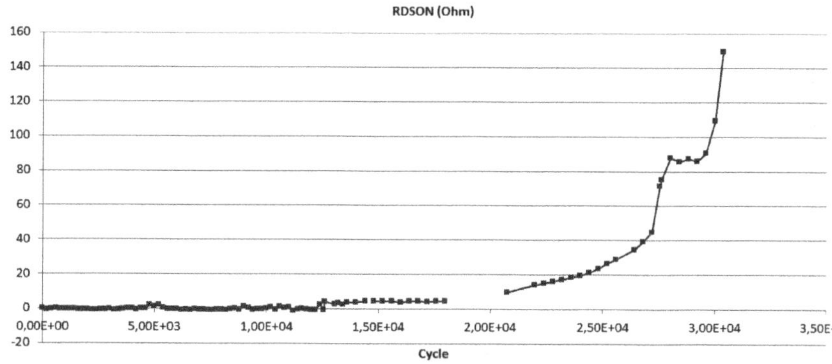

Figure 2.131 Rds_{on} versus number of cycles until failure for an accelerated aging test of a MOSFET.

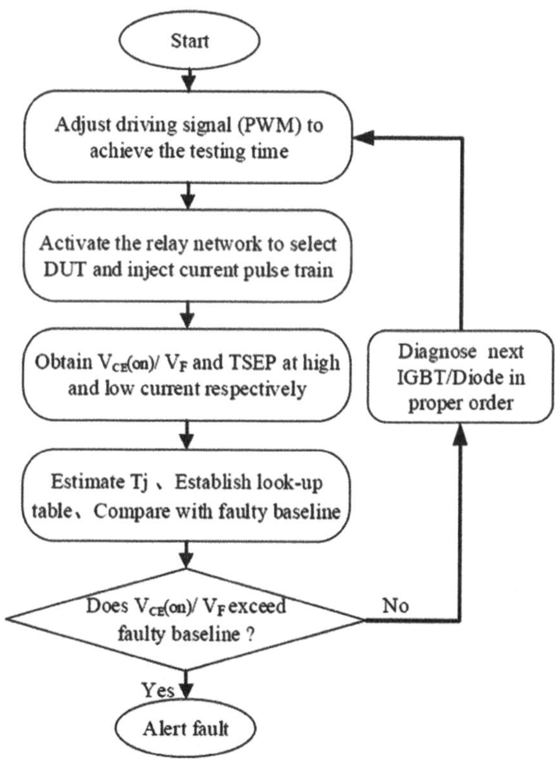

Figure 2.132 Three-level inverter diagnosis Flowchart. Courtesy of IEEE, Jinsong Han et al. (2016).

sequence of a leg. Since, at low current, Vce_{on} becomes independent of the current, it is almost linear dependent on T_j: $Vce_L(T_j)$. Therefore, Vce_L can be used as a relevant TSEP and measured at different aging levels of bond wires.

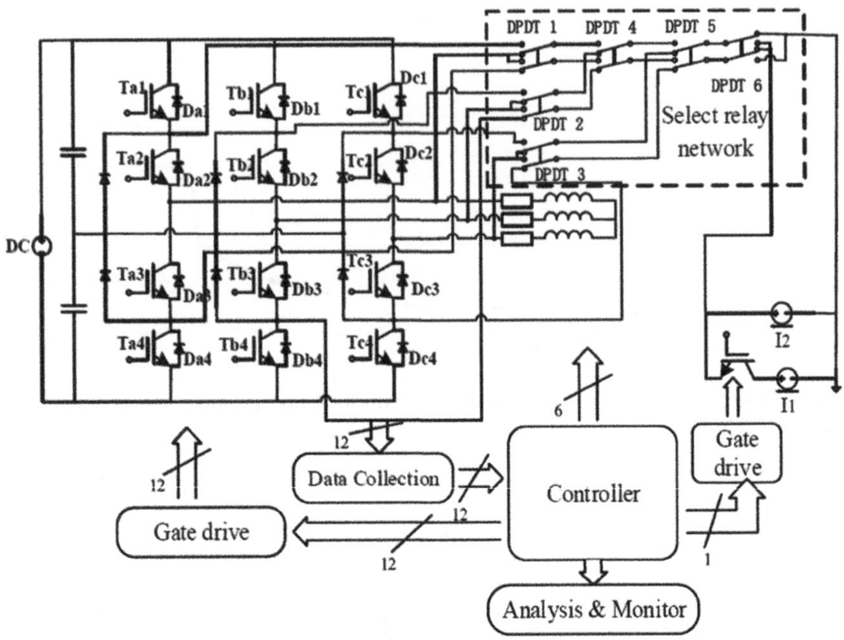

Figure 2.133 Three level inverter diagnosis system. Courtesy of IEEE.

Chen et al. (2020a) detail the localization and detection of Bond Wire Faults in Multichip IGBT Power Modules.

They first review the different signatures which have been tested to detect wire bond failures:

- $V_{CE_{on}}$;
- ΔV_{CE} at turn off;
- R_{CE};
- V_{G-AE};
- I_G;
- sense point at Kelvin emitter KE.

The approach of Chen et al. (2020a) is based on the on-state voltage measurements of $V_{CKE_{on}}$ across the collector and the Kelvin emitter and the voltage $V_{CE_{on}}$ across the collector and the emitter (see Figure 2.135). It can detect early bond wire lift-offs. In case of such a failure, $V_{CKE_{on}}$ decreases and $V_{CE_{on}}$ increases.

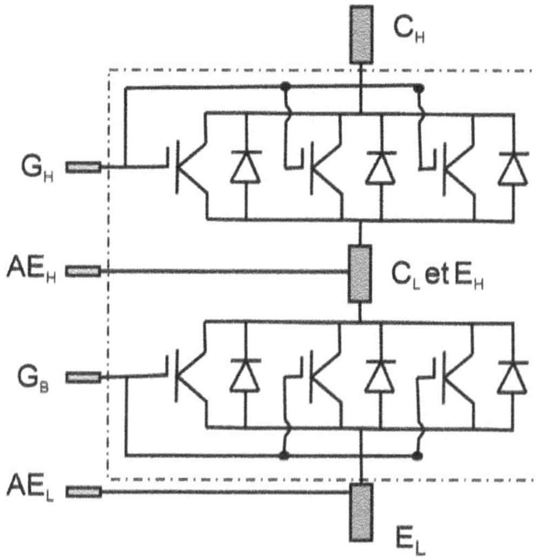

Figure 2.134 Multichip Terminal illustration. Chen et al. (2020a).

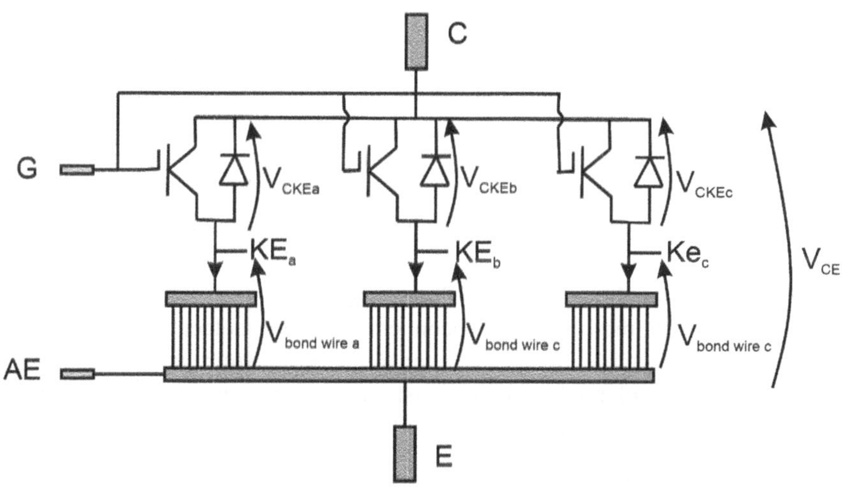

Figure 2.135 Schematic of the on-state voltage measurement. Chen et al. (2020a).

(6) Diagnosis based on a thermal model

As TSEPs are difficult to use and often need calibration, junction temperature can be given by a thermal model (Gachovska et al. (2015)). In this paper, a RC Cauer thermal model (see 2.6.2.4) is built (fig 2.136).

Its entries are the collector emitter current and voltage (I_C and V_{CE}) and the temperature of the case T_C. The thermal resistance R_{th} and capacitance C_{th} are given by the following equations:

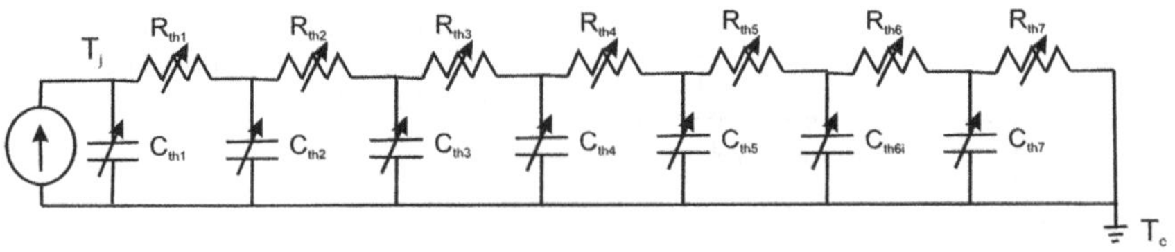

Figure 2.136 Thermal Model.

$$\begin{cases} R_{th_i} = \frac{d_i}{k_i A_i} \\ C_{th_i} = c_i \rho_i V_i \end{cases} \tag{2.119}$$

where

- d_i is the thickness of the layer i;
- A_i is the effective cross-section area of the layer i;
- k_i is the thermal conductivity of the layer i;
- c_i is the constant pressure specific heat capacity of the layer i;
- V_i is the effective volume of the layer i;
- ρ_i is the mass density of the layer i.

In this model, the two coefficients, i.e., the thermal conductivity k_i and the constant pressure specific heat capacity c_i are temperature dependent and they are fitted by polynomials. Then, the estimated junction temperature can be used to detect the thermal failures due to the cycling of power modules or any abnormal thermal behavior. It also can be used to compute their RUL.

2.6.6.3.c Diagnosis based on Signal Processing

Signal processing methods have been employed to detect and locate the different power module faults. However, since they require more computing time, they are not adapted if we need fast online treatment. In this part, three methods will be presented:

- frequency analysis (fft, ...);
- Instantaneous frequency method;
- Wavelet analysis.

(7) Frequency analysis

Zisi Tian and Ge (2016) present a dual IGBTs open-circuit fault online diagnostic method for AC-DC converters (see Figure 2.137) used in railway electric traction.

The authors studied the frequency harmonics content of the normalized current $\frac{i_s}{i_{dc}}$ in presence of a single fault or a double fault.

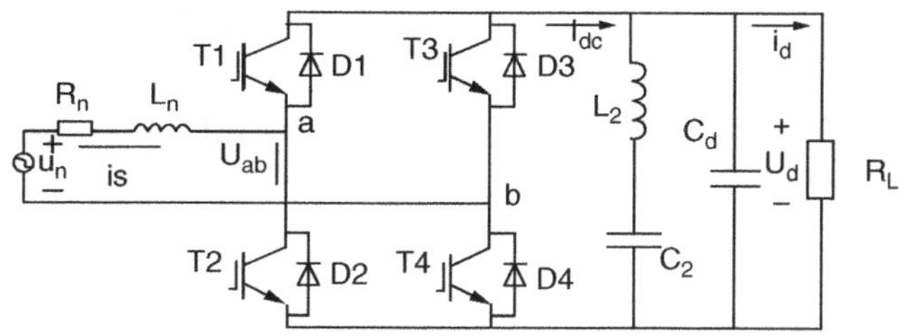

Figure 2.137 Single-phase rectifier.Zisi Tian and Ge (2016) with permission of IEEE.

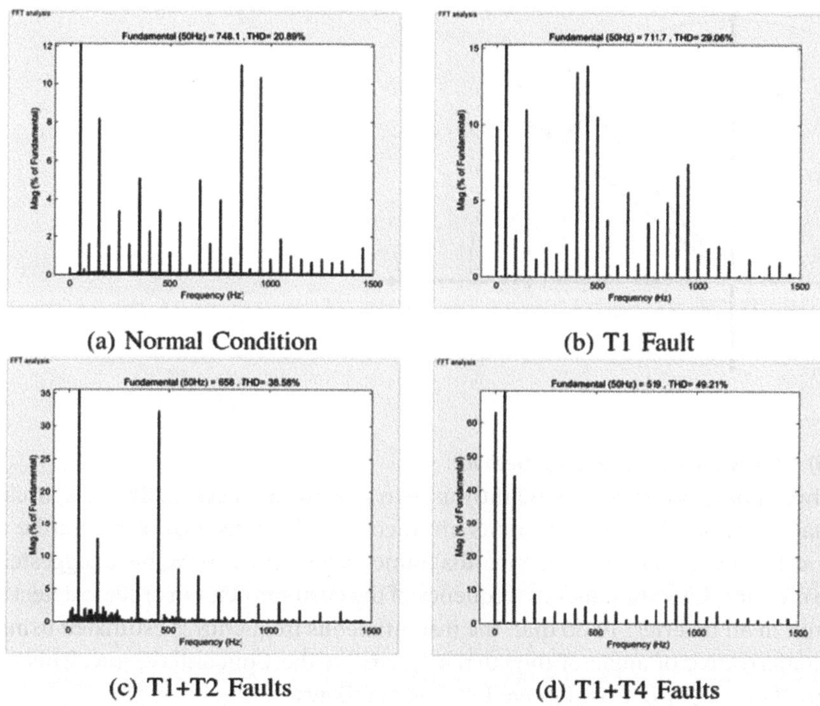

Figure 2.138 FFT of currents.Zisi Tian and Ge (2016) with permission of IEEE.

The single fault diagnosis is carried out first, followed by the double fault. The fault can be diagnosed by analyzing the harmonics of the normailzed AC current around the switching frequency 450Hz (Figure 2.139).

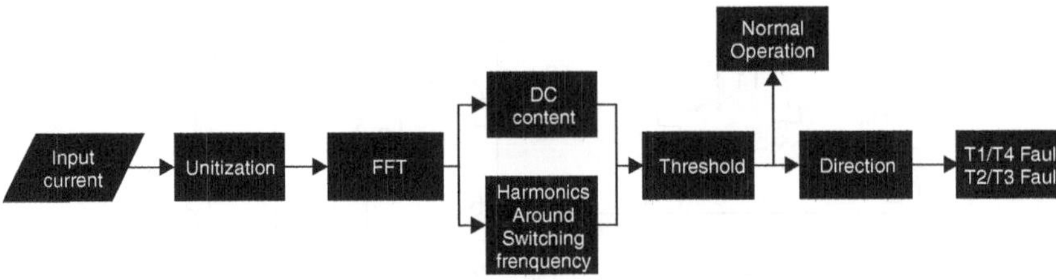

Figure 2.139 Input current frequency domain diagnosis method. Zisi Tian and Ge (2016) with permission of IEEE.

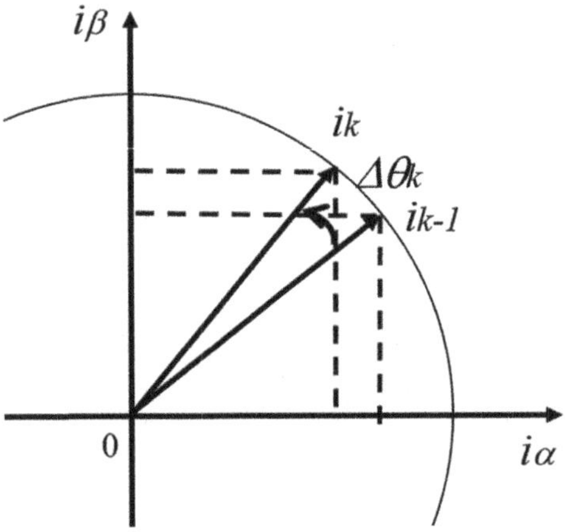

Figure 2.140 Current vector.

(8) Instantaneous frequency method:

Three-phase inverter is abundantly used in industrial process to drive electrical machines. So, the phase currents are theoretically sinusoidal and describe a circle in the $\alpha-\beta$ frame or Concordia frame. Peuget et al. (1998) have suggested to monitor the instantaneous frequency of the current vector in order to detect a fault in an inverter. To do that, the instantaneous frequency is estimated using the derivative of angle of the current vector in the Concordia frame. This is, based on $\vec{i_s}(t) = i_\alpha(t) + ji_\beta(t) = Ie^{j\omega_e t}$, is as follows:

$$f = \frac{1}{2\pi}\frac{d\theta}{dt} \tag{2.120}$$

with ω_e the angular supply frequency ($2\pi f$). A sampling frequency much higher than the maximum motor supply frequency must be used. Therefore, the angular displacement of the current vector is considered to be small. Thus, the estimation of the frequency becomes:

$$f \cong \frac{1}{2\pi}\frac{\sin\Delta\theta}{T} \tag{2.121}$$

with T the sampling time and $\Delta\theta = \theta_{kT} - \theta_{(k-1)T}$. So, it becomes: $\sin\Delta\theta = \sin\theta_{kT}\cos\theta_{(k-1)T} - \sin\theta_{(k-1)T}\cos\theta_{kT}$. From the Figure 2.140, one can deduce that:

$$\begin{aligned} i_{\alpha_{kT}} &= i_k\cos\theta_{kT} \\ i_{\beta_{kT}} &= i_k\sin\theta_{kT} \end{aligned} \tag{2.122}$$

Consequently, the instantaneous frequency can be expressed as follows:

$$f = \frac{1}{2\pi T}\frac{i_{\beta_{kT}}i_{\alpha_{(k-1)T}} - i_{\beta_{(k-1)T}}i_{\alpha_{kT}}}{i_{kT}i_{(k-1)T}} \tag{2.123}$$

Peuget et al. (1998) proposed to compare this estimate to the desired supply frequency from the driver in order to monitor the inverter fault. If the absolute value of the difference is great, the inverter operates under fault. Fuchs (2003) wrote that a direct detection is possible.

(9) Wavelet analysis:

In order to validate the ability of wavelet to inform the operator of a problem of an intermittent gate misfiring in an inverter, we will show results coming from experiments. A three-phase induction motor fed by a two-level inverter operating at 20kHz was used (see Razik et al. (2011)).

The monitoring is made thanks to the current circulating in the induction motor operating at full load (15Nm, 3kW).

Figure 2.141 illustrates the gate signals for the switch q_1, q_2, and q_3. In this Figure, one can see that the switch q_1 has a misfiring during several periods.

Figure 2.142 shows the current flowing in the phase a and b in the induction motor.

As we can see, we are not able to determine visually if the motor operates under healthy or faulty condition. However, monitoring is possible only with the existence of a singularity of the current observed. In this case, the switch should conduct but a misfiring occurs. Figure 2.143 shows the current in the phase a of the induction motor and Figure 2.144, in phase b. The gate q_1 has a problem of misfiring during $300\mu s$. The evolution of wavelet coefficients changes when the fault occurs.

One can see that the wavelet coefficients quickly raise significantly. This appears when the intermittent misfiring fault occurs. This approach is perfect to inform the operator of the appearance of misfiring in the inverter. Knowing that the sum of the three-phase currents which is equal to zero, it allows to monitor the inverter using only one phase. In conclusion, the detection of gate

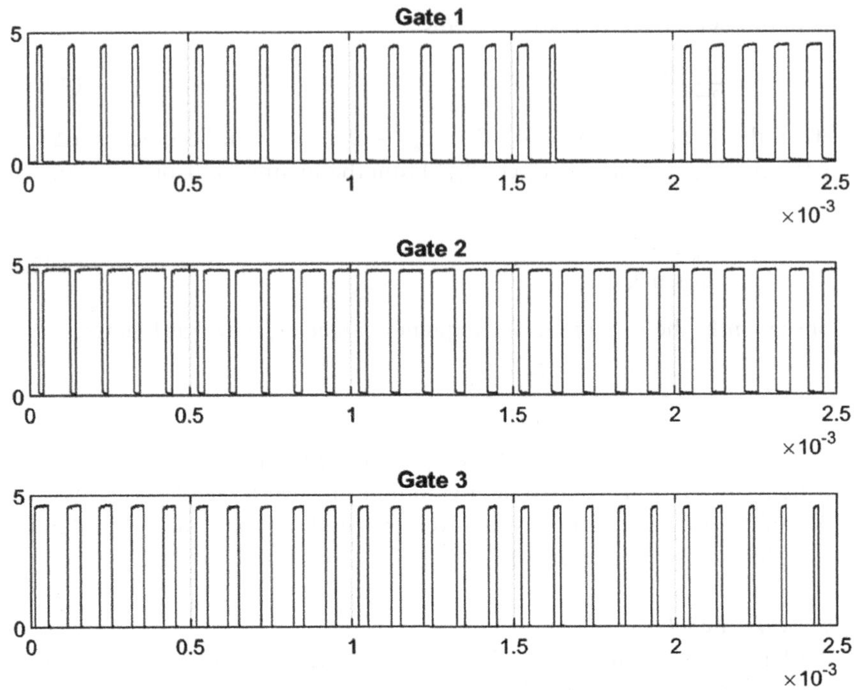

Figure 2.141 Gates signal: q_1, q_2 and q_3. Experimental results with a misfiring during $300\mu s$ for q_1.

intermittent misfiring is easily using the CWT. This approach is pertinent and convincing.

2.6.6.3.d Diagnosis based on Clustering

Clustering methods allow to diagnose multiple faults in converters. They can be based in neural network, pattern recognition, support vector classifier. The general approach is described in Figure 2.145.

In Delpha et al. (2007), the detection of an open circuit in a three-phase inverter relies on the deviation of the current vector trajectory. A pattern recognition technique is used to diagnose the fault. This one is based on a Principal Component Analysis and a linear discriminant classifier. The isolation procedure is made in two steps:

- First step: identification of the faulty leg with a three–class classifier ;
- Second step: discrimination of the faulty power switch.

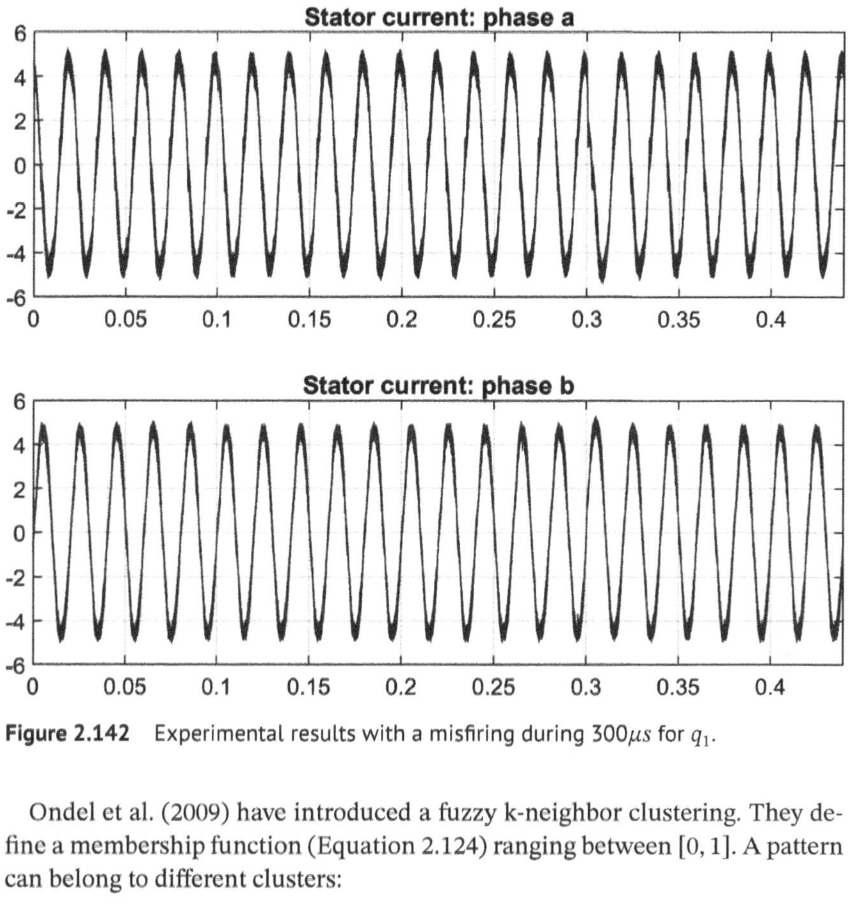

Figure 2.142 Experimental results with a misfiring during $300\mu s$ for q_1.

Ondel et al. (2009) have introduced a fuzzy k-neighbor clustering. They define a membership function (Equation 2.124) ranging between $[0, 1]$. A pattern can belong to different clusters:

$$
\begin{aligned}
&\text{if } d_{s_c} \leq d(X_u, m_c) \leq d_{s_c} \\
&\mu_{\Omega_c}(X_u) = \left(\frac{1}{N_c} - (\frac{1}{N_c} - \eta) \left(\frac{d(X_u, m_c)}{d_{s_c}} \right)^g \right) .k_c \\
&\text{if } d_{seuil_c} \leq d(X_u, m_c) \leq \beta.d_{seuil_c} \\
&\mu_{\Omega_c}(X_u) = \eta \left(\frac{\beta}{\beta - 1} \right)^g \left(\frac{\beta.d_{s_c} - d(X_u, m_c)}{\beta.d_{s_c}} \right)^g .k_c \\
&0 \text{ otherwise}
\end{aligned}
\qquad (2.124)
$$

with:

- $d(X_u, m_c)$ is the distance between an observation Xu and the center of gravity m_c of the class Ω_c;
- d_{s_c} is the parameter which defines the position of the inflection points;
- n_c is the number of samples contained in the class Ω_c ;

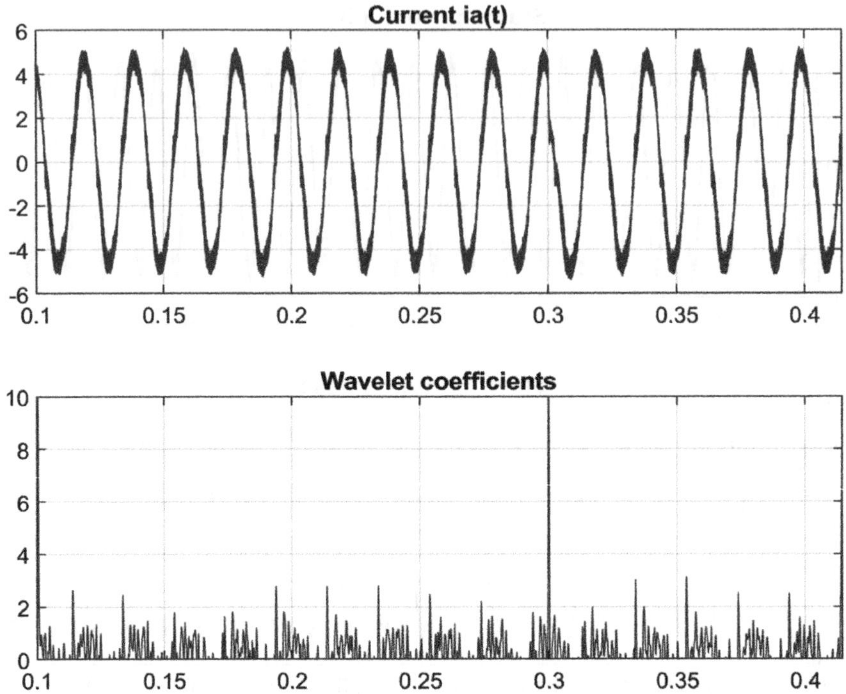

Figure 2.143 Experimental results with a misfiring during $300\mu s$ for q_1. On the top: the current i_a. On the bottom: the Wavelet coefficients.

- k_c is the number of k nearest neighbors resulting from the class; Ω_c;
- η is the parameter being used to fix the constant part of the membership function;
- β is a parameter which allows defining the bandwidth of the membership function;
- g is a parameter allowing to adjust the slope of the function.

Figure 2.146 shows the difference between exclusive and fuzzy membership functions.

Many other clustering techniques have been used. For instance:

- Pham et al. (2019) propose an auto-adaptive and dynamical clustering for double open-circuit IGBT fault diagnosis of power inverters. It is based on a Neural Network using a new Gaussian Densities Merge Mechanism and a relevant pattern vector.

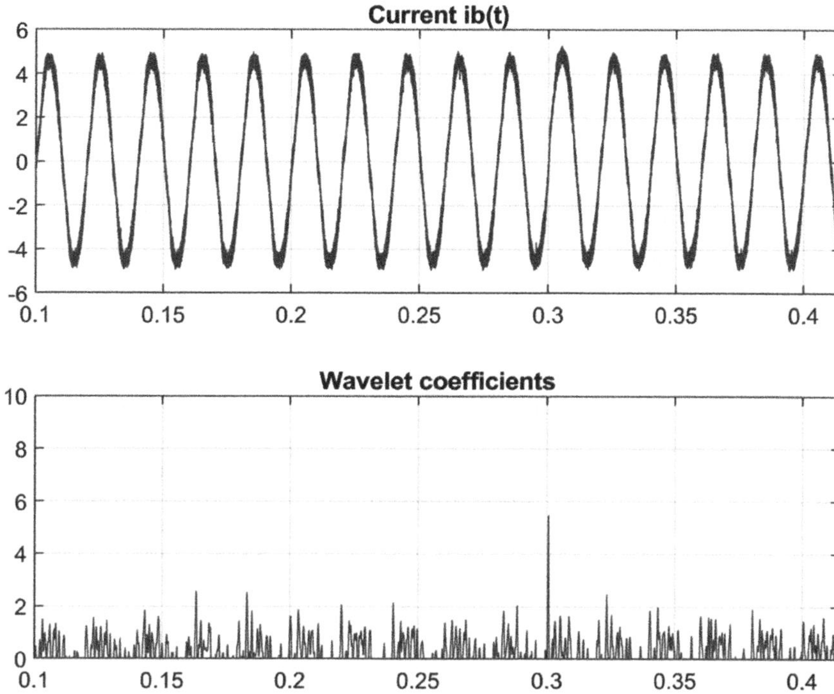

Figure 2.144 Experimental results with a misfiring during $300\mu s$ for q_1. On the top: the current i_b. On the bottom: the Wavelet coefficients.

- Bendiabdellah and Cherif (2017) dedicated a Clark decomposition of the three-phase voltages and a support vector machine classifier to diagnose an open-circuit fault at one of the IGBT of a three-phase inverter.
- Bowen and Wei (2019) deal with a method based on wavelet decomposition and support vector machine classifier to diagnose open-circuit switch faults of inverter.
- Qiu et al. (2017) use a complete ensemble empirical mode decomposition with adaptive noise algorithm and a support vector machine classifier to diagnose faults in Three-Phase Inverters of PMSM Drive Diagnosis.

2.6.6.3.e Diagnosis based on Neural Network

Neural network has been less used in scientific literature than other techniques. We can find some applications for open circuit IGBT failure diagnosis in Charfi et al. (2006) and Masrur et al. (2007) or for diode bridge failure detection Mohagheghi et al. (2009). Neural network can be used to detect a deviation of healthy behavior. For instance, the aim of the neural network

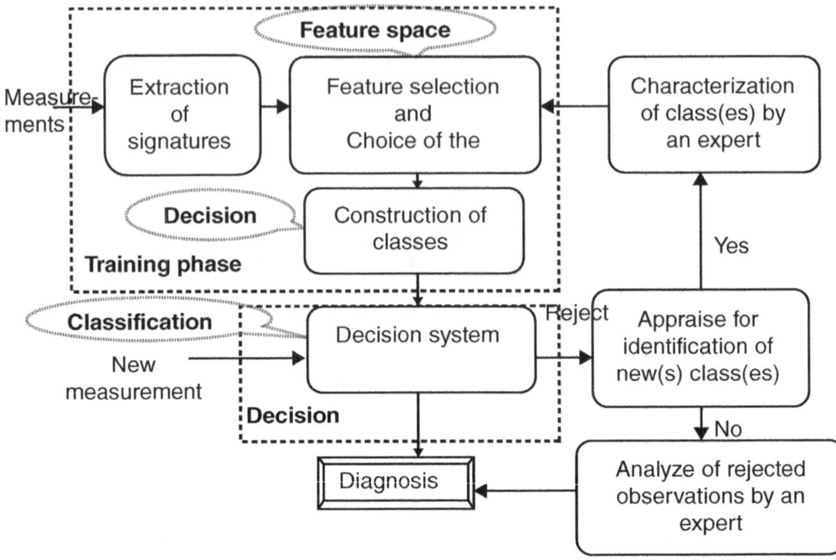

Figure 2.145 Diagnosis based on clustering Ondel et al. (2009) with permission of IEEE.

in Mohagheghi et al. (2009) is to detect if the performance properties of one or more components in the rectifie have changed from the healthy case (fig 2.147).

The input of the neural network in this case, is the vector $X = [1, v_i(t), i_i(t), i_o(t)]$ and the output is the voltage $v_o(t)$. The weights are adapted for minimizing the error between the actual output and the estimated one. The learning scheme is described in Figure 2.148.

Masrur et al. (2007) deal with the application of Neural Network to the diagnostic of open-circuit failure in an inverter. The signal is segmented, and the following parameters are computed within them:

- the maximum magnitude of the signal;
- the minimum magnitude of the signal;
- the median of the signal;
- the standard deviation;
- the mean of the signal;
- the zero-frequency component of the power spectrum.

These parameters are used by the Neural classifier in order to discriminate the different open-circuit faults. The general process is described in Figure 2.149.

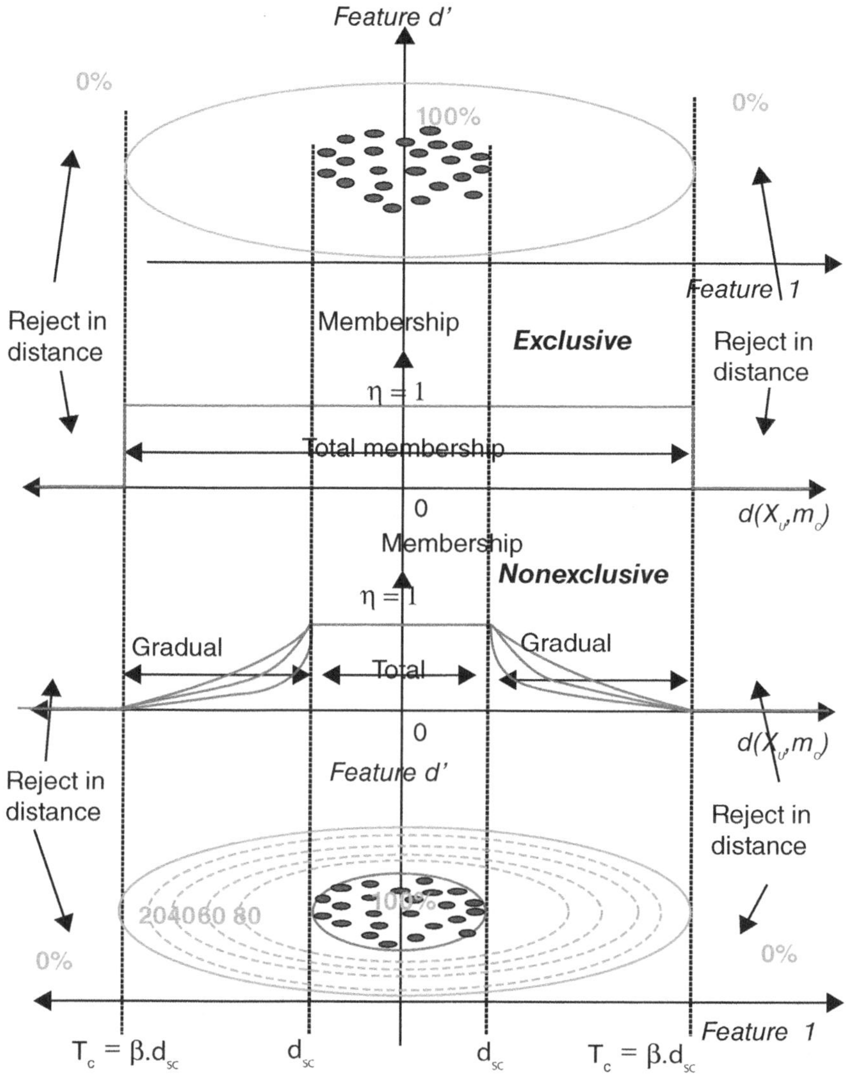

Figure 2.146 Membership function in exclusive and non-exclusive case, Ondel et al. (2009) with permission of IEEE.

A machine learning algorithm selects the operating points of the drive from the torque and the speed. A digital twin generates the training data based on these operating points for the different healthy and faulty conditions (Figure 2.150). The trained neural network is used to classify the different faults.

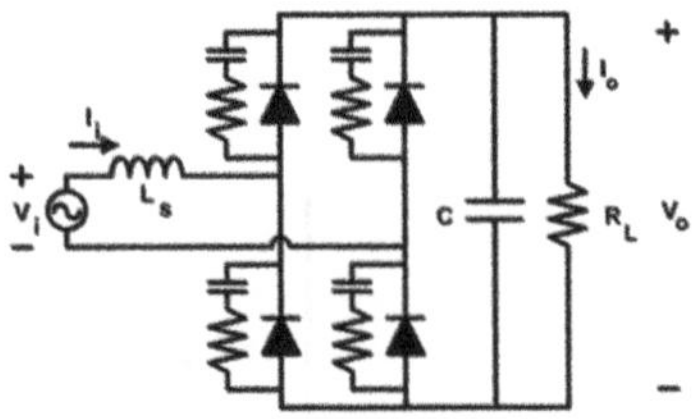

Figure 2.147 Full Bridge Rectifier (Mohagheghi et al. (2009)). Courtesy of IEEE.

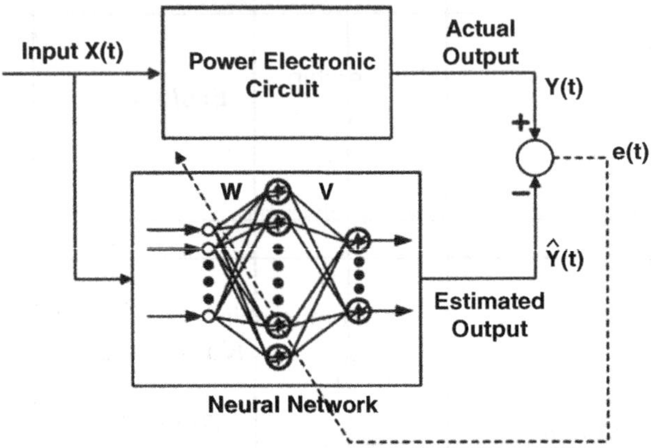

Figure 2.148 Training of the Neural Network (Mohagheghi et al. (2009)). Courtesy of IEEE.

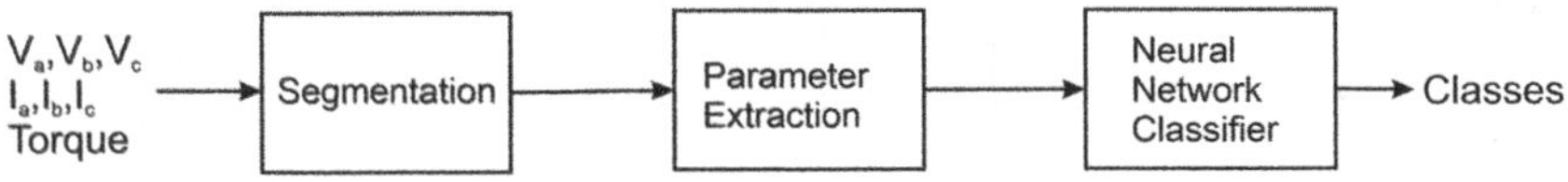

Figure 2.149 General Detection Process (Masrur et al. (2007)).

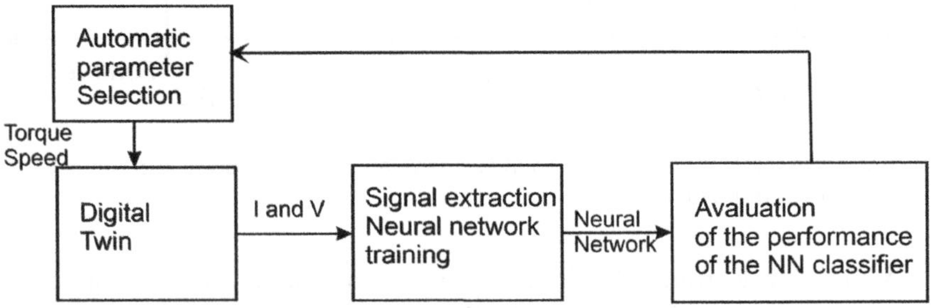

Figure 2.150 Control Parameter Selection (Masrur et al. (2007)).

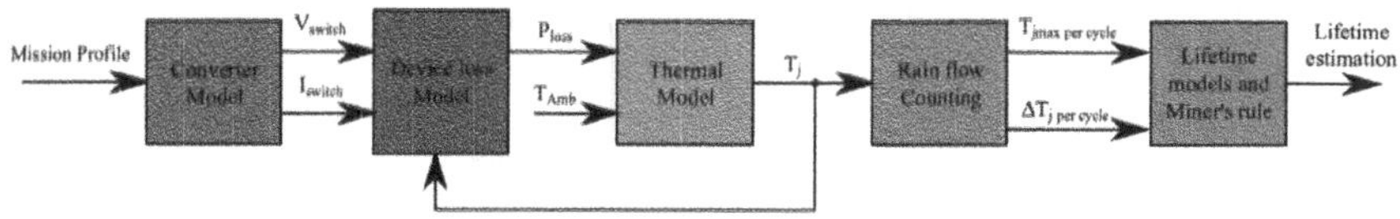

Figure 2.151 Failure mechanism based prognosis process.

Neural network could be associated with preprocessing steps. In Charfi et al. (2006), the neural network is fed by the approximation coefficients of a wavelet decomposition from the three-phase currents of an inverter.

2.6.6.3.f Synthesis

Table 2.12 gives an overview of the different diagnosis methods of power MOS and IGBT. It present a brief description of the method, the pattern vector (detection parameters), the type of failure which is addressed and the reliability of this approach (ref Lu and Sharma (2009)).

2.6.6.4 Prognosis of Failure Modes

The prognosis for power module failure is still in its early stages. It is the subject of much academic and industrial researches. It depends very strongly on the nature and the technology of the components and the field of application. However, two main classes of methods emerge (Hanif et al. (2019)).

- The first one deals with failure mechanisms. It could be based on physical of failure (PoF) models, empirical ones, or test data. They mainly provide the number of cycles to failure as a function of parameters such as the minimum or maximum junction temperature, the junction temperature swing ΔT_j, the ambient temperature, etc.
- The second one is based on failure precursor parameters such as TSEPs and extrapolation methods such as regressors or neuronal networks, etc. It mainly provides the remaining useful time in terms of real-time units.

2.6.6.4.a Prognosis based on Failure Mechanism and Statistical Data

The number of cycles is computed from the thermal profile of the devices: the ambient temperature T_a, the junction temperature T_j, the temperature swing ΔT_j etc. These parameters can be measured of provided from a thermal model and the estimated losses of the device. In case of a variable operating profile, a rain-flow algorithm can used to evaluate the contribution of the different ranges of losses (Musallam et al. (2010)).

First, we will consider the different ways to obtain the number of cycles from the thermal profiles. The Table 2.13 from (Hanif et al. (2019)) gives the advantages and the drawbacks of these different approaches.

Table 2.12 Synthesis of the different diagnosis methods for IGBT (Lu and Sharma (2009)).

Description	Detection Parameters	Type of failures	Reliability	Reference
		Direct analysis of the current		
Park's vector method	3-phase currents	Open Circuit Short Circuit	Not suitable for low current	Mendes and Marques Cardoso (1999) de Araujo Ribeiro et al. (2003a) Razik et al. (2011)
Average Current	3-phase currents	Open and Short circuit	Medium	Mendes and Marques Cardoso (1999)
Slope of current	3-phase currents	Open circuit	Not relevant at low current	Peuget et al. (1998)
Fuzzy technique	3-phase current	Open circuit	Fuzzy rules must be carefully designed	de Araujo Ribeiro et al. (2003b) Estima and Marques Cardoso (2011)
		Diagnosis based on the direct or indirect analysis of junction temperature		
Based on TSEP V_{CE}	V_{CE}, I_C	Bond wire lift-off	Medium	Jinsong Han et al. (2016) Wang et al. (2020)
Based on TSEPs $V_{CE_{on}}$ and $V_{CKE_{on}}$	$V_{CE_{on}}$, $V_{CKE_{on}}, I_C$	Bond wire lift-off	Accurate	Chen et al. (2020b)
Thermal Model	V_{CE}, I_{CE} and T_{Case}	Therma failure due to cycling	Good	Gachovska et al. (2015)
		Signal Processing		
Frequency Analysis	3-phase currents	Open Circuit	Can locate double fault	Zisi Tian and Ge (2016)
Instantaneous frequency method	3-phase currents	Open Circuit	Cannot locate faulty device	Peuget et al. (1998)

(Continued)

Table 2.12 *(Continued)*

Description	Detection Parameters	Type of failures	Reliability	Reference
Wavelet analysis	3-phasecurrents	Open Circuit Misfiring	Good.	Saleh and Rahman (2005) Torrence and Compo (1998)
		Clustering techniques		
PCA and linear discriminant classifier	3-phase currents	Open Circuit Misfiring	For only a single fault	Delpha et al. (2007)
Fuzzy k-neighbor clustering	3-phase currents and voltages	Multiple faults inverter and motor	Good	Ondel et al. (2009)
Clark decomposition and Support Vector Machine	3-phase voltages	Open Circuit Misfiring	Good	Bendiabdellah and Cherif (2017)
		Neural network		
Model-based ANN Techniques	Three phase currents,voltages and torque	Open circuit	Medium	Masrur et al. (2007) Chen et al. (2020b)
Wavelet-Neural Network Method	Three phase currents	Open Circuit	Accurate	Charfi et al. (2006)

The life time model can be empirical. For instance Bayerer et al. (2008) propose the following law for bond-wire failures:

$$N_f = K.\Delta T_j^{\beta_1}.\exp\left(\frac{\beta_2}{T_j + 273}\right).t_{on}^{\beta_3} I^{\beta_4} V^{\beta_5} D^{\beta_6} \tag{2.125}$$

where T_j the junction temperature in °C, t_{on} is the turn-on time, I is the current-per-wire, V is the blocking voltage, and D is the bond wire diameter. β_1, β_2, β_3 β_4 β_5 β_1 are identified empirical parameters.

Table 2.13 Comparison of failure mechanism-based methods (Hanif et al. (2019)).

Method	Advantages	Limitations
Empirical methods	Accurate failure rate Good indicator for reliability	Must be regularly actualized Difficult to have reliable data Difficult to construct the causality chain
Physical of failure	Could model specific mechanism of failure (Bonding, etc...) Could predict End of Life for a specific failure	Doesn't provide field reliability Complex to be applied Cannot model defect driven failures Not suitable to the whole system
Data based methods	Accurate approach for the actual system Collecting data could be made before the commissioning	Requires accelerated test Introduces uncertainties

Norris and Landzberg (1969) and Xuhua Huang et al. (2012) propose the following law for solder fatigue:

$$N_f = A(f)^{-\alpha}(\Delta T)^{-\beta} \exp\left(\frac{E_a}{k}\left(\frac{1}{T_{\max}}\right)\right) \tag{2.126}$$

where α, β are empirical parameters, f is the frequency (cycle/day), ΔT_j is the absolute temperature range in a temperature cycle, E_a is the activation energy of the failure mechanism in eV and K is the Boltzman's Constant.

The laws can be also derived from the physics of failure. Table 2.14 gives some examples:

All these data can be represented by a collection of curves, Figure 2.152 as a function of temperature, current, length of cycles, etc.

These number of cycles highly depends of the junction temperature profiles. These latter can be obtained from the TSEP, measured by optical fiber or determined by thermal modeling. In this case, the temperature is computed with the power losses and a Cauer or Foster thermal model (see Section 2.6.2.4). Power losses are provided by the voltages and currents across the different components during the time-varying mission profile (Figure 2.153).

Then, rain flow counting allows organizing the temperature profile into thermal cycles for using them as inputs of lifetime models (Figure 2.154).

Lifetime models given by the manufacturer or the previously described laws are used for performing the lifetime estimation. This estimation gives the number of cycles to failure under a given condition.

Table 2.14 Comparison of failure mechanism based methods.

Object	Law	Sources
Plastic strain-based model (Coffin-Manson law)	$N_f = C_1(\Delta\varepsilon_p)^{-C_2}$ $\Delta\varepsilon_p$ is the elastic strain amplitude. C_1 and C_2 are two parameters.	Coffin (1954)
Bond-wire lift off	$N_f = C_1(\Delta T)^{-C_2}$ ΔT is the temperature swing. C_1 and C_2 are two parameters.	Ciappa (2002) Ciappa et al. (2002)
Solder fatigue	$N_f = 0.5\left(\frac{L.\Delta\alpha.\Delta T}{\gamma.x}\right)^{1/c}$ where L is the lateral size of the solder joint	
$\Delta\alpha$ is the CTE mismatch between the upper and the lower plate, ΔT is the temperature swing, c is the fatigue exponent, x and γ are the thickness and the ductility factor of the solder.	Ciappa (2002) Ciappa et al. (2002)	

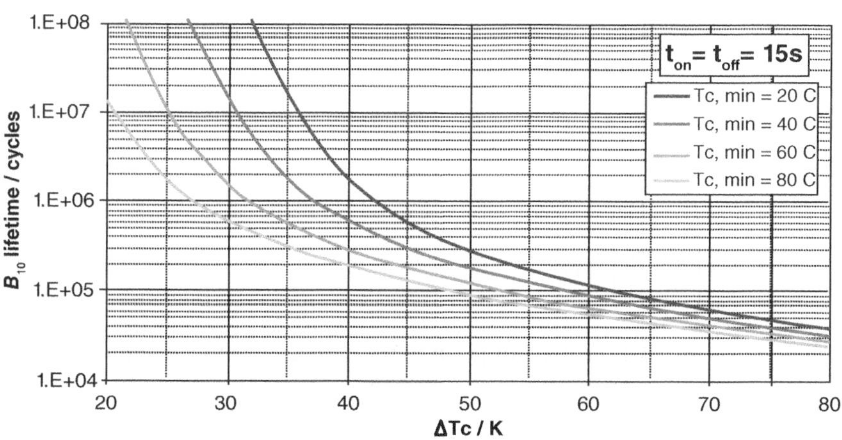

Figure 2.152 Example of B10 lifetime curves of the solder joints of the conductor leads and substrates for t_{cycle} equal to 30s. process (ABB (2004)). Courtesy of ABB Power Grids Switzerland Ltd., Semiconductors.

Each cycle or half cycle gives a number of consumed time life cycles. All these cycles are aggregated with the Miner's rule (Miner (1945)), which supposes the damage effect can be linearly accumulated.

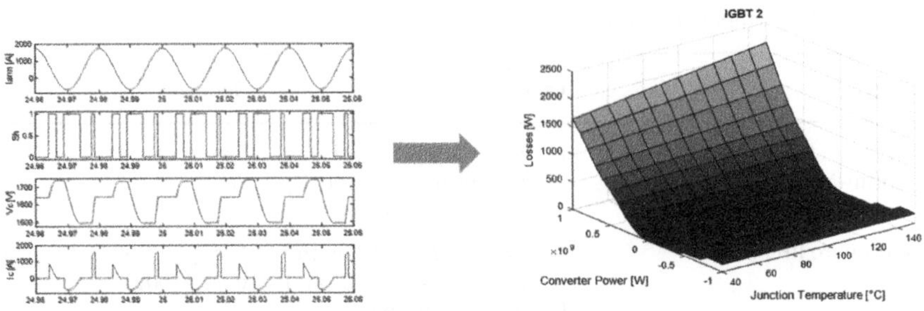

Figure 2.153 Temperature profile modeling.

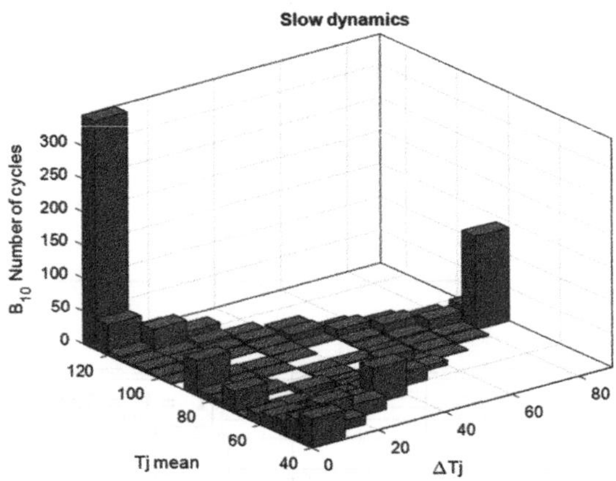

Figure 2.154 Rain flow counting.

$$LC = \left(\sum_{i=1}^{N_T} \frac{n_i}{N_i} \right) .100\% \tag{2.127}$$

where LC is the lifetime consumption of the device, n_i is the actual number of cycle for a given operating point and N_i is the number of cycles until failure for this operating point.

2.6.6.4.b Prognosis based on Failure Precursors

These methods are based on the following steps (Figure 2.155):

- Some accelerated aging tests are carried out in a highly controlled environment until failure or the required function is no longer ensured. The tests must be representative of the conditions under which the equipment in operation will evolve.

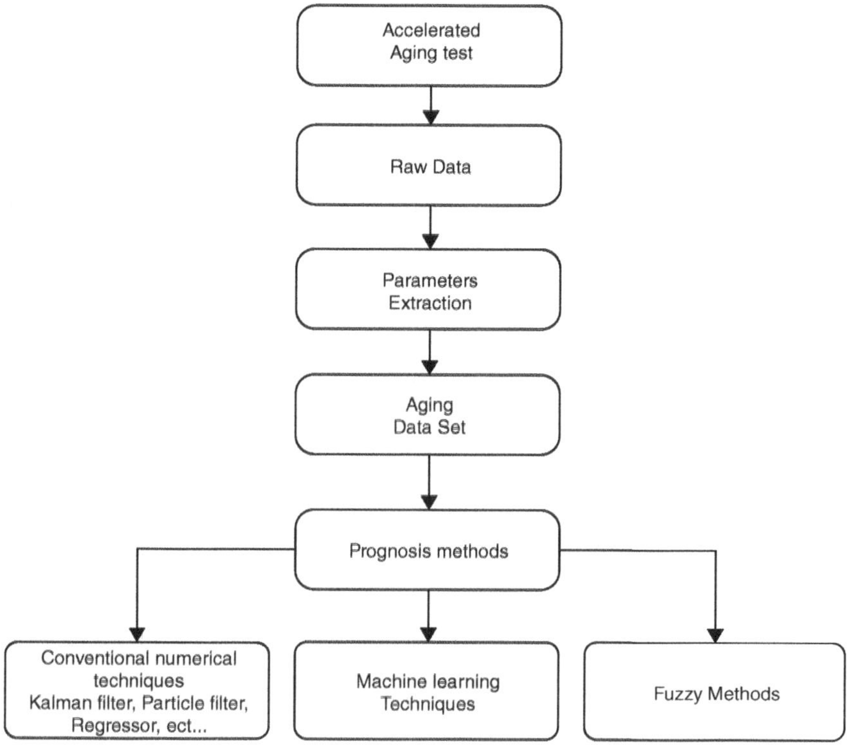

Figure 2.155 Prognosis process based on Failure Precursors. *Source:* Fang et al. (2018).

- Some relevant parameters which are correlated to the aging process are extracted. They are mainly linked to TSEP (see Section 2.6.6.2). The monitoring of these parameters provides aging data set (or they make it possible to identify the aging laws previously described).
- Data driven prognostic algorithms, regressors or fuzzy techniques give an estimate of the remaining useful life based on these parameters.

The evaluation of the uncertainties is inherent to the forecasting activity itself (Sankararaman and Goebel (2013)), see Figure 2.156.

However, in the current state of research, this aspect is not mature in the field of power electronics. Therefore, we will not deal with this subject in the context of this chapter. Readers can refer to the works of NASA's prognostic Center of Excellence team, which have greatly contributed to putting the notion of uncertainty back at the center of the prognosis (Sankararaman and Goebel (2013) Sankararaman and Goebel (2015) Celaya et al. (2012))(Goebel et al. (2017)). In

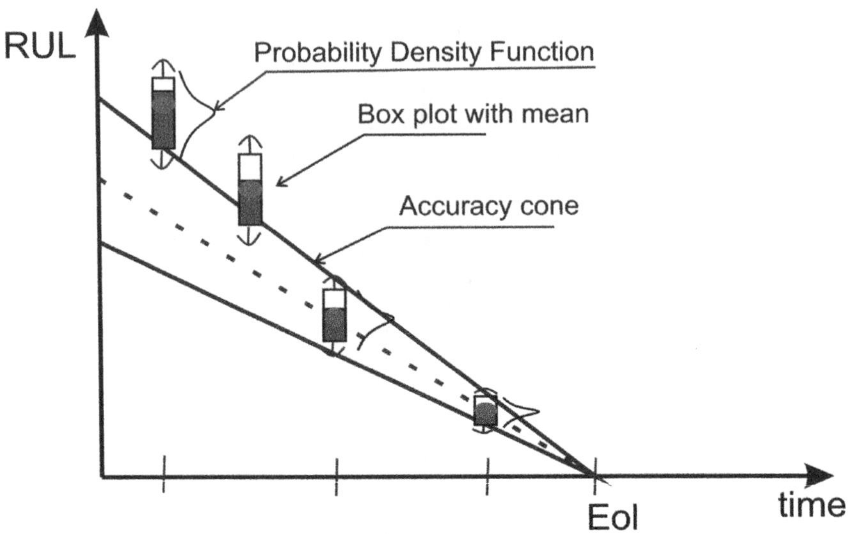

Figure 2.156 Prognosis uncertainty and performance. *Source:* Saxena et al. (2010).

addition, (Saxena et al. (2008) Saxena et al. (2010) Saxena et al. (2014)) give some metrics to evaluate the prognostic performance.

(1) Conventional numerical techniques: Regressors, Kalman filters, Particle filter, etc.

(1)1 General description These methods use algorithms such as linear regression, Kalman filter, Particle filter (PF) in order to identify the future state from past values. The different model can be expressed as recursive laws, cumulative contributions, discrete or continuous solutions of recursive equations.

The prediction of the future state is based on a sequence of past states. In general, it can be expressed in the form of the Equation 2.128:

$$\hat{X}_{t+1\rightarrow t+h_p} = f_p(t_{t-n_0\rightarrow t}, \theta) \tag{2.128}$$

with

- $f_p(t_{t-n_0\rightarrow t})$ the sequence of n_o observations of signature values up to t,
- f_p: the prediction model,
- θ the parameters of the prediction model,
- $\hat{X}_{t+1\rightarrow t+h_p}$: the estimated sequence up to the prediction horizon h_p.

The prediction method can be iterative. This method consists in learning a model to make predictions at $t + 1$. Then, the same model is cascaded $(h_p - 1)$ times, in order to make a prediction at $t + h_p$. This method is quite

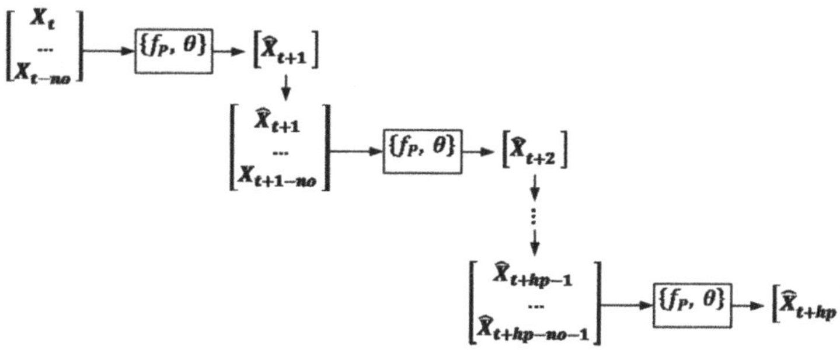

Figure 2.157 Illustration of an iterative prediction. *Source:* Breuneval (2017).

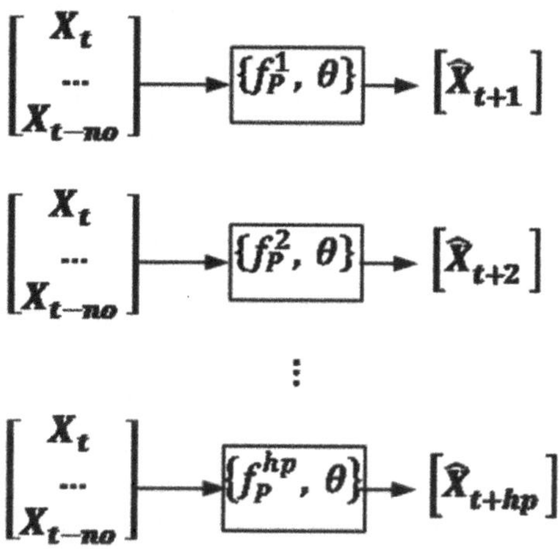

Figure 2.158 Illustration of a set of direct, multi-horizon predictions. *Source:* Breuneval (2017).

easy to implement but the error accumulates as the predictions are made (see Figure 2.157).

The prediction can be straightforward. This method consists in training a model to forecast the state from $t + 1$ to $t + h_p$ with the a historical sequence ending at t. It gives a good accuracy but needs a big learning base and it is difficult to implement since there are many models (see Figure 2.158).

The multi-step prediction can be made in parallel. It consists in using an extrapolation model with several outputs (multi inputs multi outputs: MIMO). This is more accurate for short term prediction but less precise for a long horizon (see Figure 2.159).

These different model can generate the RUL or a probability of failure.

(1)2 Gaussian process regression (GPR) Ali et al. (2018) apply GPR to the failure prognosis of an IGBT. This paper considers a linear model

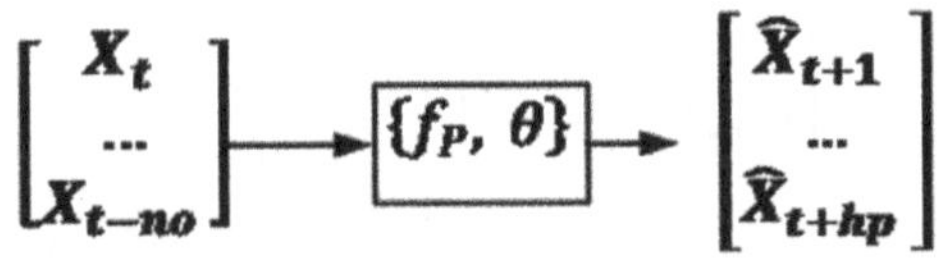

Figure 2.159 Illustration of a parallel prediction. *Source:* Breuneval (2017).

of the variation of $V_{CE_{on}}$ as a function of time and vector of parameters (eq 2.129):

$$y = x^t.w + \varepsilon \tag{2.129}$$

where $y = V_{CE_{on}}$, w is the weight vector that corresponds to the parameters of the model, ε is a measurement Gaussian noise $N(0, \sigma_n{}^2)$, $X = [x1, x2, ..., xn]^t$ and $y = [y1, y2, ..., yn]^t$ are matrix representation of the training set.

The learning base is built with aging accelerated tests. The probability distribution function of unknown parameters is identified by Bayesian inference (eq. 2.130). The GPR provides the variance around the mean value to describe the uncertainty of the model.

$$w_{opt} = \arg\max_{w}(p(y|X, w).p(w)) \tag{2.130}$$

The RUL is given by the intersection of the fitted GPR with a failure threshold value V_f.

Hong et al. (2013) deals also with the failure prognosis of an IGBT. It is based on the junction temperature. This one is predicted by a Gaussian Bayesian process. Assuming a centered Gaussian process, the output predictive mean and variance is given by equation (eq 2.131):

$$\begin{cases} \hat{y} = k^T.K.t \\ \sigma_{\hat{y}}{}^2 = C(x, x) - k^T(x)K^{-1}k(x) \end{cases} \tag{2.131}$$

where

- y is the scalar output (junction temperature T_j),
- x are the latent variables. $x \in R^d$,
- $\{(x^{(i)}, y^{(i)})\}$ the set of training points with $i \in [1\ n]$,
- n the number of training points,
- K is the covariance matrix of the training set: $K_{ij} = C(x^{(i)}, x^{(j)})$,
- C is a covariance function. In this case, it has been chosen: $C(x^{(i)}, x^{(j)}) = v_0 \exp\{-\frac{1}{2}\sum_{l=1}^{d} w_l(x_l{}^{(i)} - x_l^{(j)})\} + a_0 + a_1 \sum_{l=1}^{d} x_l{}^{(i)} x_l^{(j)} + v_1\delta(i, j)$,
- $k(x) = \left[C(x, x^{(1)}), ..., C(x, x^{(n)})\right]^T$,
- $t = \left[t^{(1)}, ..., t^{(n)}\right]$.

The hyper-parameters $\theta = [a_0, a_1, w_1, ..., w_d, v_0, v_1]$ is estimated by maximizing the log-like hood of the data on the learning base.

(1)3 Cumulative law For instance, Brown et al. (2006) use the TSEP V_{CE} and compute the probability of failure f_1 of a switch mode power supply with IGBT, according to Equation 2.132.

$$f_1 = t_0 \sum_{i=1}^{M} \left(e^{\frac{Ea}{k.T_i}} \Delta t_i \right) \tag{2.132}$$

where E_a, k, T, and t_0 represent the activation energy, the Boltzmann's constant, the junction temperature and a constant of proportionality and Δt_i the operating times.

(1)4 Kalman filter Eleffendi and Johnson (2016) use a Kalman filter in order to fulfill an adaptive estimate of the junction temperature of an IGBT from V_{CE}, the load current, and a thermal Foster model. The residual signal from the Kalman filter can be used to detect changes in the thermal model parameters, thus gives an indicator of the thermal path degradation.

(1)5 Particle filter Saha et al. (2009) use a particle filter (PF) applied on a decay model (eq. 2.133)

$$\theta(t) = \mathrm{C}e^{-\lambda t} \tag{2.133}$$

The current I_{CE} is approximated by a polynomial estimator (eq. 2.134). The PF identifies one of these parameters p_1. This latter is assumed to follow the aging law (eq. 2.134).

$$I_{CE}(t) = e^{p_1.t^3 + p_2.t^2 + p_3.t^1 + p_4} \tag{2.134}$$

p_1 is used as an aging indicator (see Figure 2.160).

Pugalenthi and Raghavan (2018) use the particle filter algorithm to identify the aging process of a MOSFET. PF is applied to track the on-state resistance R_{DSon} assuming the evolution law (2.135):

$$R_{DSon}(t) = \alpha.e^{\beta.t} + R_{DSon_0} \tag{2.135}$$

The end of life is characterized by an increase of about 10% of R_{DSon}.

Hu et al. (2020a) also applies a particle filter to realize a bond-wire IGBT failure prognosis. This paper presents a fusion method of a model-driven and a data-driven approach. Finite-element analysis provides a suitable parameter to track the crack propagation processes of the wire bonds (V_{CE}).

The model of the process is established from the physical fracture mechanism theory. A state space equation describes the crack propagation, and an output equation provides the relation between V_{CE} and a crack propagation

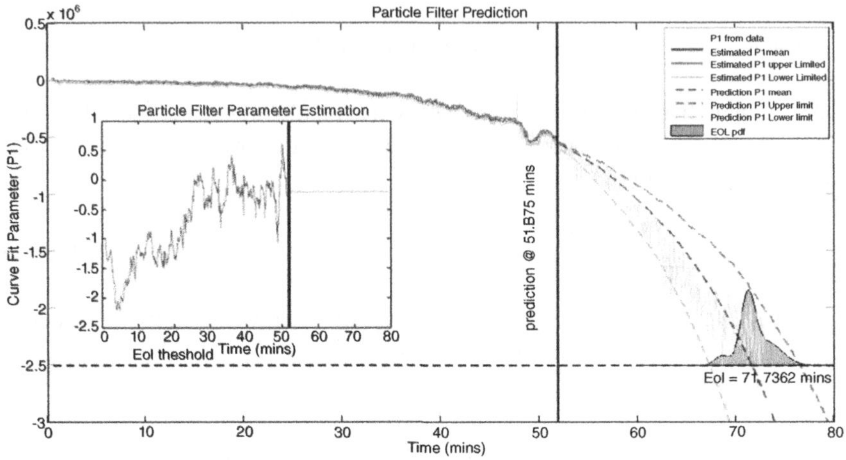

Figure 2.160 Particle Filter Tracking and Prediction. *Source:* Saha et al. (2009). Courtesy of IEEE

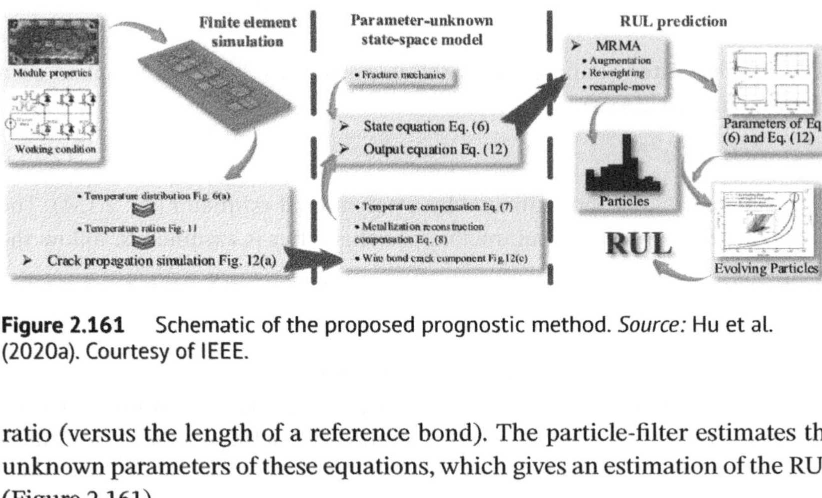

Figure 2.161 Schematic of the proposed prognostic method. *Source:* Hu et al. (2020a). Courtesy of IEEE.

ratio (versus the length of a reference bond). The particle-filter estimates the unknown parameters of these equations, which gives an estimation of the RUL (Figure 2.161).

(2) Machine learning techniques

(2)1 General description Two approaches are mainly used.

- using classical clustering techniques such as pattern recognition, PCA, etc.
- using neural network to extrapolate time series.

(2)2 Clustering techniques Unsupervised clustering time series and Supervised clustering of unknown operating points

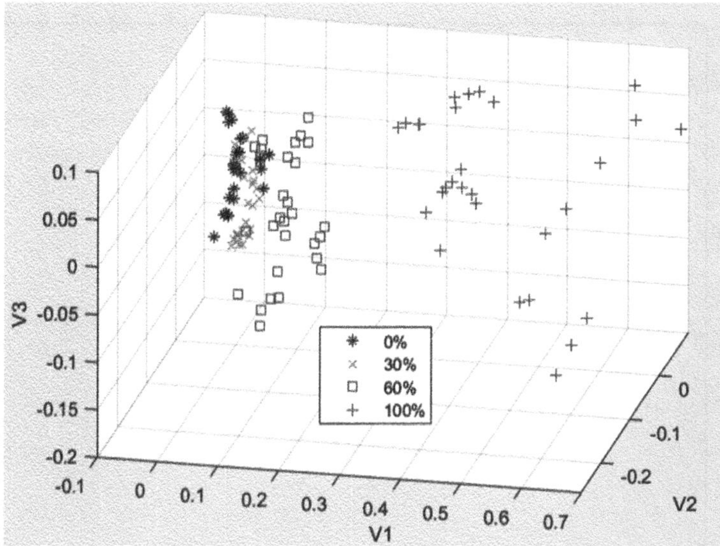

Figure 2.162 3-D representation of the 120 samples of the signature according to the 4 learning classes. *Source:* Hologne (2018).

In a first step, the learning base with the selected TSEPs is partitioned, by an unsupervised classifier (kmeans, self organizing map, etc.), into disjointed areas according to aging state. Accelerated aging test provides this learning base. These different classes can represent, for instance, a percentage of aging (see Figure 2.162).

In a second step, a supervised classifier (k Near Neighbor, Support vector machine, Neural network classifier ...) assigns any new operating point to one of this areas and consequently to a aging state (see Figure 2.163).

(2)3 Neural network Extrapolation of times series by Neural network

The Neural network is used to extrapolate time series such the method describe in the paragraph (1)1. For instance, Wu et al. compute the IGBT junction temperature with a Back Propagation (BP) Neural Network. The inputs of the neural network are the TSEP V_{CE} and I_C (see Figure 2.164).

A comparison was made with a polynomial fitting scheme is presented. BP neural network seems to be more relevant in this study.

Hong et al. (2013) compare the prognosis of the junction temperature of an IGBT with Gaussian process regression and with a wavelet neural network. In this work, GPR seems to be more relevant.

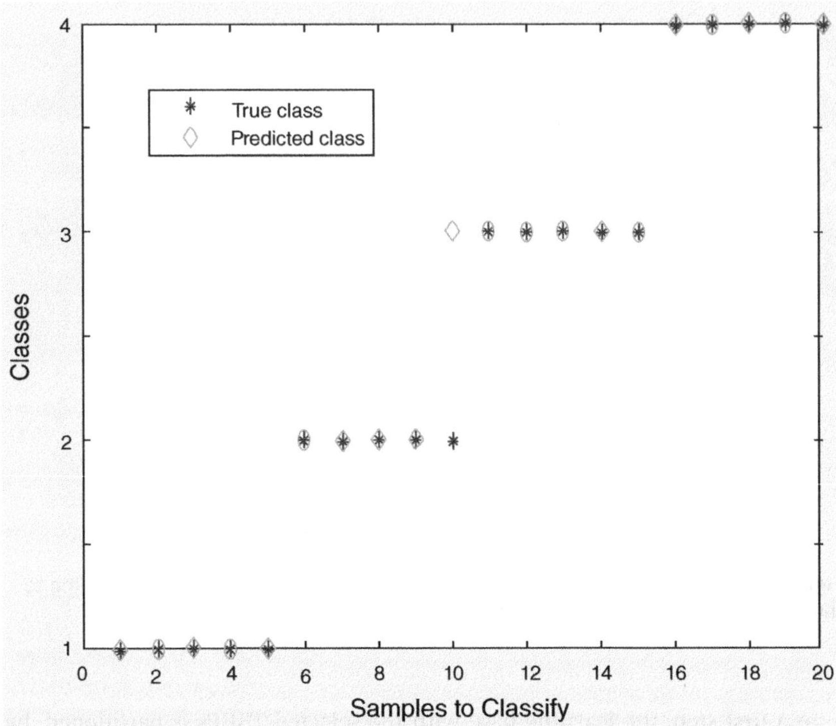

Figure 2.163 Application of a Neural Network Classifier. *Source:* Hologne (2018).

This method is currently scarcely used for the prognosis of static converter failure in the scientific literature, but it seems promising.

The paper Ahsan et al. (2016) deals with an application of the Neural Network to predict the RUL of an IGBT. Data come from the IGBT accelerated aging experimental platform developed by NASA AMES laboratory (Celaya et al. (2009)), using thermal over-stress aging condition. In this application, the collector-emitter voltage V_{CE} provides the pattern and duration of different phases in the applied voltage load (see Figure 2.165). The degradation phase durations feed the inputs of the neural network. The only output is the RUL.

A comparison has been made with an ANFIS network (see Figure 2.166).

(3) Fuzzy methods

In this case, the RUL is built from a fuzzy logic estimator.

The process is described by Alghassi et al. (2015) (see Figure 2.167b).

Ten successive steps of the TSEP V_{CE} are stored in a FIFO (First-In First-Out stack), which is divided in two parts. Each of them gives a mean value and

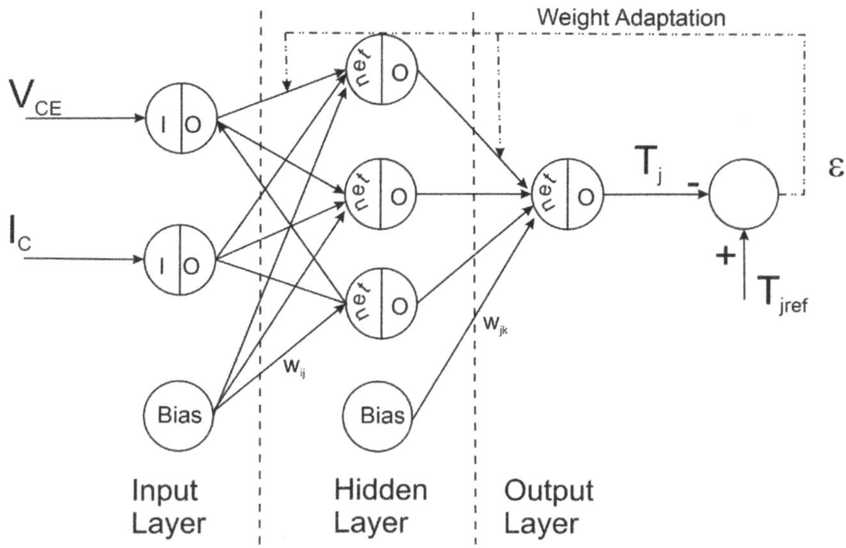

Figure 2.164 Structure of BP neural network. *Source:* Wu et al..

the difference ΔV_{CE}. The mean value of V_{CE} and ΔV_{CE} feeds the fuzzification stage.

In the fuzzification stage, nine fuzzy rules are built from linguistic variables (Low, Medium and High) and the knowledge of expert. They provide a set of outputs: very low, not quite low, low, somewhat medium, medium, indeed medium, slightly high, very high.

- If input1 is Low and input2 is Low then output = very low,
- If input1 is Low and input2 is Medium then output = not quite low,
- If input1 is Low and input2 is High then output = low,
- If input1 is Medium and input2 is Low then output = somewhat medium,
- If input1 is Medium and input2 is Medium then output = medium,
- If input1 is Medium and input2 is High then output = indeed medium,
- If input1 is High and input2 is Low then output = slightly high,
- If input1 is High and input2 is Medium then output = indeed high,
- If input1 is High and input2 is High then output = very high.

The IGBT life model is learnt from the Mamdani fuzzy inference system (FIS). The output was made fuzzy by Gaussain membership function.

In this case, the aggregation and defuzzification procedure considers the memberships function which have the maximum values. Thus, the output

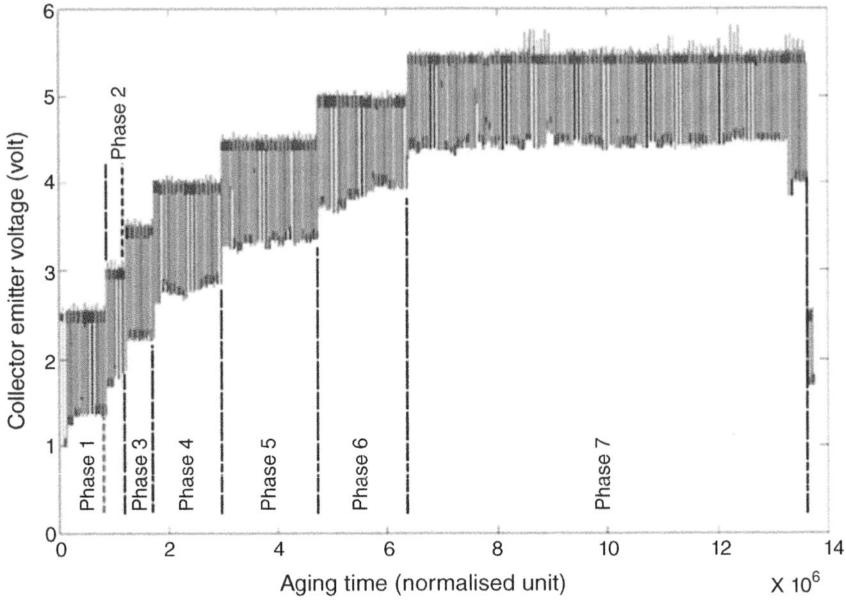

Figure 2.165 Degradation profile of collector-emitter voltage for one IGBT. *Source:* Wu et al.. Courtesy of IEEE.

fuzzy sets are combined into a single fuzzy set using the aggregation method of the FIS.

Finally, this method was tested with a simulated data set by Alghassi et al. (2015) (see Figure 2.167b). This method is accurate, but the design of the fuzzy rules requires the intervention of an expert.

Samie et al. (2015) propose an adaptation of the previous scheme, see Figure 2.168. This is a hybrid method. The RUL is computed by an adaptive neuro-fuzzy inference system (ANFIS). This structure combines neural networks and fuzzy logic principles.

This method needs a learning phase. The inputs of this estimator are the operating condition (OC) and TSEP V_{CE}. In this case, the RUL is given in per unit (%) rather in time.

2.6.7 Diagnosis and Prognosis Techniques for Capacitors

2.6.7.1 Fault Diagnosis Techniques

Catastrophic failures can result from manufacturer or user failures, or are simply the result of the natural aging process. Fault diagnosis techniques seek to avoid catastrophic failures resulting from a natural aging process, which are the

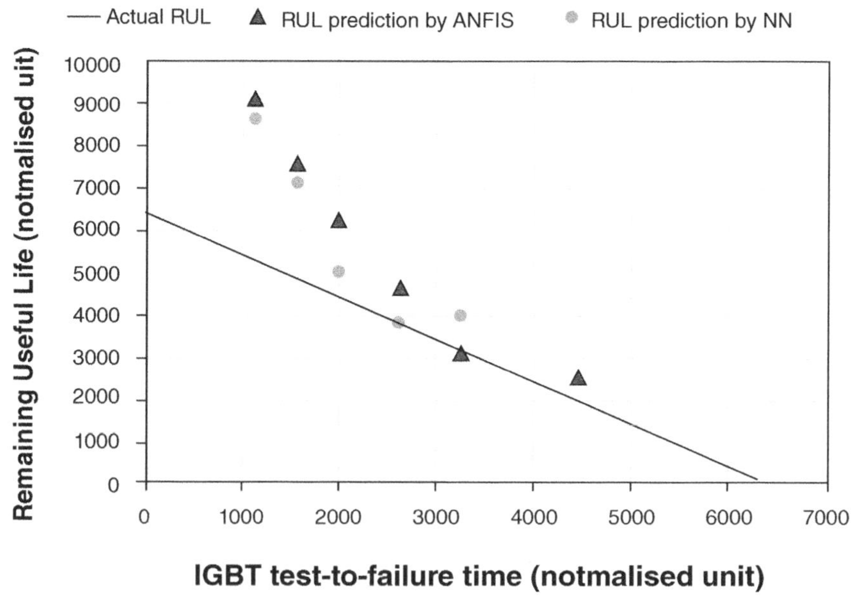

Figure 2.166 Comparing RUL prediction between ANFIS and Neural network with actual RUL. *Source:* Wu et al.. Courtesy of IEEE.

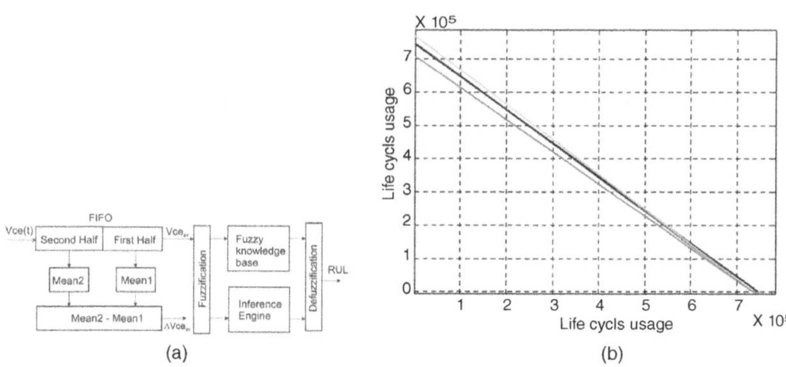

Figure 2.167 Fuzzy prognosis method. (a) Flowchart diagram of fuzzy logic. (b) RUL Simulation with the defuzzied model. *Source:* Alghassi et al. (2015) Courtesy of IEEE.

result of a parametric failure not detected in due time. For electrolytic capacitors, the main failure mechanism is the evaporation of the electrolyte solution, which is accelerated by an increase in temperature during operation and by the ripple of the current. This situation leads to a decrease in capacity (C) and an increase in equivalent series resistance (ESR) which can further increase the

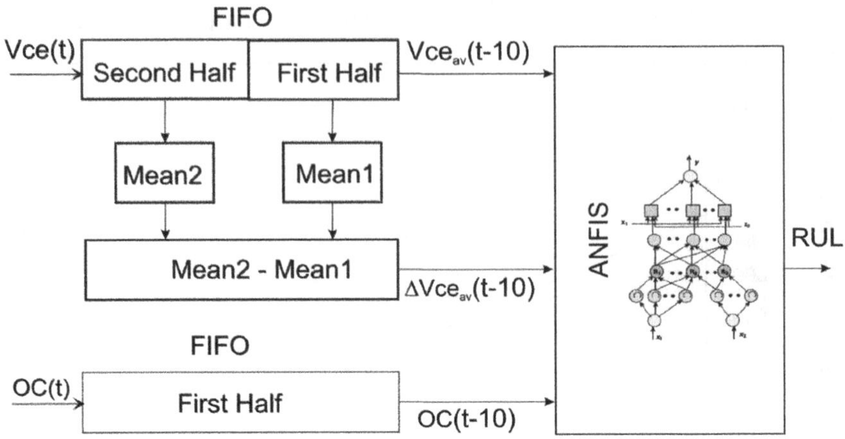

Figure 2.168 Prognosis incertitude and performance. *Source:* Samie et al. (2015).

temperature (Cardoso (2018)). In some applications, it may be necessary to use other criteria, depending on the performance of the converter. Aging of the capacitors increases the ESR of the capacitor, which in turn leads to an increase in the voltage ripple in the intermediate circuit. Increased ripple can affect the efficiency of the converter or even damage other components. For example, in photovoltaic solar inverters, increasing the ripple of the intermediate circuit voltage decreases the extracted power and can simultaneously damage the semiconductor switches due to an overvoltage (Cardoso (2018)). In these cases, it is necessary to introduce a condition that protects the converter, such as a maximum voltage ripple limit. The technique for diagnosing capacitor failures can be divided into three types:

- Off-line;
- online;
- quasi-online.

Some work is reported in the literature for monitoring the state of electrolytic capacitors in electronic power circuits, and some of them are summarized below. The main failure mechanism is the evaporation of the electrolyte solution, which is accelerated by an increase in temperature during operation and by the ripple of the current. This situation leads to a decrease in capacity (C) and an increase in equivalent series resistance (ESR) which can further increase the temperature. The standards suggest that the capacitor should be considered as failing if there is an increase of about twice the initial ESR value and a decrease of 20% in the capacity value (Abdennadher (2010) and Lahyani et al. (1998)).

This proves that the estimation of these two parameters can provide a good indicator of a potential failure of the capacitor. Since electrolytic capacitors are one of the most sensitive components of power electronics, the development of fault detection techniques is essential to assess their condition and thus avoid total failure or equipment malfunction. In recent years, several papers have been published on monitoring the condition of electrolytic capacitors, and various online techniques have been proposed for the estimation of C and ESR.

In Amaral and Cardoso (2012), an online fault detection technique is proposed to avoid structural failures in aluminum electrolytic capacitors used in the output filter of step-down DC-DC converters. A simple relationship between the slopes of the input current and the output voltage, during the conduction stage, has been developed to estimate the ESR value of the capacitor. The effect of temperature was also considered using a mathematical model of ESR as a function of temperature. It has been shown that no additional current sensor is required, which makes this technique simple and inexpensive. In Amaral and Cardoso (2009), the current and voltage waveforms of the capacitor on the primary side of an ATX power supply were analyzed using the discrete Fourier transform (DFT) at the switching frequency of the MOSFETs, to estimate the ESR value. Although DFT is a quick and easy method, there are some drawbacks that greatly reduce its practical effectiveness. Indeed, the extraction of harmonic amplitudes using DFT does not make sense if the harmonics are non-stationary. In addition, good frequency resolution requires storing long samples of data, which is not always possible due to the limitation of system memory. On the other hand, the steady state of the signals cannot be guaranteed for a long period of acquisition. To overcome these drawbacks, the use of STFT has been proposed. The principle of STFT consists in applying the DFT algorithm on a sliding window of short duration; this allows a time-amplitude representation for the harmonics sought. However, the frequency resolution always depends on the length of the sliding window. In fact, a very short window provides good time resolution with poor frequency resolution, while a long window gives high-frequency resolution with poor time frequency resolution.

In Yao et al. (2016), another technique was successfully proposed for the online estimation of ESR and C, when it is applied to a boost power factor correction converter. It has been shown that this method does not require additional current sensors; only two capacitor voltage values at particular times in a line cycle are required. The same technique was also applied for a step-down converter with continuous conduction mode (CCM) operating under different conditions (Yao et al. (2015)). This technique is well suited to DC-DC converters operating in CCM, but this mode cannot be guaranteed under all operating conditions.

In Sun et al. (2017), an interesting method for online monitoring of the conditions of IGBT modules and electrolytic capacitors with DC link, in a single-phase DC-AC converter, was proposed. This method is based on the introduction of a controlled short circuit in a bridge arm of a power converter during a test time of 2-5μs. The short circuit current resulting from the IGBT module will pass through the DC link capacitor, causing a significant change in the ripple of the capacitor current. The authors developed an expression for estimating the ESR using the short-circuit current of the IGBT and the step voltage of the DC-link capacitor.

In Farjah et al. (2017), the application of a new type of Rogowski coil sensor has been reported. It has been shown that the ESR can be successfully estimated using the ripple of the output voltage and the output voltage of the Rogowski coil. It has also been indicated that the implementation of this technique is relatively simple, even for converters already manufactured. The optimization method based on biogeography (OBB) was proposed in (Li and Low (2016)) as a new approach for the online estimation of the capacitor parameters of a continuous-continuous step-down converter. This approach is characterized by a low sampling rate and uses an averaged model. OBB can be considered as a new evolutionary algorithm which can give very good results compared to other optimization algorithms such as genetic algorithms (GA). The process of estimating the parameters requires the injection of a perturbation of the pseudo-random binary sequence which produces the ripple voltage and current. Despite the interesting results obtained by this approach, the high computational load remains its main drawback.

In Yao et al. (2017), a non-invasive method has been proposed to estimate the parameters of the output capacitor of a flyback converter. A trip circuit and an isolated amplification circuit have been designed to measure the output voltage ripples at certain instants of the switching cycle. The measured values are sent to a digital signal processor for the calculation of ESR and C. In Ahmad et al. (2017), the authors suggested monitoring the electrolytic capacitor by evaluating its impedance at twice the network frequency (100 Hz) . This technique was applied to a single-phase photovoltaic system connected to the network where the power of the network fluctuates with twice the fundamental frequency of the network. The available PV and inductor currents were used to estimate the current of the capacitor and therefore no additional current sensor was required. The online capacity estimate was also presented in Ahmad et al. (2018). This parameter has been shown to be a better indicator for assessing the health of ACS. In Soualhi et al. (2018), a new online approach was proposed to predict C and ESR of capacitors and super-capacitors. The proposed approach is based on the combination of fuzzy logic and neural networks. Another method, based on the analysis of the response of the output voltage step with respect to a

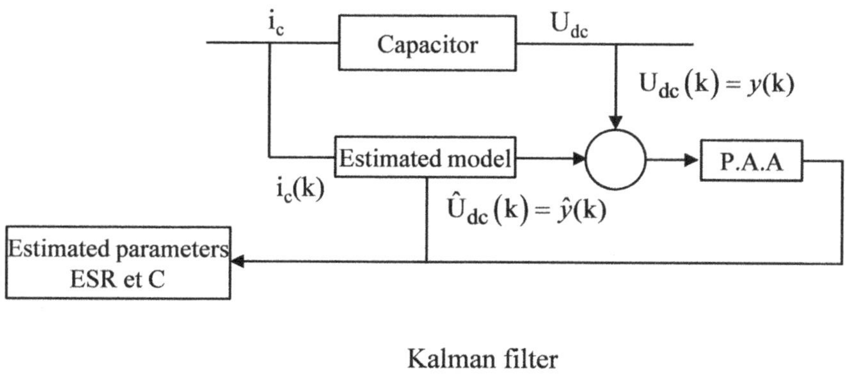

Figure 2.169 Schematic diagram of the recursive estimation by recursive least squares (RLS).

decrease in capacitance, has been proposed to detect the wear of the capacitors on the output stage of a DC-DC step-down converter (Hannonen et al. (2015)).

In Ahmad et al. (2018), an online technique for ESR estimation has been proposed to assess the health of the electrolytic capacitor in a DC system based on solar PV. This method uses PV voltage and current. A technique for online evaluation of the degradation of electrolytic capacitors has been proposed in Nakao et al. (2017).

One of the algorithms used for identifying ESR and C parameters is the parameter estimation algorithm based on the Kalman filter. Indeed, knowing the input and the output of the system namely respectively the current and the voltage across the capacitor and the order of the system (given by a validated electric model) we can use the identification of the parameters by the method of recursive least squares given by the diagram in Figure 2.169. A Parameter Adaptation Algorithms (PAA) can adapt at each iteration (k) the system parameters (ESR and C in our case). The goal is that the error between the actual output and the estimated output is as low as possible.

The problem of estimation is now formalized by the search for a parameter vector $\widehat{\theta}$, minimizing a quadratic criterion given by the following equation:

$$J_k(\widehat{\theta}) = \frac{1}{k}\sum_{\ell=0}^{k}\left[y(k) - \widehat{y}(k,\widehat{\theta})\right]^2 \tag{2.136}$$

where θ is the vector of the parameters to be identified. The electrolytic capacitor can be modeled by an equivalent electrical diagram represented by the following figure:

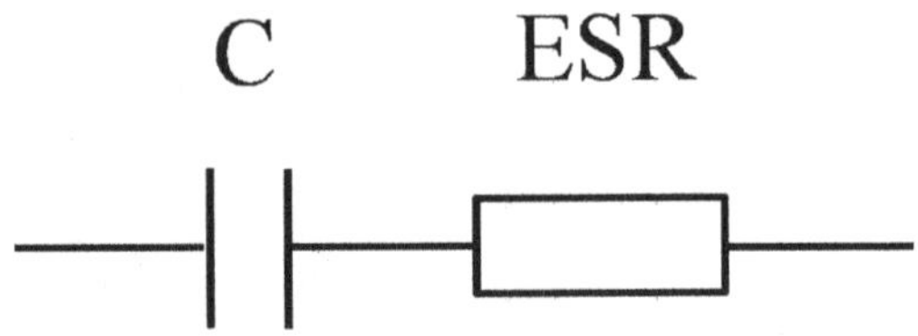

Figure 2.170 Electrical diagram of a capacitor.

The equivalent transfer function is given by:

$$H(s) = \frac{I_c(s)}{U_{dc}(s)} = \frac{ESR.C.s + 1}{C.s} \tag{2.137}$$

with s the Laplace Variable.

For power converters we can use sampled measurements. For this, we consider the transfer function corresponding to equation (2.137), using the z-transform using the zero order blocker (ZOB). This gives the following transfer function:

$$\mathrm{H}(\mathrm{z}^{-1}) = 1 - \mathrm{z}^{-1}.\mathrm{TFz}\left[\frac{\mathrm{ESR}}{\mathrm{p}} + \frac{1}{\mathrm{Cp}^2}\right] \tag{2.138}$$

$$\mathrm{H}(\mathrm{z}^{-1}) = 1 - \mathrm{z}^{-1}\left[\mathrm{ESR}\frac{1}{1-\mathrm{z}^{-1}} + \frac{1}{\mathrm{C}}\frac{\mathrm{Te.z}^{-1}}{(1-\mathrm{z})^2}\right] \tag{2.139}$$

$$\mathrm{H}(\mathrm{z}^{-1}) = \frac{\mathrm{b}_0 + \mathrm{b}_1.\mathrm{z}^{-1}}{1-\mathrm{z}^{-1}} = \frac{\mathrm{ESR} + \left(\frac{\mathrm{Te}}{\mathrm{C}} - \mathrm{ESR}\right).\mathrm{z}^{-1}}{1-\mathrm{z}^{-1}} \tag{2.140}$$

The bilinear transform (called Tustin's transform) checks the equation (2.141):

$$\mathrm{S} = \frac{2}{\mathrm{T}}\frac{1\text{-}\mathrm{z}^{-1}}{1+\mathrm{z}^{-1}} \tag{2.141}$$

By applying this equation (2.141) to equation (2.137) we obtain the following transfer function:

$$\mathrm{H}(\mathrm{z}^{-1}) = \frac{\mathrm{b}_0 + \mathrm{b}_1.\mathrm{z}^{-1}}{1-\mathrm{z}^{-1}} = \frac{\left(\mathrm{ESR} + \frac{\mathrm{Te}}{2.\mathrm{C}}\right) + \left(\frac{\mathrm{Te}}{2.\mathrm{C}} - \mathrm{ESR}\right).\mathrm{z}^{-1}}{1-\mathrm{z}^{-1}} \tag{2.142}$$

with T_e the sampling period. From equation (2.142), the vector of the parameters to be identified, θ, (in particular b_0 and b_1) can be formulated as follows:

$$\theta^{T} = \left[ESR+\frac{Te}{2.C} \quad \frac{Te}{2.C} - ESR \right] \tag{2.143}$$

The solution of this problem in the least squares sense gives:

$$\widehat{\theta} = \left[\sum_{\ell=0}^{k} \Phi(\ell)\Phi^{T}(\ell) \right]^{-1} \sum_{\ell=0}^{k} \Phi(\ell)y(\ell) \tag{2.144}$$

with $\Phi(\ell)$ the vector of the measurements which is identified as follows:

$$\Phi^{T}(\ell) = [i_c(\ell) \;\; i_c(\ell\text{-}1)] \tag{2.145}$$

with i_c the current absorbed by the capacitor. For our purpose here, it is important to express (2.145) in recursive form:

$$F(k) = \left[\sum_{\ell=0}^{k} \Phi(\ell)\,\Phi^{T}(\ell) \right]^{-1} \tag{2.146}$$

So

$$F^{-1}(k) = F^{-1}(k-1) + \Phi(k)\Phi^{T}(k) \tag{2.147}$$

Equation (2.144) can be rewritten in the form:

$$\widehat{\theta}(k) = F(k) \left[\sum_{\ell=0}^{k-1} \Phi(\ell)y(\ell) + \Phi_r(k)y(k) \right] \tag{2.148}$$

But we have

$$\sum_{\ell=0}^{k-1} \Phi(\ell)y(\ell) = F^{-1}(k-1)\,\widehat{\theta}(k-1) = \left[F^{-1}(k) - \Phi(k)\Phi^{T}(k)\right]\widehat{\theta}(k-1) \tag{2.149}$$

Substituting (2.149) in (2.148), we will have:

$$\widehat{\theta}(k) = \widehat{\theta}(k-1) - F(k)\Phi(k)\Phi^{T}(k)\widehat{\theta}(k-1) + F(k)\Phi(k)y(k) \tag{2.150}$$

equation (2.150) can be rewritten in the following form:

$$\widehat{\theta}(k) = \widehat{\theta}(k-1) + F(k)\Phi(k)\left[y(k) - \Phi^{T}(k)\widehat{\theta}(k-1)\right] \tag{2.151}$$

from equation (2.147) we will have:

$$F(k) = \left[F^{-1}(k-1) + \Phi(k)\Phi^{T}(k)\right]^{-1} \tag{2.152}$$

introduce the lemmea of matrix inversion:

$$(A+BCD)^{-1} = A^{-1} - A^{-1}B\left(C^{-1} + DA^{-1}B\right)^{-1}DA^{-1} \tag{2.153}$$

with $A = F^{-1}(k-1), B = \Phi(k), C = I, D = \Phi^T$, we obtain:

$$\begin{cases} F(k) = F(k-1) - \frac{F(k-1)\Phi(k)\Phi^T(k)F(k-1)}{1+\Phi^T(k)F(k-1)\Phi(k)} \\ \widehat{\theta}(k) = \widehat{\theta}(k-1) + \frac{F(k-1)\Phi_r(k)}{1+\Phi^T(k)F(k-1)\Phi(k)}\left[y(k) - \widehat{\theta}(k-1)\Phi(k))\right] \end{cases} \tag{2.154}$$

From numerical measurements and knowing the sampling period, it suffices to find the two parameters b_0 and b_1 of equation (2.142) to determine ESR and C as follows:

$$ESR = \frac{b_0 - b_1}{2} \tag{2.155}$$

$$C = \frac{Te}{b_0 + b_1} \tag{2.156}$$

2.6.7.2 Methods for Predicting Electrolytic Capacitor Failures

Several methods exist in the literature for detecting the aging of electrolytic capacitors (Abdennadher et al. (2010); Gupta et al. (2018); Gasperi (1996); Renwick et al. (2015); Jano and Pitica (2012); Celaya et al. (2011); Zhou et al. (2017); Yu et al. (2012); Sun et al. (2016); Naikan and Rathore (2016); Kul (2010); Pang and Bryan (2010); Venet et al. (2002); Ertl et al. (2006)). The most significant are summarized below:

Aging Factors:

- Temperature
  - A failure physics model to predict the change in ESR by evaluating the loss of electrolyte was proposed. The lifetime of capacitors at different temperatures was calculated according to Arrhenius' law and the results are compared with the lifetimes obtained via an approximation law used in industry (Gasperi (1996));
  - The method proposed by Venet et al. (2002) makes it possible to predict the lifetime of an electrolytic capacitor via an estimate of the ESR in the current state and a comparison with its value in new condition at operating temperature;
  - For a fixed ambient temperature of the capacitor, the total lifetime is given by the Arrhenius equation. The operating time of the capacitor is obtained using regression equations characterizing the temporal variations of the

capacitance C and the ESR as a function of the temperature. The remaining life is estimated by subtracting the total life from the operating time obtained by the regression equations (Abdennadher et al. (2010); Yu et al. (2012));
  - An approach to modeling degradation is proposed in Sun et al. (2016), in which the parameters of the regression equations characterizing the capacity and the ESR over time follow Arrhenius' law. The increase in current ripple in the capacitors is considered to be a failure criterion. The delay before failure can be calculated under variable temperature operating conditions.
- Temperature + Voltage
  - A method for predicting the lifetime of capacitors is introduced in Jano and Pitica (2012); Naikan and Rathore (2016). The service life obtained is more precise than the previous models because the capacitors are affected by the applied voltages.
- Temperature + Current
  - A method proposed by Pang and Bryan (2010) makes it possible to predict the lifetime of an electrolytic capacitor via an estimate of the ESR and of the current flowing through the capacitor.
- Voltage
  - A methodology for predicting the remaining useful life for electrolytic capacitors based on a Kalman filter is presented Celaya et al. (2011).
- Nominal conditions
  - Under nominal conditions, the drift of the ESR as a function of time is determined by the Arrhenius model (exponential model). A third degree regression model is proposed for the degradation of average capacity as a function of time (Kul (2010)).

The references cited above require the availability of capacitor degradation models in real time, taking into account the operating conditions (ambient temperature, voltage, ripple current, frequency, etc.) to predict the lifetime of the capacitors. Physical models and data-driven models should be combined to represent degradation of the capacitors. Regression-based degradation modeling is the most widely used of the data-based approaches. But these regression-based models require large sets of samples to increase the precision of estimation. The work carried out to date is based on accelerated aging tests to define an aging benchmark. These constitute an essential learning phase, to be carried out offline, before any online evaluation of the lifetime of the capacitors. Below is an example of a regression model based on the prediction of time series by neural networks and fuzzy logic.

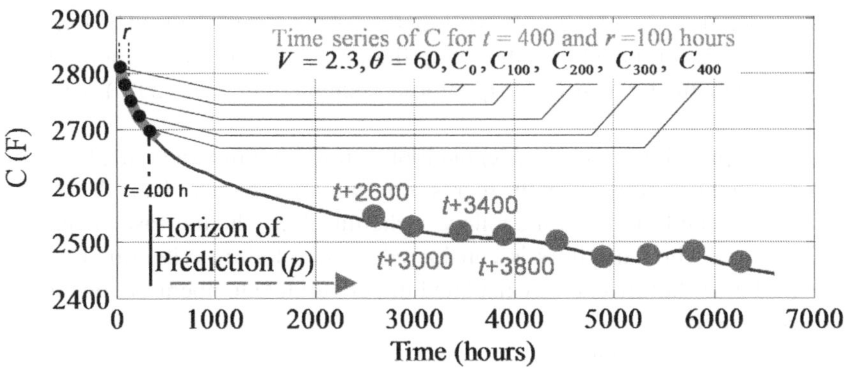

Figure 2.171 Trending prediction of C by a one-step ahead prediction strategy for accelerated calendar aging test (2.3V, 60°C) of 3000F, 2.7 V SCs.

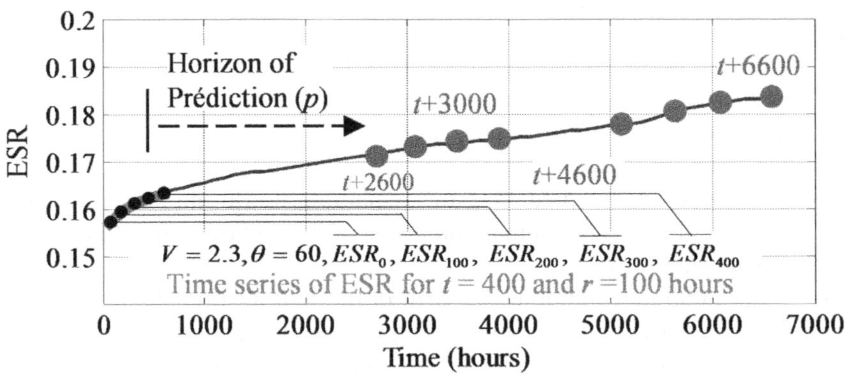

Figure 2.172 Trending prediction of ESR by a one-step ahead prediction strategy for accelerated calendar aging test (2.3V, 60°C) of 3000F, 2.7 V SCs.

Consider a time series X of parameters measured on a capacitor (SC). These measurements (also called observations) can be used to estimate an unobservable parameter noted $\hat{x}_{t+p}$ via a function $f(X)$. This function is expressed as follows:

$$\hat{x}_{t+p} = f(X) = f(V_t, \theta_t, x_{t-(n-2)r} \dots x_{t-3r}, x_{t-2r}, x_{t-r}, x_t) \tag{2.157}$$

$\hat{x}_{t+p}$ represents an estimated parameter x until an horizon of prediction noted p. $\hat{x}_t$ represents a measured parameter at time t. $\hat{x}_{t-r}$ represents a measured parameter at time $t - r$, where r is the interval between two measurements. $x_{t-(n-2)r}$ represents a measured parameter at time $t - (n - 2)r$, where where

$(n-2)$ represents the number of previous measurements. V_t, θ_t represent additional parameters measured at time t. For the SC, the additional parameters are the supply voltage V_t and the temperature θ_t. In this case, the time series of the ESR, X_{ESR}, is replaced in (2.157) to estimate the value of the ESR until the horizon p:

$$\begin{aligned} E\widehat{S}R_{t+p} &= f(X_{ESR}) \\ &= f(V_t, \theta_t,\ ESR_{t-(n-2)r}, \dots,\ ESR_{t-3r},\ ESR_{t-2r},\ ESR_{t-r},\ ESR_t) \end{aligned} \tag{2.158}$$

For the capacitance, the times series X_c is replaced in (2.157) to estimate the value of the C until the horizon p:

$$\widehat{C}_{t+p} = f(X_C) \ = f(V_t, \theta_t,\ C_{t-(n-2)r}, \dots,\ C_{t-3r},\ C_{t-2r},\ C_{t-r}, C_t) \tag{2.159}$$

When r=1, this is interpreted by a one-step ahead prediction (one interval), and when $(n-2) = 3$, this implies the use of three previous measures in addition to the current measurement to carry out an estimation of $E\widehat{S}R_{t+p}$ and $\widehat{C}_{t+p}$.

This is illustrated in Figures 2.171 and 2.172 for r=100, $(n-2)$=4 and p=2600, 3000, 3400 hours...

For this application, the predictor is the Neo-fuzzy neuron (NFN), which is considered as a black box used to represent the function $f(X)$. This model takes into account the relationship between the input in order to estimate a desired output after a training phase. The NFN is a nonlinear multi-input single-output model. Its most important advantages are the high rate of training, the computational simplicity and its characterization by a set of fuzzy if-then rules. The structure of a NFN with n-inputs and one-output is shown in Figure 2.173.

$\widehat{x}_{t+p}$ represents the desired output and $\bar{x}_{ir}$ are the normalized i^{th} inputs of the model. In this case, two NFN provide the estimation of the $E\widehat{S}R_{t+p}$ and $\widehat{C}_{t+p}$ until the horizon p. The output of the NFN is expressed as follows:

$$\widehat{x}_{t+p} = \sum_{i=0}^{n} f_i(\bar{x}_{ir}) \tag{2.160}$$

with $\bar{x}_{ir}$ a normalized observation situated in the ith position of the times series $\{\bar{x}_0, \bar{x}_r, \dots, \bar{x}_{(n-2)r}\}$. For the ESR, this times series is equivalent to $\{\overline{ESR}_t, \overline{ESR}_{t-r}, \dots, \overline{ESR}_{t-(n-2)r}\}$. The same remark is given for the capacitance with $\{\bar{x}_0, \bar{x}_r, \dots, \bar{x}_{(n-2)r}\} = \{\bar{C}_t, \bar{C}_{t-r}, \dots, \bar{C}_{t-(n-2)r}\}$. The last and before the last position of the times series represent respectively the supply voltage and the temperature of the capacitor $\bar{x}_{nr} = \bar{V}_t$, $\bar{x}_{(n-1)r} = \bar{\theta}_t$.

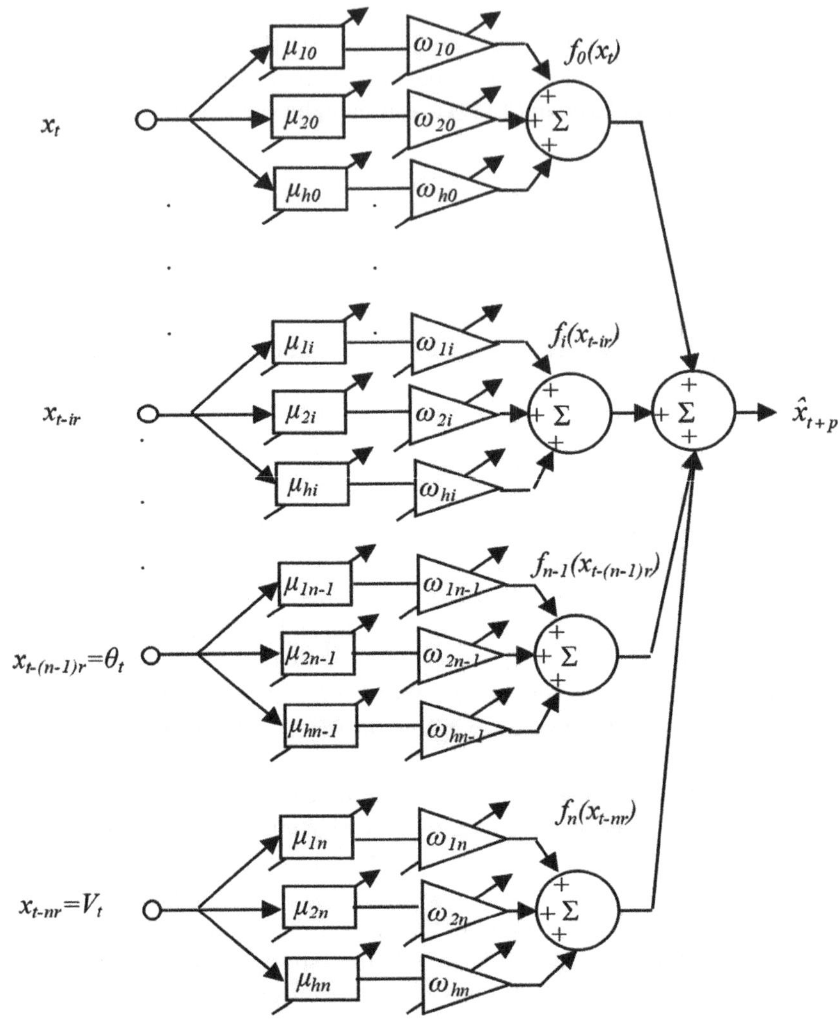

Figure 2.173 NFN architecture for the prediction of the ESR and C.

The function $f_i(\bar{x}_{ir})$ is a nonlinear synapse which performs the transformation of the ith input in the following form:

$$f_i(\bar{x}_{ir}) = \sum_{j=1}^{h} \omega_{ji} \cdot \mu_{ji}(\bar{x}_{ir}) \tag{2.161}$$

where ω_{ji} are the interconnecting weights between the ith input $\bar{x}_{ir}$ and the jth membership value $\mu_{ji}(\bar{x}_{ir})$. The membership values are obtained from

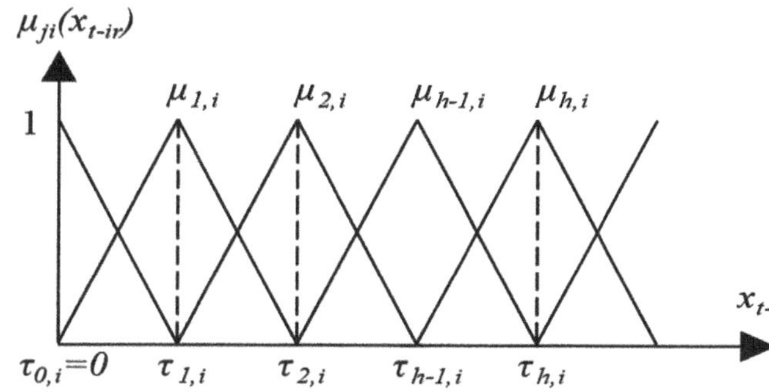

Figure 2.174 Membership value for an input according to triangular curves.

triangular functions spaced equivalently according the number h of desired membership functions (as shown in Figure 2.174). $\tau_{0,i}, \tau_{1,i}, ..., \tau_{h,i}$ are the intervals of the triangular functions with $\tau_{h,i} = 1$.

The membership value $\mu_{ji}(\bar{x}_{ir})$ is obtained according to the position of the ith input in the intervals $[\tau_{j-1,i}, \tau_{ji}]$ and $[\tau_{ji}, \tau_{j+1,i}]$. This function is expressed as follows:

$$\mu_{ji}(\bar{x}_{ir}) = \begin{cases} \frac{\bar{x}_{ir}-\tau_{j-1,i}}{\tau_{ji}-\tau_{j-1,i}} & , \bar{x}_{ir} \in [\tau_{j-1,i}, \tau_{ji}] \\ \frac{\tau_{j+1,i}-\bar{x}_{ir}}{\tau_{j+1,i}-\tau_{ji}} & , \bar{x}_{ir} \in [\tau_{ji}, \tau_{j+1,i}] \\ 0 & \text{otherwise} \end{cases} \tag{2.162}$$

Based on Figure 2.174, the output value of the nonlinear synapse $f_i(\bar{x}_{ir})$ depends only on two neighboring membership functions $(j, j+1)$. It is expressed as follows:

$$f_i(\bar{x}_{ir}) = \omega_{ji} \cdot \mu_{ji}(\bar{x}_{ir}) + \omega_{j+1,i} \cdot \mu_{j+1,i}(\bar{x}_{ir}) \tag{2.163}$$

The weight ω_{ji} is updated by the incremental updating (Stepwise Training) learning algorithm based on the following formula:

$$\omega_{ji} = -\rho \cdot (x_{t+p} - \hat{x}_{t+p}) \cdot \mu_{ji}(\bar{x}_{ir}) \tag{2.164}$$

where ρ is the learning rate and x_{t+p} is the measured output. Finally, the estimation of $\hat{x}_{t+p}$ is given by:

$$\hat{x}_{t+p} = f_0(\bar{x}_0) + f_i(\bar{x}_{ir}) + ... + f_n(\bar{x}_{nr}) \tag{2.165}$$

where t denotes current time instant, $\hat{x}_{t+p}$ is the predicted value at the horizon $t + p$. The aim of such technique relies on the development of an adaptive network in order to approximate unknown functions through training data and identify the precise values of synaptic weights.

The prediction of the *ESR* and *C* is divided into three steps: data base creation, learning and validation.

- Step1. Database creation: this step is used to acquire and to create a database composed by the *ESR* and *C* of different SCs working under different temperatures θ_t and supply voltages V_t. This step is done by an impedance spectrometer. The database of the ESR and C is divided into two parts: learning database and validation database.
- Step 2. Learning: The training set represents the aging evolution of the ESR and C under a temperature θ_t and a supply voltage Vt. This database is used to update the interconnecting weights of the model.
- Step 3. Validation: The validation step implements the second database in the NFN to predict the aging of the ESR and C. Note that the second database corresponds to the *ESR* and *C* of SCs working under a different supply voltage and/or a different temperature.

Bibliography

The Case Western Reserve University Bearing Data Center website/bearing data center test seeded fault test data. http://csegroups.case.edu/bearingdatacenter/pages/download-data-file.

Experimental studies of aging in electrolytic capacitors, Porltand, OR, 10/2010 2010. Prognostics and Health Management Society.

ABB. Application Note 5SYA 2043-0: Load-cycling capability of HiPak IGBT modules, 2004. URL https://library.e.abb.com/public/1f4fb71e0af3356883257c8d00443ca1/Load-cycling%20capability%20of%20HiPak_5SYA%202043-04.pdf.

ABB. Data Sheet, Doc. No. 5SYA 1447-00 2016-09 5SNG 0150Q170300 62Pak phase leg IGBT Module, 2016. URL https://search.abb.com/library/Download.aspx?DocumentID=5SYA%201447&LanguageCode=en&DocumentPartId=&Action=Launch.

K. Abdennadher, P. Venet, G. Rojat, J. Rétif, and C. Rosset. A real-time predictive-maintenance system of aluminum electrolytic capacitors used in uninterrupted power supplies. *IEEE Transactions on Industry Applications*, 46 (4):1644–1652, 2010.

Mohamed Karim Abdennadher. *Study and elaboration of a monitoring and predictive maintenance system for capacitors and batteries used in Uninterruptible Power Supplies (UPS).* Theses, Université Claude Bernard - Lyon I, June 2010. URL https://tel.archives-ouvertes.fr/tel-00532642.

M. W. Ahmad, N. Agarwal, P. N. Kumar, and S. Anand. Low-frequency impedance monitoring and corresponding failure criteria for aluminum electrolytic capacitors. *IEEE Transactions on Industrial Electronics*, 64(7): 5657–5666, 2017.

M. W. Ahmad, P. N. Kumar, A. Arya, and S. Anand. Noninvasive technique for dc-link capacitance estimation in single-phase inverters. *IEEE Transactions on Power Electronics*, 33(5):3693–3696, 2018.

M. Ahsan, S. Stoyanov, and C. Bailey. Data driven prognostics for predicting remaining useful life of igbt. In *2016 39th International Spring Seminar on Electronics Technology (ISSE)*, pages 273–278, 2016.

O. Alavi, M. Abdollah, and A. Hooshmand Viki. Assessment of thermal network models for estimating igbt junction temperature of a buck converter. In *2017 8th Power Electronics, Drive Systems Technologies Conference (PEDSTC)*, pages 102–107, 2017.

A. Albanna, A. Malburg, M. Anwar, A. Guta, and N. Tiwari. Performance comparison and device analysis between si igbt and sic mosfet. In *2016 IEEE Transportation Electrification Conference and Expo (ITEC)*, pages 1–6, June 2016. doi: 10.1109/ITEC.2016.7520242.

P.L. Alger. *The Nature of Induction Machines.* Gordon and Breach.

A. Alghassi, P. Soulatiantork, M. Samie, S. Perinpanayagam, and M. Faifer. Reliability enhance powertrain using fuzzy knowledge base prognostics model. In *2015 17th European Conference on Power Electronics and Applications (EPE'15 ECCE-Europe)*, pages 1–9, 2015.

M. Z. Ali, M. N. S. K. Shabbir, X. Liang, Y. Zhang, and T. Hu. Machine learning-based fault diagnosis for single- and multi-faults in induction motors using measured stator currents and vibration signals. *IEEE Transactions on Industry Applications*, 55(3):2378–2391, 2019. doi: 10.1109/TIA.2019.2895797.

S. H. Ali, M. Heydarzadeh, S. Dusmez, X. Li, A. S. Kamath, and B. Akin. Lifetime estimation of discrete IGBT devices based on gaussian process. *IEEE Transactions on Industry Applications*, 54(1):395–403, 2018.

A. M. R. Amaral and A. J. M. Cardoso. State condition estimation of aluminum electrolytic capacitors used on the primary side of atx power supplies. In *2009 35th Annual Conference of IEEE Industrial Electronics*, pages 442–447, 2009.

Acacio Amaral and A.J.M. Cardoso. online fault detection of aluminum electrolytic capacitors, in step-down dc-dc converters, using input current and output voltage ripple. *IET Power Electronics*, pages 315–322, 03 2012. doi: 10.1049/iet-pel.2011.0163.

Anastasios Amoiridis, Anup Anurag, Pramod Ghimire, Stig Munk-Nielsen, and Nick Baker. Vce-based chip temperature estimation methods for high power igbt modules during power cycling – a comparison. pages 1–9, 2015. doi: 10.1109/EPE.2015.7309449. URL https://sfx.aub.aau.dk/sfxaub?sid=pureportal&doi=10.1109/EPE.2015.7309449.

J. M. Anderson, R. W. Cox, and J. Noppakunkajorn. An online fault diagnosis method for power electronic drives. In *2011 IEEE Electric Ship Technologies Symposium*, pages 492–497, 2011.

P. Arumugam, T. Hamiti, and C. Gerada. Turn–turn short circuit fault management in permanent magnet machines. *IET Electric Power Applications*, 9(9):634–641, 2015.

Y. Avenas, L. Dupont, and Z. Khatir. Temperature measurement of power semiconductor devices by thermo-sensitive electrical parameters—a review. *IEEE Transactions on Power Electronics*, 27(6):3081–3092, June 2012. ISSN 1941-0107. doi: 10.1109/TPEL.2011.2178433.

A. S. Babel and E. G. Strangas. Condition-based monitoring and prognostic health management of electric machine stator winding insulation. In *2014 International Conference on Electrical Machines (ICEM)*, pages 1855–1861, 2014. doi: 10.1109/ICELMACH.2014.6960436.

R. Bayerer, T. Herrmann, T. Licht, J. Lutz, and M. Feller. Model for power cycling lifetime of igbt modules - various factors influencing lifetime. In *5th International Conference on Integrated Power Electronics Systems*, pages 1–6, 2008.

A. Bendiabdellah and B. D. E. Cherif. A proposed voltage technique for inverter open fault-circuit detection based on svm strategy. In *2017 IEEE 12th International Conference on Power Electronics and Drive Systems (PEDS)*, pages 8–13, 2017.

M. Bermudez, I. Gonzalez-Prieto, F. Barrero, H. Guzman, X. Kestelyn, and M. J. Duran. An experimental assessment of open-phase fault-tolerant virtual-vector-based direct torque control in five-phase induction motor drives. *IEEE Transactions on Power Electronics*, 33(3):2774–2784, 2018.

Giorgio Bertotti. *Hysteresis in Magnetism*. Academic Press, 1988.

A. Bhure, E. G. Strangas, J. Agapiou, and R. M. Lesperance. Partial discharge detection in medium voltage stators using an antenna. In *2017 IEEE 11th International Symposium on Diagnostics for Electrical Machines, Power Electronics and Drives (SDEMPED)*, pages 480–485, 2017. doi: 10.1109/DEMPED.2017.8062398.

P. Bidan, T. Lebey, and C. Neacsu. Development of a new off-line test procedure for low voltage rotating machines fed by adjustable speed drives (ASD). *IEEE Transactions on Dielectrics and Electrical Insulation*, 10(1):168–175, 2003. doi: 10.1109/TDEI.2003.1176584.

T. Boumegoura, J. Marques, Hamed Yahoui, Guy Clerc, and Hassan Hammouri. Rotor induction machine failure: Analysis and diagnosis. *European Transactions on Electrical Power*, 14(2):71–84, April 2004. doi: 10.1002/etep.6. URL https://hal.archives-ouvertes.fr/hal-00140572. Complete paper: http://www3.interscience.wiley.com/cgi-bin/fulltext/108061601/PDFSTART.

C. Bowen and T. Wei. Switch open-circuit faults diagnosis of inverter based on wavelet and support vector machine. In *2019 14th IEEE International Conference on Electronic Measurement Instruments (ICEMI)*, pages 1178–1184, 2019.

G. Bower, C. Rogan, J. Kozlowski, and M. Zugger. Sic power electronics packaging prognostics. In *2008 IEEE Aerospace Conference*, pages 1–12, March 2008. doi: 10.1109/AERO.2008.4526605.

R. Breuneval, G. Clerc, B. Nahid-Mobarakeh, and B. Mansouri. Hybrid diagnosis of intern-turn short-circuit for aircraft applications using svm-mbf. In *2017 IEEE International Conference on Fuzzy Systems (FUZZ-IEEE)*, pages 1–6, 2017. doi: 10.1109/FUZZ-IEEE.2017.8015588.

Romain Breuneval. *Surveillance de l'état de santé des actionneurs électromécaniques: application à l'aéronautique*. PhD thesis, Université de Lyon, Université Claude Bernard Lyon 1, Villeurbanne, 2017.

F. Briz, M. W. Degner, A. B. Diez, and J. M. Guerrero. Online diagnostics in inverter-fed induction machines using high-frequency signal injection. *IEEE Transactions on Industry Applications*, 40(4):1153–1161, 2004.

D. Brown, P. Kalgren, M. Roemer, and T. Dabney. Electronic prognostics - a case study using switched-mode power supplies (smps). In *2006 IEEE Autotestcon*, pages 636–642, 2006.

G. Capolino, R. Romary, H. Hénao, and R. Pusca. State of the art on stray flux analysis in faulted electrical machines. In *2019 IEEE Workshop on Electrical Machines Design, Control and Diagnosis (WEMDCD)*, volume 1, pages 181–187, 2019. doi: 10.1109/WEMDCD.2019.8887805.

A.J.M. Cardoso. *Diagnosis and Fault Tolerance of Electrical Machines, Power Electronics and Drives*. 10 2018. ISBN 978-1-78561-531-3.

A. Ceban, R. Pusca, and R. Romary. Study of rotor faults in induction motors using external magnetic field analysis. *IEEE Transactions on Industrial Electronics*, 59(5):2082–2093, 2012.

J. Celaya, P. F. Wysocki, and K. Goebel. IGBT Accelerated Aging Data Set, 2009. URL https://ti.arc.nasa.gov/tech/dash/groups/pcoe/prognostic-data-repository/.

Jose Celaya, Chetan Kulkarni, Gautam Biswas, and Kai Goebel. A model-based prognostics methodology for electrolytic capacitorsbased on electrical overstress accelerated aging. 09 2011.

Jose Celaya, Abhinav Saxena, and Kai Goebel. Uncertainty representation and interpretation in model-based prognostics algorithms based on kalman filter estimation. In *Annual Conference of the Prognostics and Health Management Society 2012*, 09 2012.

C. Chakraborty and V. Verma. Speed and current sensor fault detection and isolation technique for induction motor drive using axes transformation. *IEEE*

Transactions on Industrial Electronics, 62(3):1943–1954, 2015. doi: 10.1109/TIE.2014.2345337.

C. Chakraborty, A. V. R. Teja, S. Maiti, and Y. Hori. A new v×i based adaptive speed sensorless four quadrant vector controlled induction motor drive. In *The 2010 International Power Electronics Conference - ECCE ASIA -*, pages 3041–3048, 2010. doi: 10.1109/IPEC.2010.5543736.

M. Chapman, N. Frost, and R. Bruetsch. Insulation systems for rotating low-voltage machines. In *Conference Record of the 2008 IEEE International Symposium on Electrical Insulation*, pages 257–260, 2008. doi: 10.1109/ELINSL.2008.4570323.

F. Charfi, F. Sellami, and K. Al-Haddad. Fault diagnostic in power system using wavelet transforms and neural networks. In *2006 IEEE International Symposium on Industrial Electronics*, volume 2, pages 1143–1148, 2006.

C. Chen, V. Pickert, M. Al-Greer, C. Jia, and C. Ng. Localization and detection of bond wire faults in multichip igbt power modules. *IEEE Transactions on Power Electronics*, 35(8):7804–7815, 2020a.

W. Chen, L. Zhang, K. Pattipati, A. M. Bazzi, S. Joshi, and E. M. Dede. Data-driven approach for fault prognosis of sic mosfets. *IEEE Transactions on Power Electronics*, 35(4):4048–4062, 2020b.

X. Chen, W. Xu, Y. Liu, and M. R. Islam. Bearing corrosion failure diagnosis of doubly fed induction generator in wind turbines based on stator current analysis. *IEEE Transactions on Industrial Electronics*, 67(5):3419–3430, 2020c.

Sung-Dong Cho and Kyung-Wook Paik. Study on the amorphous ta2o5 thin film capacitors deposited by dc magnetron reactive sputtering for multichip module applications. *Materials Science and Engineering: B*, 67(3):108 – 112, 1999. ISSN 0921-5107.

U. Choi, F. Blaabjerg, and K. Lee. Study and Handling Methods of Power IGBT Module Failures in Power Electronic Converter Systems. *IEEE Transactions on Power Electronics*, 30(5):2517–2533, May 2015. ISSN 1941-0107. doi: 10.1109/TPEL.2014.2373390.

M. Ciappa, F. Carbognani, P. Cova, and W. Fichtner. A novel thermomechanics-based lifetime prediction model for cycle fatigue failure mechanisms in power semiconductors. *Microelectronics Reliability*, 42(9):1653 – 1658, 2002. ISSN 0026-2714. doi: https://doi-org-s.bibliopam.ec-lyon.fr/10.1016/S0026-2714(02)00206-8. URL https://www-sciencedirect-com.bibliopam.ec-lyon.fr/science/article/pii/S0026271402002068.

Mauro Ciappa. Selected failure mechanisms of modern power modules. *Microelectronics Reliability*, 42(4):653 – 667, 2002. ISSN 0026-2714.

doi: https://doi-org-s.bibliopam.ec-lyon.fr/10.1016/S0026-2714(02)00042-2. URL https://www-sciencedirect-com.bibliopam.ec-lyon.fr/science/article/pii/S0026271402000422.

J. G. Cintron-Rivera, S. N. Foster, and E. G. Strangas. Mitigation of turn-to-turn faults in fault tolerant permanent magnet synchronous motors. *IEEE Transactions on Energy Conversion*, 30(2):465–475, 2015.

V. Climente-Alarcon, J. A. Antonino-Daviu, A. Haavisto, and A. Arkkio. Diagnosis of induction motors under varying speed operation by principal slot harmonic tracking. *IEEE Transactions on Industry Applications*, 51(5):3591–3599, 2015a. doi: 10.1109/TIA.2015.2413963.

V. Climente-Alarcon, J. A. Antonino-Daviu, E. G. Strangas, and M. Riera-Guasp. Rotor-bar breakage mechanism and prognosis in an induction motor. *IEEE Transactions on Industrial Electronics*, 62(3):1814–1825, 2015b.

Jr Coffin, L F. A study of the effects of cyclic thermal stresses on a ductile metal. *ASME*, 76(6):931–950, 1954.

M. Correvon. Les semiconducteurs de puissance Troisième partie: L'IGBT, 2020a. URL https://www.chireux.fr/mp/cours/electronique_puissance/Chap8_IGBT.pdf.

M. Correvon. Les semiconducteurs de puissance Deuxième partie: Le MOSFET, 2020b. URL https://www.chireux.fr/mp/cours/electronique_puissance/Chap7_Mosfet.pdf.

S. M. A. Cruz, H. A. Toliyat, and A. J. M. Cardoso. DSP implementation of the multiple reference frames theory for the diagnosis of stator faults in a DTC induction motor drive. *IEEE Transactions on Energy Conversion*, 20(2):329–335, 2005.

R. L. de Araujo Ribeiro, C. B. Jacobina, E. R. C. da Silva, and A. M. N. Lima. Fault detection of open-switch damage in voltage-fed pwm motor drive systems. *IEEE Transactions on Power Electronics*, 18(2):587–593, 2003a.

R. L. de Araujo Ribeiro, C. B. Jacobina, E. R. C. da Silva, and A. M. N. Lima. Fault detection of open-switch damage in voltage-fed pwm motor drive systems. *IEEE Transactions on Power Electronics*, 18(2):587–593, 2003b.

C. Delpha, D. Diallo, M. E. H. Benbouzid, and C. Marchand. Pattern recognition for diagnosis of inverter fed induction machine drive: A step toward reliability. In *2007 IET Colloquium on Reliability in Electromagnetic Systems*, pages 1–5, 2007.

S. Diao, Z. Makni, J. Bisson, D. Diallo, and C. Marchand. Sensor fault diagnosis for improving the availability of electrical drives. In *IECON 2013 - 39th Annual Conference of the IEEE Industrial Electronics Society*, pages 3108–3113, 2013. doi: 10.1109/IECON.2013.6699625.

M. Drif and A. J. M. Cardoso. Stator fault diagnostics in squirrel cage three-phase induction motor drives using the instantaneous active and reactive power signature analyses. *IEEE Transactions on Industrial Informatics*, 10(2): 1348–1360, 2014.

C. Durand, M. Klingler, D. Coutellier, and H. Naceur. Confrontation of failure mechanisms observed during active power cycling tests with finite element analyze performed on a mosfet power module. In *2013 14th International Conference on Thermal, Mechanical and Multi-Physics Simulation and Experiments in Microelectronics and Microsystems (EuroSimE)*, pages 1–4, April 2013. doi: 10.1109/EuroSimE.2013.6529900.

H. T. Eickhoff, R. Seebacher, A. Muetze, and E. G. Strangas. Post-fault operation strategy for single switch open-circuit faults in electric drives. *IEEE Transactions on Industry Applications*, 54(3):2381–2391, 2018.

H. H. Eldeeb, A. Berzoy, and O. Mohammed. Comprehensive investigation of harmonic signatures resulting from inter- turn short-circuit faults in DTC driven IM operating in harsh environments. In *2018 XIII International Conference on Electrical Machines (ICEM)*, pages 2579–2585, 2018.

M. A. Eleffendi and C. M. Johnson. Application of kalman filter to estimate junction temperature in igbt power modules. *IEEE Transactions on Power Electronics*, 31(2):1576–1587, 2016.

Mohd Amir Eleffendi. *In-service estimation of state of health of power modules*. PhD thesis, University of Nottingham., July 2016. URL https://www.researchgate.net/publication/305348999_In-service_estimation_of_state_of_health_of_power_modules.

J. Ertl, K. Edelmoser, F. Zach, and J.W. Kolar. A novel method for online monitoring and managing of electrolytic capacitors of dc voltage link pwm converters. In *Official proceedings of the International Conference Power Electronics, Intelligent Motion, Power Quality*, pages 295 – 302. ZM Communications GmbH, 2006.

J. O. Estima and A. J. Marques Cardoso. A new approach for real-time multiple open-circuit fault diagnosis in voltage-source inverters. *IEEE Transactions on Industry Applications*, 47(6):2487–2494, 2011.

Fairchild. AN-9010 MOSFET Basics, 2000. URL https://www.ele.uri.edu/courses/ele447/Slides/AN-9010.pdf.

X. Fang, S. Lin, X. Huang, F. Lin, Z. Yang, and S. Igarashi. A review of data-driven prognostic for igbt remaining useful life. *Chinese Journal of Electrical Engineering*, 4(3):73–79, 2018.

E. Farjah, H. Givi, and T. Ghanbari. Application of an efficient rogowski coil sensor for switch fault diagnosis and capacitor esr monitoring in nonisolated single-switch dc–dc converters. *IEEE Transactions on Power Electronics*, 32(2): 1442–1456, 2017.

Allianz France. Présentation des avaries. Machines électriques tournantes, cahier de prévention, CP2, 1988.

N. M. A. Freire, J. O. Estima, and A. J. M. Cardoso. A voltage-based approach without extra hardware for open-circuit fault diagnosis in closed-loop PWM AC regenerative drives. *IEEE Transactions on Industrial Electronics*, 61(9): 4960–4970, 2014.

L. Frosini, M. Minervini, L. Ciceri, and A. Albini. Multiple faults detection in low voltage inverter-fed induction motors. In *2019 IEEE 12th International Symposium on Diagnostics for Electrical Machines, Power Electronics and Drives (SDEMPED)*, pages 323–329, 2019.

F. W. Fuchs. Some diagnosis methods for voltage source inverters in variable speed drives with induction machines - a survey. In *IECON'03. 29th Annual Conference of the IEEE Industrial Electronics Society (IEEE Cat. No.03CH37468)*, volume 2, pages 1378–1385 Vol.2, 2003.

T. K. Gachovska, B. Tian, J. L. Hudgins, W. Qiao, and J. F. Donlon. A real-time thermal model for monitoring of power semiconductor devices. *IEEE Transactions on Industry Applications*, 51(4):3361–3367, 2015.

M. L. Gasperi. Life prediction model for aluminum electrolytic capacitors. In *IAS '96. Conference Record of the 1996 IEEE Industry Applications Conference Thirty-First IAS Annual Meeting*, volume 3, pages 1347–1351 vol.3, 1996.

P. Ghimire, K. B. Pedersen, A. R. d. Vega, B. Rannestad, S. Munk-Nielsen, and P. B. Thogersen. A real time measurement of junction temperature variation in high power igbt modules for wind power converter application. In *CIPS 2014; 8th International Conference on Integrated Power Electronics Systems*, pages 1–6, 2014.

E. Ghosh, A. Mollaeian, S. Kim, J. Tjong, and N. C. Kar. Intelligent flux predictive control through online stator inter-turn fault detection for fault-tolerant control of induction motor. In *2017 IEEE International Conference on Industrial Technology (ICIT)*, pages 306–311, 2017.

R. Ghosh, P. Seri, R. E. Hebner, and G. C. Montanari. Noise rejection and detection of partial discharges under repetitive impulse supply voltage. *IEEE Transactions on Industrial Electronics*, 67(5):4144–4151, 2020.

Kai Goebel, Jose Celaya, Shankar Sankararaman, Indranil Roychoudhury, Matthew Daigle, and Abhinav Saxena. *Prognostics: The Science of Making Predictions*. Createspace Independent Publishing Platform, 04 2017. ISBN ISBN-10: 1539074838 ISBN-13: 978-1539074830.

T. Goktas, M. Zafarani, and B. Akin. Discernment of broken magnet and static eccentricity faults in permanent magnet synchronous motors. *IEEE Transactions on Energy Conversion*, 31(2):578–587, 2016.

Guy Grellet and Guy Clerc. *Actionneurs électriques*. éditions Eyrolles, January 1997. URL https://hal.archives-ouvertes.fr/hal-01926904.

A. S. Guedes and S. M. Silva. Insulation failures prognosis in electric machines: preventive detection and time to failure forecast. *IET Electric Power Applications*, 14(6):1108–1117, 2020.

M. Gunther, K. Wolter, M. Rittner, and W. Nuchter. Failure mechanisms of direct copper bonding substrates (dcb). In *2006 1st Electronic Systemintegration Technology Conference*, volume 2, pages 714–718, Sep. 2006. doi: 10.1109/ESTC.2006.280090.

Anunay Gupta, Om Prakash Yadav, Douglas DeVoto, and Joshua Major. A review of degradation behavior and modeling of capacitors. page V001T04A004, 08 2018. doi: 10.1115/IPACK2018-8262.

K. N. Gyftakis and A. J. Marques Cardoso. Reliable detection of stator inter-turn faults of very low severity level in induction motors. *IEEE Transactions on Industrial Electronics*, pages 1–1, 2020.

R. Z. Haddad, C. A. Lopez, S. N. Foster, and E. G. Strangas. A voltage-based approach for fault detection and separation in Permanent Magnet Synchronous Machines. *IEEE Transactions on Industry Applications*, 53(6):5305–5314, 2017.

N. Haje Obeid, A. Battiston, T. Boileau, and B. Nahid-Mobarakeh. Early intermittent interturn fault detection and localization for a Permanent Magnet Synchronous Motor of electrical vehicles using wavelet transform. *IEEE Transactions on Transportation Electrification*, 3(3):694–702, 2017.

T. J. Å. Hammarström. Partial discharge characteristics within motor insulation exposed to multi-level PWM waveforms. *IEEE Transactions on Dielectrics and Electrical Insulation*, 25(2):559–567, 2018.

A. Hanif, Y. Yu, D. DeVoto, and F. Khan. A comprehensive review toward the state-of-the-art in failure and lifetime predictions of power electronic devices. *IEEE Transactions on Power Electronics*, 34(5):4729–4746, 2019.

J. Hannonen, J. Honkanen, J. Ström, T. Kärkkäinen, P. Silventoinen, and S. Räisänen. Capacitor aging detection in dc-dc converter output stage. In *2015 IEEE Energy Conversion Congress and Exposition (ECCE)*, pages 5538–5545, 2015.

S. Hiller, M. Beier-Moebius, S. Frankeser, and J. Lutz. Using the zth(t) - power pulse measurement to detect a degradation in the module structure. In *Proceedings of PCIM Europe 2015; International Exhibition and Conference for Power Electronics, Intelligent Motion, Renewable Energy and Energy Management*, pages 1–7, 2015.

M. Hoeer, M. Meissner, F. Filsecker, and S. Bernet. Online temperature estimation of a high-power 4.5 kv igbt module based on the gate-emitter threshold voltage. In *2015 17th European Conference on Power Electronics and Applications (EPE'15 ECCE-Europe)*, pages 1–8, 2015.

Malorie Hologne. *Contribution to condition monitoring of Silicon Carbide MOSFET based Power Module*. Theses, Université de Lyon, December 2018. URL https://tel.archives-ouvertes.fr/tel-02061648.

J. Hong, D. Hyun, S. B. Lee, and C. Kral. Offline Monitoring of Airgap Eccentricity for Inverter-Fed Induction Motors Based on the Differential Inductance. *IEEE Transactions on Industry Applications*, 49(6):2533–2542, 2013.

S. Hong, Z. Zhou, C. Lv, and H. Guo. Prognosis for insulated gate bipolar transistor based on Gaussian Process Regression. In *2013 IEEE conference on prognostics and health management (PHM)*, pages 1–5, 2013.

Ian Howard. *A Review of rolling element bearing vibration: detection, diagnosis and prognosis*. DSTO Aeronautical and Maritime Research Laboratory, Oct. 1994.

K. Hu, Z. Liu, H. Du, L. Ceccarelli, F. Iannuzzo, F. Blaabjerg, and I. A. Tasiu. Cost-effective prognostics of igbt bond wires with consideration of temperature swing. *IEEE Transactions on Power Electronics*, 35(7):6773–6784, 2020a.

K. Hu, Z. Liu, I. A. Tasiu, and T. Chen. Fault diagnosis and tolerance with low torque ripple for open-switch fault of IM drives. *IEEE Transactions on Transportation Electrification*, pages 1–1, 2020b.

R. Hu, J. Wang, A. R. Mills, E. Chong, and Z. Sun. PWM ripple currents based turn fault detection for 3-phase permanent magnet machines. In *2017 IEEE International Electric Machines and Drives Conference (IEMDC)*, pages 1–7, 2017. doi: 10.1109/IEMDC.2017.8002165.

S. Huang, A. Aggarwal, E. Strangas, K. Li, F. Niu, and X. Huang. Robust stator winding faults detection in PMSMs with respect to current controller bandwidth. *IEEE Transactions on Power Electronics*, pages 1–1, 2020.

R. Jano and D. Pitica. Accelerated aging tests of aluminum electrolytic capacitors for evaluating lifetime prediction models. *Acta Tech. Napoc*, 53(2):36, 2012.

W. R. Jensen, E. G. Strangas, and S. N. Foster. A method for online stator insulation prognosis for inverter-driven machines. *IEEE Transactions on Industry Applications*, 54(6):5897–5906, 2018.

B. Ji, X. Song, W. Cao, V. Pickert, Y. Hu, J. W. Mackersie, and G. Pierce. In situ diagnostics and prognostics of solder fatigue in igbt modules for electric vehicle drives. *IEEE Transactions on Power Electronics*, 30(3):1535–1543, 2015.

C. Jiang, S. Li, and T. G. Habetler. A review of condition monitoring of induction motors based on stray flux. In *2017 IEEE Energy Conversion Congress and Exposition (ECCE)*, pages 5424–5430, 2017.

Y. Jiang, Z. Zhang, W. Jiang, W. Geng, and J. Huang. Three-phase current injection method for mitigating turn-to-turn short-circuit fault in concentrated-winding permanent magnet aircraft starter generator. *IET Electric Power Applications*, 12 (4):566–574, 2018.

Jinsong Han, Mingyao Ma, Kaiqi Chu, Xing Zhang, and Zhemin Lin. In-situ diagnostics and prognostics of wire bonding faults in igbt modules of three-level neutral-point-clamped inverters. In *2016 IEEE 8th International Power Electronics and Motion Control Conference (IPEMC-ECCE Asia)*, pages 3262–3267, 2016.

Z. Khatir. Junction temperature investigations based on a general semi-analytical formulation of forward voltage of power diodes. *IEEE Transactions on Electron Devices*, 59(6):1716–1722, 2012.

Hack-Eun Kim, Andy C.C Tan, Joseph Mathew, and Byeong-Keun Choi. Bearing fault prognosis based on health state probability estimation. *Expert systems with applications*, 39, 2012.

Y. Kim, H. Kim, J. Moon, Y. Kim, and S. Jung. A study on the estimation of bearing life of electric motor using ISO 281 and accelerated life test. In *2017 2nd International Conference on System Reliability and Safety (ICSRS)*, pages 223–226, 2017.

H. Kuhn and A. Mertens. online junction temperature measurement of igbts based on temperature sensitive electrical parameters. In *2009 13th European Conference on Power Electronics and Applications*, pages 1–10, 2009.

A. Lahyani, P. Venet, G. Grellet, and P. Viverge. Failure prediction of electrolytic capacitors during operation of a switchmode power supply. *IEEE Transactions on Power Electronics*, 13(6):1199–1207, 1998.

A. Lebaroud and G. Clerc. Classification of induction machine faults by optimal time–frequency representations. *IEEE Transactions on Industrial Electronics*, 55 (12):4290–4298, 2008. doi: 10.1109/TIE.2008.2004666.

S. S. Lee, W. K. Tham, and C. K. Ang. Igbt collector-emitter failure mechanism. In *2016 IEEE 23rd International Symposium on the Physical and Failure Analysis of Integrated Circuits (IPFA)*, pages 180–184, July 2016. doi: 10.1109/IPFA.2016.7564276.

E. Levi. Multiphase electric machines for variable-speed applications. *IEEE Transactions on Industrial Electronics*, 55(5):1893–1909, 2008.

B. X. Li and K. S. Low. Low sampling rate online parameters monitoring of dc–dc converters for predictive-maintenance using biogeography-based optimization. *IEEE Transactions on Power Electronics*, 31(4):2870–2879, 2016.

Q. Li, X. Ding, Q. He, W. Huang, and Y. Shao. Manifold sensing-based convolution sparse self-learning for defective bearing morphological feature extraction. *IEEE Transactions on Industrial Informatics*, pages 1–1, 2020.

H. Liu, Y. Sun, J. Huang, Z. Hou, and T. Wang. Inter-turn fault detection for the inverter-fed induction motor based on the Teager-Kaiser energy operation of switching voltage harmonics. In *2015 18th International Conference on Electrical Machines and Systems (ICEMS)*, pages 1214–1219, 2015.

J. Liu, G. Zhang, Q. Chen, L. Qi, Y. Geng, and J. Wang. In situ condition monitoring of igbts based on the miller plateau duration. *IEEE Transactions on Power Electronics*, 34(1):769–782, 2019.

B. Lu and S. K. Sharma. A literature review of igbt fault diagnostic and protection methods for power inverters. *IEEE Transactions on Industry Applications*, 45(5): 1770–1777, 2009.

H. Luo, Y. Chen, W. Li, and X. He. Online high-power p-i-n diode junction temperature extraction with reverse recovery fall storage charge. *IEEE Transactions on Power Electronics*, 32(4):2558–2567, 2017.

H. Luo, W. Li, F. Iannuzzo, X. He, and F. Blaabjerg. Enabling junction temperature estimation via collector-side thermo-sensitive electrical parameters through emitter stray inductance in high-power igbt modules. *IEEE Transactions on Industrial Electronics*, 65(6):4724–4738, 2018.

Y. Ma, X. Jia, Q. Hu, H. Bai, C. Guo, and S. Wang. A new state recognition and prognosis method based on a sparse representation feature and the Hidden Semi-Markov Model. *IEEE Access*, 8:119405–119420, 2020.

V. Madonna, P. Giangrande, W. Zhao, G. Buticchi, H. Zhang, C. Gerada, and M. Galea. Reliability vs. performances of electrical machines: Partial discharges issue. In *2019 IEEE Workshop on Electrical Machines Design, Control and Diagnosis (WEMDCD)*, volume 1, pages 77–82, 2019.

R. Mandeya, C. Chen, V. Pickert, and R. T. Naayagi. Prethreshold voltage as a low-component count temperature sensitive electrical parameter without self-heating. *IEEE Transactions on Power Electronics*, 33(4):2787–2791, 2018.

M. A. Masrur, Z. Chen, B. Zhang, and Y. Lu Murphey. Model-based fault diagnosis in electric drive inverters using artificial neural network. In *2007 IEEE Power Engineering Society General Meeting*, pages 1–7, 2007.

P. Maussion, A. Picot, M. Chabert, and D. Malec. Lifespan and aging modeling methods for insulation systems in electrical machines: A survey. In *2015 IEEE Workshop on Electrical Machines Design, Control and Diagnosis (WEMDCD)*, pages 279–288, 2015.

A. Medoued, A. Lebaroud, A. Boukadoum, T. Boukra, and G. Clerc. Back propagation neural network for classification of induction machine faults. In *8th IEEE Symposium on Diagnostics for Electrical Machines, Power Electronics Drives*, pages 525–528, 2011. doi: 10.1109/DEMPED.2011.6063673.

A. M. S. Mendes and A. J. Marques Cardoso. Voltage source inverter fault diagnosis in variable speed ac drives, by the average current park's vector approach. In *IEEE International Electric Machines and Drives Conference. IEMDC'99. Proceedings (Cat. No.99EX272)*, pages 704–706, 1999.

A. M. S. Mendes and A. J. Marques Cardoso. Fault-tolerant operating strategies applied to three-phase induction-motor drives. *IEEE Transactions on Industrial Electronics*, 53(6):1807–1817, 2006.

J. Millán, P. Godignon, X. Perpiñà, A. Pérez-Tomás, and J. Rebollo. A survey of wide bandgap power semiconductor devices. *IEEE Transactions on Power*

Electronics, 29(5):2155–2163, May 2014. ISSN 1941-0107. doi: 10.1109/TPEL.2013.2268900.

M.A. Miner. Cumulative Damage in Fatigue. *Journal of Applied Mechanics*, 12(3): 159–164, 1945.

G. Mirzaeva and K. I. Saad. Advanced diagnosis of rotor faults and eccentricity in induction motors based on internal flux measurement. *IEEE Transactions on Industry Applications*, 54(3):2981–2991, 2018a.

G. Mirzaeva and K. I. Saad. Advanced diagnosis of stator turn-to-turn faults and static eccentricity in induction motors based on internal flux measurement. *IEEE Transactions on Industry Applications*, 54(4):3961–3970, 2018b.

A. J. Mitcham, G. Antonopoulos, and J. J. A. Cullen. Implications of shorted turn faults in bar wound PM machines. *IEE Proceedings - Electric Power Applications*, 151(6):651–657, 2004.

Mitsubishi. Power Module reliability, December 2019. URL http://www.mitsubishielectric.com/semiconductors/products/pdf/reliability/powermodule_reliability_e.pdf.

S. Mohagheghi, R. G. Harley, T. G. Habetler, and D. Divan. Condition monitoring of power electronic circuits using artificial neural networks. *IEEE Transactions on Power Electronics*, 24(10):2363–2367, 2009.

A. Mohammadpour and L. Parsa. Post-fault control technique for multi-phase PM motor drives under short-circuit faults. In *2013 Twenty-Eighth Annual IEEE Applied Power Electronics Conference and Exposition (APEC)*, pages 817–822, 2013.

A. Mohammadpour, S. Sadeghi, and L. Parsa. A generalized fault-tolerant control strategy for Five-Phase PM motor drives considering star, pentagon, and pentacle connections of stator windings. *IEEE Transactions on Industrial Electronics*, 61(1):63–75, 2014.

A. Muetze and A. Binder. Practical Rules for Assessment of Inverter-Induced Bearing Currents in Inverter-Fed AC Motors up to 500 kW. *IEEE Transactions on Industrial Electronics*, 54(3):1614–1622, 2007. doi: 10.1109/TIE.2007.894698.

A. Muetze, V. Niskanen, and J. Ahola. On radio-frequency-based detection of high-frequency circulating bearing current flow. *IEEE Transactions on Industry Applications*, 50(4):2592–2601, 2014. doi: 10.1109/TIA.2013.2296626.

K. Murthy and R. Bedford. Transformation between foster and cauer equivalent networks. *IEEE Transactions on Circuits and Systems*, 25(4):238–239, 1978.

M. Musallam, C. M. Johnson, C. Yin, C. Bailey, and M. Mermet-Guyennet. Real-time life consumption power modules prognosis using online rainflow algorithm in metro applications. In *2010 IEEE Energy Conversion Congress and Exposition*, pages 970–977, Sep. 2010. doi: 10.1109/ECCE.2010.5617883.

V. Naikan and Arvind Rathore. Accelerated temperature and voltage life tests on aluminium electrolytic capacitors. *International Journal of Quality & Reliability Management*, 33:120–139, 01 2016. doi: 10.1108/IJQRM-12-2014-0201.

Hiroshi Nakao, Yu Yonezawa, Takahiko Sugawara, Yoshiyasu Nakashima, and Fujio Kurokawa. Online evaluation method of electrolytic capacitor degradation for digitally controlled smps failure prediction. *IEEE Transactions on Power Electronics*, PP:1–1, 04 2017. doi: 10.1109/TPEL.2017.2691048.

S. Nandi, S. Ahmed, and H. A. Toliyat. Detection of rotor slot and other eccentricity related harmonics in a three phase induction motor with different rotor cages. *IEEE Transactions on Energy Conversion*, 16(3):253–260, 2001.

C. Neacsu, P. Bidan, and T. Lebey. Off line measurements of partial discharges in turn insulation of low voltage rotating machines. In *Conference Record of the 2000 IEEE International Symposium on Electrical Insulation (Cat. No.00CH37075)*, pages 235–238, 2000.

Patrick Nectoux, Rafael Gouriveau, Kamal Medjaher, Emmanuel Ramasso, Brigitte Chebel-Morello, and et al. PRONOSTIA: An experimental platform for bearings accelerated degradation tests. In *IEEE International Conference on Prognostics and Health Management, PHM'12*, pages 1–8, June 2012.

V. Nguyen, J. Seshadrinath, D. Wang, S. Nadarajan, and V. Vaiyapuri. Model-Based Diagnosis and RUL Estimation of Induction Machines Under Interturn Fault. *IEEE Transactions on Industry Applications*, 53(3):2690–2701, 2017. doi: 10.1109/TIA.2017.2669195.

Atsushi Nishino. Capacitors: operating principles, current market and technical trends. *Journal of Power Sources*, 60(2):137 – 147, 1996. ISSN 0378-7753.

F. Niu, B. Wang, A. S. Babel, K. Li, and E. G. Strangas. Comparative evaluation of direct torque control strategies for permanent magnet synchronous machines. *IEEE Transactions on Power Electronics*, 31(2):1408–1424, 2016.

K. C. Norris and A. H. Landzberg. Reliability of controlled collapse interconnections. *IBM Journal of Research and Development*, 13(3):266–271, 1969.

P. Nussbaumer, M. A. Vogelsberger, and T. M. Wolbank. Induction machine insulation health state monitoring based on online switching transient exploitation. *IEEE Transactions on Industrial Electronics*, 62(3):1835–1845, 2015.

Y. Nyanteh, L. Graber, C. Edrington, S. Srivastava, and D. Cartes. Overview of simulation models for partial discharge and electrical treeing to determine feasibility for estimation of remaining life of machine insulation systems. In *2011 Electrical Insulation Conference (EIC).*, pages 327–332, 2011. doi: 10.1109/EIC.2011.5996172.

O. Ondel, G. Clerc, E. Boutleux, and E. Blanco. Fault detection and diagnosis in a set "inverter–induction machine" through multidimensional membership function and pattern recognition. *IEEE Transactions on Energy Conversion*, 24 (2):431–441, 2009.

R. Ouaida, M. Berthou, J. León, X. Perpiñà, S. Oge, P. Brosselard, and C. Joubert. Gate oxide degradation of sic mosfet in switching conditions. *IEEE Electron Device Letters*, 35(12):1284–1286, Dec 2014. ISSN 1558-0563. doi: 10.1109/LED.2014.2361674.

Rémy Ouaida. *Vieillissement et mécanismes de dégradation sur des composants de puissance en carbure de silicium (SIC) pour des applications haute température.* PhD thesis, 2014. URL http://www.theses.fr/2014LYO10228. Thèse de doctorat dirigée par Joubert, CharlesBrosselard, Pierre et Oge, Sébastien Génie électrique Lyon 1 2014.

Rob O'Connor. Understanding polymer and hybrid capacitors - electronic products. *Panasonic - Electronic Products*, 56(4), 2014.

P. A. Panagiotou, I. Arvanitakis, N. Lophitis, and K. N. Gyftakis. Fem study of induction machines suffering from rotor electrical faults using stray flux signature analysis. In *2018 XIII International Conference on Electrical Machines (ICEM)*, pages 1861–1867, 2018.

H.M. Pang and Pong Bryan. A life prediction scheme for electrolytic capacitors in power converters without current sensor. pages 973 – 979, 03 2010. doi: 10.1109/APEC.2010.5433384.

Y. Park, C. Yang, S. B. Lee, D. Lee, D. Fernandez, D. Reigosa, and F. Briz. Online detection and classification of rotor and load defects in PMSMs based on hall sensor measurements. *IEEE Transactions on Industry Applications*, 55(4): 3803–3812, 2019. doi: 10.1109/TIA.2019.2911252.

Abhijit Pathak. MOSFET/IGBT drivers theory and applications, 2001. URL http://www.ixys.com/Documents/AppNotes/IXAN0010.pdf.

Frédéric Perisse. *ETUDE ET ANALYSE DES MODES DE DEFAILLANCES DES CONDENSATEURS ELECTROLYTIQUES A L'ALUMINIUM ET DES THYRISTORS, APPLIQUEES AU SYSTEME DE PROTECTION DU LHC (LARGE HADRON COLLIDER)*. Theses, Université Claude Bernard - Lyon I, July 2003.

R. Peuget, S. Courtine, and J. . Rognon. Fault detection and isolation on a pwm inverter by knowledge-based model. *IEEE Transactions on Industry Applications*, 34(6):1318–1326, 1998.

S. Peyghami, Z. Wang, and F. Blaabjerg. Reliability modeling of power electronic converters: A general approach. In *2019 20th Workshop on Control and Modeling for Power Electronics (COMPEL)*, pages 1–7, June 2019. doi: 10.1109/COMPEL.2019.8769685.

T. H. Pham, S. Lefteriu, E. Duviella, and S. Lecoeuche. Auto-adaptive and dynamical clustering for double open-circuit fault diagnosis of power inverters.

In *2019 4th Conference on Control and Fault Tolerant Systems (SysTol)*, pages 306–311, 2019.

K. Pugalenthi and N. Raghavan. Roughening particle filter based prognosis on power mosfets using on-resistance variation. In *2018 Prognostics and System Health Management Conference (PHM-Chongqing)*, pages 1170–1175, 2018.

Juha Pyrh'onen, Valeria Hrabovcov'a, and Tapani Jokinen. *Design of Rotating Electrical Machines*. Chichester: Wiley, second edition, 2014.

Y. Qi, E. Bostanci, M. Zafarani, and B. Akin. Severity estimation of interturn short circuit fault for PMSM. *IEEE Transactions on Industrial Electronics*, 66(9): 7260–7269, 2019.

L. Qiu, Q. Peng, Y. Yang, W. Xu, and Q. Liang. Using ceemdan algorithm and svm fault diagnosis methodology in three-phase inverters of pmsm drive systems. In *2017 36th Chinese Control Conference (CCC)*, pages 7127–7132, 2017.

H. Razik, T. M. Oliveira, M. B. d. R. Corrêa, C. B. Jacobina, and E. R. C. da Silva. Analysis and identification of furtive misfiring in converter using wavelet. *8th IEEE Symposium on Diagnostics for Electrical Machines, Power Electronics Drives*, pages 651–656, 2011.

Jason D. Renwick, Chetan S. Kulkarni, and José R. Celaya. Analysis of electrolytic capacitor degradation under electrical overstress for prognostic studies. In *Annual conference of the prognostics and health management society*, 2015.

M. Rocchi, F. Mosciaro, F. Grottesi, M. Scortichini, A. Giantomassi, M. Pirro, M. Grisostomi, and G. Ippoliti. Fault prognosis for rotating electrical machines monitoring using recursive least square. In *2014 6th European Embedded Design in Education and Research Conference (EDERC)*, pages 269–273, 2014. doi: 10.1109/EDERC.2014.6924402.

Alfréd Rényi. On measures of entropy and information. In *Proceedings of the Fourth Berkeley Symposium on Mathematical Statistics and Probability, Volume 1: Contributions to the Theory of Statistics*, pages 547–561, Berkeley, Calif., 1961. University of California Press. URL https://projecteuclid.org/euclid.bsmsp/1200512181.

B. Saha, J. R. Celaya, P. F. Wysocki, and K. F. Goebel. Towards prognostics for electronics components. In *2009 IEEE Aerospace conference*, pages 1–7, 2009.

S. A. Saleh and M. A. Rahman. Modeling and protection of and three-phase power transformer using wavelet packet transform. *IEEE Transactions on Power Delivery*, 20(2):1273–1282, Apr. 2005.

M. Samie, A. Alghassi, and S. Perinpanayagam. Unified igbt prognostic using natural computation. In *2015 IEEE International Conference on Digital Signal Processing (DSP)*, pages 698–702, 2015.

Shankar Sankararaman and Kai Goebel. Why is the remaining useful life prediction uncertain? *PHM 2013 - Proceedings of the Annual Conference of the Prognostics and Health Management Society 2013*, pages 337–349, 01 2013.

Shankar Sankararaman and Kai Goebel. Uncertainty in prognostics and systems health management. *International Journal of Prognostics and Health Management*, 6, 07 2015.

M. Sathik, T. K. Jet, C. J. Gajanayake, R. Simanjorang, and A. K. Gupta. Comparison of power cycling and thermal cycling effects on the thermal impedance degradation in igbt modules. In *IECON 2015 - 41st Annual Conference of the IEEE Industrial Electronics Society*, pages 001170–001175, 2015.

A. Saxena, J. Celaya, E. Balaban, K. Goebel, B. Saha, S. Saha, and M. Schwabacher. Metrics for evaluating performance of prognostic techniques. In *2008 international conference on prognostics and health management*, 2008.

Abhinav Saxena, Jose Celaya, Bhaskar Saha, Sankalita Saha, and Kai Goebel. Metrics for offline evaluation of prognostic performance. *International Journal of Prognostics and Health Management*, 1, January 2010.

Abhinav Saxena, Shankar Sankararaman, and Kai Goebel. Performance evaluation for fleet-based and unit-based prognostic methods. *Second european conference of the prognostics and health management society 2014*, 2014.

R. K. Singleton, E. G. Strangas, and S. Aviyente. Extended Kalman filtering for remaining-useful-life estimation of bearings. *IEEE Transactions on Industrial Electronics*, 62(3):1781–1790, 2015.

A. Soualhi, H. Razik, G. Clerc, and D. D. Doan. Prognosis of bearing failures using hidden markov models and the adaptive neuro-fuzzy inference system. *IEEE Transactions on Industrial Electronics*, 61(6):2864–2874, 2014.

A. Soualhi, M. Makdessi, R. German, F. R. Echeverría, H. Razik, A. Sari, P. Venet, and G. Clerc. Heath monitoring of capacitors and supercapacitors using the neo-fuzzy neural approach. *IEEE Transactions on Industrial Informatics*, 14(1): 24–34, 2018.

J. R. Stack, T. G. Habetler, and R. G. Harley. Fault classification and fault signature production for rolling element bearings in electric machines. *IEEE Transactions on Industry Applications*, 40(3):735–739, 2004.

M. K. W. Stranges, G. C. Stone, and D. L. Bogh. Voltage endurance testing. *IEEE Industry Applications Magazine*, 15(6):12–18, 2009. doi: 10.1109/MIAS.2009.934439.

B. Sun, X. Fan, C. Qian, and G. Zhang. Pof-simulation-assisted reliability prediction for electrolytic capacitor in led drivers. *IEEE Transactions on Industrial Electronics*, 63(11):6726–6735, 2016.

P. Sun, C. Gong, X. Du, Y. Peng, B. Wang, and L. Zhou. Condition monitoring igbt module bond wires fatigue using short-circuit current identification. *IEEE Transactions on Power Electronics*, 32(5):3777–3786, 2017.

V. Sundaramoorthy, E. Bianda, R. Bloch, I. Nistor, G. Knapp, and A. Heinemann. Online estimation of igbt junction temperature (tj) using gate-emitter voltage (vge) at turn-off. In *2013 15th European Conference on Power Electronics and Applications (EPE)*, pages 1–10, 2013.

A. Tani, M. Mengoni, L. Zarri, G. Serra, and D. Casadei. Control of multiphase induction motors with an odd number of phases under open-circuit phase faults. *IEEE Transactions on Power Electronics*, 27(2):565–577, 2012.

O. V. Thorsen and M. Dalva. A survey of faults on induction motors in offshore oil industry, petrochemical industry, gas terminals, and oil refineries. *IEEE Transactions on Industry Applications*, 31(5):1186–1196, September 1995. ISSN 1939-9367. doi: 10.1109/28.464536.

C. Torrence and G. P. Compo. A practival guide to wavelet analysis. *Bulletin of the American Metereological Society*, 79(1):61–78, Jan. 1998.

I. Tsyokhla, A. Griffo, and J. Wang. Online condition monitoring for diagnosis and prognosis of insulation degradation of inverter-fed machines. *IEEE Transactions on Industrial Electronics*, 66(10):8126–8135, 2019.

Z. Ullah and J. Hur. Irreversible demagnetization fault prognosis in a permanent magnet type machines. In *2020 IEEE Energy Conversion Congress and Exposition (ECCE)*, pages 742–747, 2020. doi: 10.1109/ECCE44975.2020.9235942.

T. v. Essen, M. A. Ras, S. Stegmeier, and G. Mitic. Investigation of the influence of aging processes on thermal characteristics of an igbt power module by means of transient thermal analysis. In *PCIM Europe 2016; International Exhibition and Conference for Power Electronics, Intelligent Motion, Renewable Energy and Energy Management*, pages 1–5, 2016.

C. H. van der Broeck, T. Polom, R. D. Lorenz, and R. W. De Doncker. Thermal monitoring of power electronic modules using device self-sensing. In *2018 IEEE Energy Conversion Congress and Exposition (ECCE)*, pages 4699–4706, 2018.

C. H. Van der Broeck, T. Polom, R. D. Lorenz, and R. W. De Doncker. Thermal monitoring of power electronic modules using device self-sensing. In *2018 IEEE Energy Conversion Congress and Exposition (ECCE)*, pages 4699–4706, 2018.

Christophe Van der Broeck. *Methodology for Thermal Modeling, Monitoring and Control of Power Electronics*. PhD thesis, Rheinisch-Westfalischen Technischen Hochschule Aachen University, November 2018. URL https://www.researchgate.net/publication/331261496_Methodology_for_Thermal_Modeling_Monitoring_and_Control_of_Power_Electronics.

P. Venet, F. Perisse, M. H. El-Husseini, and G. Rojat. Realization of a smart electrolytic capacitor circuit. *IEEE Industry Applications Magazine*, 8(1):16–20, 2002.

X. Wang, P. Sun, L. Sun, Q. Luo, and X. Du. Online condition monitoring for bond wire degradation of igbt modules in three-level neutral-point-clamped converters. *IEEE Transactions on Industrial Electronics*, pages 1–1, 2020.

K. Wei, W. Wang, Z. Hu, and M. Du. Condition monitoring of igbt modules based on changes of thermal characteristics. *IEEE Access*, 7:47525–47534, 2019.

B. A. Welchko, J. Wai, T. M. Jahns, and T. A. Lipo. Magnet-flux-nulling control of interior PM machine drives for improved steady-state response to short-circuit faults. *IEEE Transactions on Industry Applications*, 42:113–120, 2006.

Rolf Winter. Handbook for Robustness Validation of Automotive Electrical/Electronic Modules, June 2008. URL https://www.zvei.org/fileadmin/user_upload/Presse_und_Medien/Publikationen/2013/Oktober/Handbook_for_Robustness_Validation_of_Automotive_Electrical/Robust-Validation-EE-Modules.pdf.

Junke Wu, Luowei Zhou, and Xiong Du. Junction Temperature Prediction of IGBT Power Module Based on BP Neural Network. *Journal of Electrical Engineering and Technology*, 9(3).

L. Xu, Y. Zhou, and S. Liu. Dbc substrate in si- and sic-based power electronics modules: Design, fabrication and failure analysis. In *2013 IEEE 63rd Electronic Components and Technology Conference*, pages 1341–1345, May 2013. doi: 10.1109/ECTC.2013.6575747.

Xuhua Huang, Wen-Fang Wu, and P. Chou. Fatigue life and reliability prediction of electronic packages under thermal cycling conditions through fem analysis and acceleration models. In *2012 14th International Conference on Electronic Materials and Packaging (EMAP)*, pages 1–6, 2012.

H. Yan, Y. Xu, F. Cai, H. Zhang, W. Zhao, and C. Gerada. PWM-VSI fault diagnosis for a PMSM drive based on the fuzzy logic approach. *IEEE Transactions on Power Electronics*, 34(1):759–768, 2019.

K. Yao, W. Tang, W. Hu, and J. Lyu. A current-sensorless online esr and c identification method for output capacitor of buck converter. *IEEE Transactions on Power Electronics*, 30(12):6993–7005, 2015.

K. Yao, W. Tang, X. Bi, and J. Lyu. An online monitoring scheme of dc-link capacitor's esr and c for a boost pfc converter. *IEEE Transactions on Power Electronics*, 31(8):5944–5951, 2016.

K. Yao, C. Cao, and S. Yang. Noninvasive online condition monitoring of output capacitor's esr and c for a flyback converter. *IEEE Transactions on Instrumentation and Measurement*, 66(12):3190–3199, 2017.

H. Yilmaz, J. L. Benjamin, R. F. Dyer, L. S. Chen, W. R. Van Dell, and G. C. Pifer. Comparison of the punch-through and non-punch-through IGT structures. *IEEE Transactions on Industry Applications*, IA-22(3):466–470, May 1986. ISSN 1939-9367. doi: 10.1109/TIA.1986.4504744. Citation Key:.

K. Younsi, P. Neti, M. Shah, J. Y. Zhou, J. Krahn, K. Weeber, and C. D. Whitefield. online capacitance and dissipation factor monitoring of AC stator insulation. *IEEE Transactions on Dielectrics and Electrical Insulation*, 17(5):1441–1452, 2010.

Y. Yu, T. Zhou, M. Zhu, and D. Xu. Fault diagnosis and life prediction of dc-link aluminum electrolytic capacitors used in three-phase ac/dc/ac converters. In

2012 Second International Conference on Instrumentation, Measurement, Computer, Communication and Control, pages 825–830, 2012.

Y. Yu, Y. Zhao, B. Wang, X. Huang, and D. Xu. Current sensor fault diagnosis and tolerant control for vsi-based induction motor drives. *IEEE Transactions on Power Electronics*, 33(5):4238–4248, 2018. doi: 10.1109/TPEL.2017.2713482.

I. Zamudio-Ramirez, J. A. Antonino-Daviu, R. A. Osornio-Rios, R. de Jesus Romero-Troncoso, and H. Razik. Detection of winding asymmetries in wound-rotor induction motors via transient analysis of the external magnetic field. *IEEE Transactions on Industrial Electronics*, 67(6):5050–5059, 2020.

I. Zamudio-Ramirez, R. A. Osornio-Rios, J. A. Antonino-Daviu, H. Razik, and R. de Jesus Romero-Troncoso. Magnetic flux analysis for the condition monitoring of electric machines: A review. *IEEE Transactions on Industrial Electronics*, 2021. doi: 10.1109/TII.2021.3070581.

W. G. Zanardelli, E. G. Strangas, and S. Aviyente. Identification of intermittent electrical and mechanical faults in Permanent-Magnet AC drives based on time–frequency analysis. *IEEE Transactions on Industry Applications*, 43(4): 971–980, 2007.

M. Zeng and Z. Chen. SOSO boosting of the K-SVD denoising algorithm for enhancing fault-induced impulse responses of rolling element bearings. *IEEE Transactions on Industrial Electronics*, 67(2):1282–1292, 2020.

Bo Zhang, Wentong Zhang, Ming Qiao, Zhenya Zhan, and Zhaoji Li. Concept and design of super junction devices. *Journal of Semiconductors*, 39(2):021001, February 2018. doi: 10.1088/1674-4926/39/2/021001.

J. Zhang, M. Salman, W. Zanardelli, S. Ballal, and B. Cao. An integrated fault isolation and prognosis method for electric drive systems of battery electric vehicles. *IEEE Transactions on Transportation Electrification*, pages 1–1, 2020. doi: 10.1109/TTE.2020.3025107.

Ludi Zhang. *Etude de fiabilité des modules d'électronique de puissance à base de composant SiC pour applications hautes températures*. Theses, Université Sciences et Technologies - Bordeaux I, January 2012. URL https://tel.archives-ouvertes.fr/tel-00988235.

W. Zhang, D. Xu, P. N. Enjeti, H. Li, J. T. Hawke, and H. S. Krishnamoorthy. Survey on fault-tolerant techniques for power electronic converters. *IEEE Transactions on Power Electronics*, 29(12):6319–6331, 2014.

D. Zhou, H. Wang, F. Blaabjerg, S. K. Kær, and D. Blom-Hansen. Degradation effect on reliability evaluation of aluminum electrolytic capacitor in backup power converter. In *2017 IEEE 3rd International Future Energy Electronics Conference and ECCE Asia (IFEEC 2017 - ECCE Asia)*, pages 202–207, 2017.

Zisi Tian and X. Ge. An online fault diagnostic method based on frequency-domain analysis for igbts in traction pwm rectifiers. In *2016 IEEE 8th*

International Power Electronics and Motion Control Conference (IPEMC-ECCE Asia), pages 3403–3407, 2016.

C. Zoeller, M. A. Vogelsberger, R. Fasching, W. Grubelnik, and T. M. Wolbank. Evaluation and current-response-based identification ofinsulation degradation for high utilized electrical machines in railway application. *IEEE Transactions on Industry Applications*, 53(3):2679–2689, 2017. doi: 10.1109/TIA.2017.2661718.

Clemens Zoeller, Markus A. Vogelsberger, Thomas M. Wolbank1, and Hans Ertl. Impact of SiC semiconductors switching transition speed on insulation health state monitoring of traction machines. *IET Power Electron.*, page 2769–2775, 2016.

3

Fault Diagnosis and Prognosis for Reliability Enhancement

3.1 Introduction

Reliability has been a human attribute before it became a technical one. The original definition, which remains essentially the same to date is the inverse of the number of accidents per hour of operation (Song and Wang (2013)). The classic one has been restated as the probability that an item (component, subsystem, or system) performs the required functions that its design defines, for the intended period of time, under predetermined environmental and operational conditions. The **reliability function**, $R(t)$, represents the probability that the system will operate without failures over a time interval $[0, t]$.

The reliability of drives has become dramatically more important as electrical drives are replacing an increasing portion of mechanical actuators, and also finding new and even more demanding and critical applications. A drives designer, as usual, has to balance demands with conflicting requirements: the drive has to be affordable, to satisfy performance and weight, to respect volume requirements, and to perform well without failure during its expected lifetime and operational profile. The tools that the designer can use to satisfy the last requirement (i.e., high reliability) are: selection of the drive topology, which may include internal and external redundancies, determination of component stresses and selection of components based on their reliability for the expected use, and the design of a controller that can handle redundancies and operational changes. A less desirable part of the managing of faults, the decision system, may incorporate reconfigurations that will decrease the performance of the drive, and even modify the mission profile to avoid catastrophic failure.

In this chapter, we discuss how the diagnosis and prognosis tools introduced earlier can be integrated with the design and operation of a drive for high

Fault Diagnosis, Prognosis, and Reliability for Electrical Machines and Drives. First Edition.
Elias G. Strangas, Guy Clerc, Hubert Razik, and Abdenour Soualhi.

reliability. We discuss the sources of reliability information about components, how this information can be included in the reliability estimation, and how prognosis can lead to a more reliable drive.

3.2 Fundamentals

As the drive operates, the stresses on the components together with possible unidentified internal weaknesses result in faults. These faults themselves, in their early stages, do not interrupt and may do not even affect the operation of the drive. But they can be considered as precursors or early manifestations of impending failures. To avoid failures as a fault develops, a number of actions are possible: Redundancies, partial or full, either in hardware or in the control algorithm, which have been installed before operation starts, can be utilized to allow continued operation at full or reduced performance. The redundancy itself, along with the identification of the fault and its severity, is designed to prevent interruption of operation and extend the estimate of the remaining useful life of the part or the drive.

Having in place a system that includes diagnosis, prognosis, and mitigation as part of its design, before its commissioning, should result in a drive design with a longer anticipated time to fail and higher reliability, compared to a similar drive where some of these components and actions are not included.

Determining the reliability of each component in order to calculate the reliability of the drive, even in the simplest case, is not an easy task. Manuals and manufacturers' data provide accurate values of the reliability for many components, especially for a given operating environment and condition. Although grossly inaccurate, the bathtub curve (Figure 3.1) gives a first approach to a more accurate method of comparing simple drives and indicates that early faults have to be addressed at commissioning.

Not included in the classic definition are idle times, where the component is not aging, with some exceptions, for example, insulation collecting humidity or aging due to surrounding heat sources and false brinelling of bearings under static load. It is then significant to develop metrics that have to do with the specifics of the operation of the drive and its components in terms of *a priori* estimates. These can be specific to a drive type, the mode of operation, and operational demands. As a drive operates, its components consume part of their life through stresses, and an accurate reliability estimate accounts for this fact. The main use of the reliability calculation is, in the design stage, the selection of components and the system architecture for a specific application and to define maintenance schedules and processes. It is important to note that the reliability

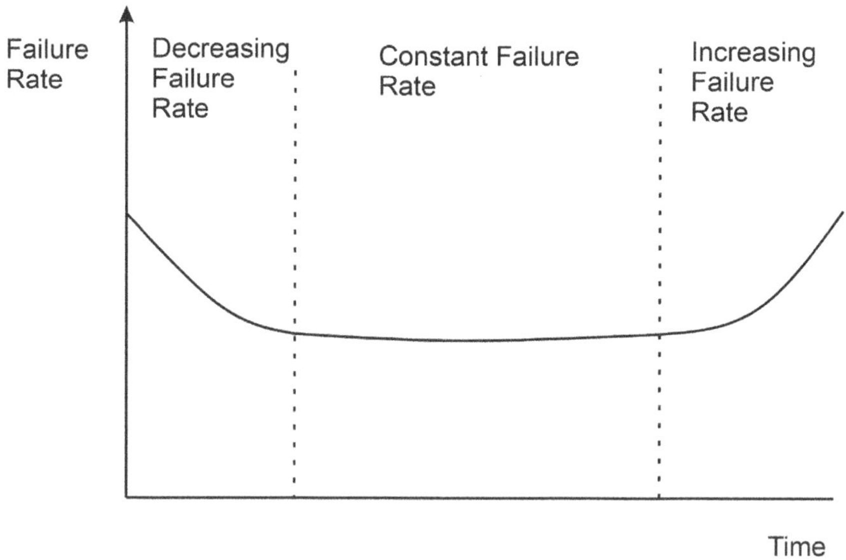

Figure 3.1 Bath Curve.

value is a characteristic of a class of components and a design rather than of a specific drive. The better the specifics of the operation and of the components are known, the more accurate is the reliability computation. Not all possible operating characteristics can be included in the original estimate of the mean time to failure, nor can this estimate be absolutely accurate.

A simple way to determine the reliability of a drive system is to start from the reliability of every component in it: the power source, the batteries and capacitor, the inverter (switches and gate drivers), the motor or generator, the sensors. Any component failure in this series connection will precipitate the failure of the total drive, and the drive reliability Equation 3.1):

$$R_{\text{drive}} = 1/\sum_{i}(1/r_i) \tag{3.1}$$

where r_i, $i = 1, 2, 3, ...$ is the reliability of each component in series. In a parallel connection, a failure of one component may result in an overload of the rest, but will allow the continuation of operation. Equations 3.2 give the reliability for three typical block connections in Figure 3.2.

$$\begin{aligned} R_s &= r_1 r_2 r_3 \\ R_p &= 1 - (1 - r_1)(1 - r_2)(1 - r_3) \\ R_{sp} &= [1 - (1 - r_1)(1 - r_2)][1 - (1 - r_2)(1 - r_3)(1 - r_4] \end{aligned} \tag{3.2}$$

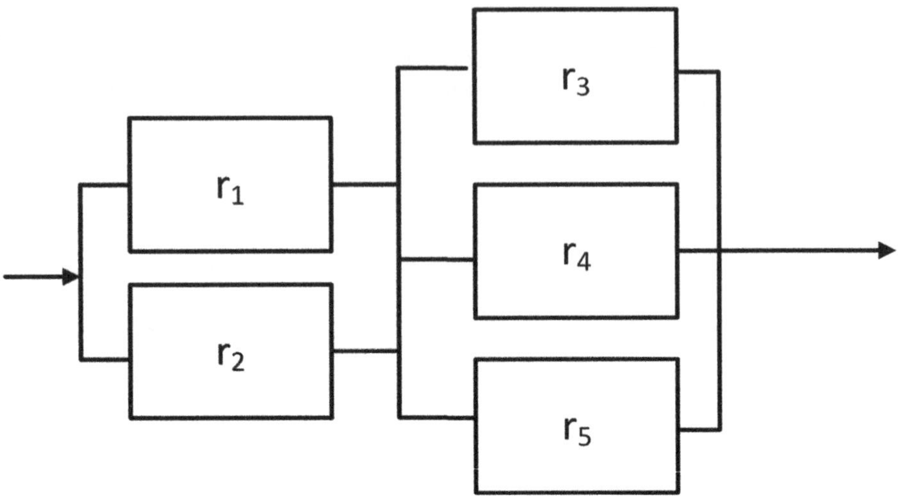

Figure 3.2 Elementary block connections for reliability calculation.

Together with the definition of reliability, other definitions and metrics are used, see for example Song and Wang (2013). The **failure rate** of an item is an indication of its 'proneness to failure' after time t has elapsed. A typical failure rate curve, that is, the number of failures per unit time, as a function of time, is commonly described as the bathtub curve. The shape of the curve suggests that the life cycle of an item can be divided into three elementary periods: the burn-in period, the useful life period, and the wear-out period. Extensive testing at commissioning or end-of-line, and/or burn-in testing aims to decrease that high failure rate at the beginning of life. The failure rate tends to stabilize at a level where it remains relatively constant for a certain period of time, before the item begins to wear-out. The published or calculated failure rate is usually the one during the useful life of the part, while the wear-out period is meant to characterize operation beyond the designed useful life.

The failure rate, $\lambda(t)$, is related to the reliability function, $R(t)$, by:

$$\lambda(t) = \lim_{\Delta t \to 0} \frac{R(t) - R(t + \Delta t)}{R(t)\Delta t} \tag{3.3}$$

$$= -\frac{1}{R(t)} \frac{dR(t)}{dt} \tag{3.4}$$

where Δt is a time interval, $\Delta t > 0$. The reliability $R(t)$ is determined from the failure rate $\lambda(t)$ with the consideration of $R(0) = 1$, that is, the item is fully functional at the initial state

$$R(t) = \exp\left(-\int_0^t \lambda(\tau)d\tau\right). \tag{3.5}$$

Mean Time to Failure (MTTF) is the expected time before a failure occurs. It gives the average time in which an item (component, system) operates without failing. MTTF is a widely quoted performance metric for comparison of various system designs. This indicator reflects the life distribution of an item. As the mission time may be shorter than the MTTF, this definition does not provide adequate meaning, and does not convey the associated high reliability in this case.

The relationship between MTTF and reliability function is described by

$$\text{MTTF} = \int_0^{+\infty} R(t)dt, \tag{3.6}$$

where $R(t)$ is the reliability function. When the failure rate $\lambda(t)$ is constant, the expression for MTTF is:

$$\text{MTTF} = \frac{1}{\lambda} \tag{3.7}$$

Mean Time to Repair (MTTR) is the mean repair time that it takes to eliminate a failure and to restore the system to a specified state.

Availability is the probability that a system will be functioning at a given time.

As reliability $R(t)$ represents the percentage of a group of samples (or the probability of one sample) that can properly function at operating time t, the unreliability or probability of failure $F(t)$ can be defined as percentage of a group of samples that fail at operating time t. The plot of $F(t)$ against the operating time t is referred to as the 'cumulative distribution function' (CDF) curve, which increases with time and can be represented by an analytical function developed in 1951 by Weibull, discussed later.

Compared to the metrics of constants, including MTTF or lifetime, which are widely used in the existing approaches for reliability prediction of power electronics, the metric of variable **cumulative distribution function** (CDF) provides more information about the reliability performance, especially the time-varying and probability-distributed characteristics. Therefore, the CDF of a power electronics component/system is set as the prediction target. CDF is normally generated based on statistics of failed components/products, and it is used to describe the behaviors of failures which have already happened. Of greater interest here, CDF is used for prediction of failures that have not yet happened, and the CDF is generated based on a given mission profile of drive, as well as a series of models and designed tests.

For a more insightful concept, Bazzi et al. (2012) define the concept of fault coverage in response to the fact that it may not be possible to forecast with complete certainty whether a system will be able to deliver its function after a fault. **Fault coverage**, denoted by c, can be interpreted as the conditional probability, given that a fault has occurred altering the system structure, that the system recovers and keeps delivering its intended function, that is, $c = Pr[\text{system recovers}|\text{system fails}]$. For example, the response of a system to a component fault might differ, depending on the operating conditions. In a dynamic system, operating conditions are governed by reference set points and external disturbances (a load in a motor drive is actually an external disturbance for the motor-drive dynamics). Thus, without loss of generality, assume that a system has an input and a fault occurs at different values of the input; assume that the system survives for 80% of the possible inputs, so the coverage for that fault is 80%. This is the probability that the system survives under that fault for different inputs.

Section 3.9.1 discusses additional metrics based on the costs associated with the operation and management of a fault, and provides new metrics of reliability. In that case, cost is not easily tangible, as it may include issues of performance, safety, and so on.

3.2.1 The Pattern of Failures with Time for Non-Repairable Items

Source: O'Connor and Kleyner (2011)

By appreciating the way the hazard rate behaves in time, a lot of information can be collected about the causes of failure and about the reliability of the item.

Decreasing hazard rates are observed in items, which become less likely to fail as their survival time increases. This is often observed in electronic equipment and parts. 'Burn-in' of electronic parts is a good example of the way in which knowledge of a decreasing hazard rate is used to generate an improvement in reliability. The parts are operated under failure-provoking stress conditions for a time before delivery. As substandard parts fail and are rejected, the hazard rate decreases and the surviving population is more reliable.

A constant hazard rate is characteristic of failures, which are caused by the application of loads in excess of the design strength, at a constant average rate. For example, overstress failures due to accidental or transient circuit overload, or maintenance-induced failures of mechanical equipment, typically occur randomly and at a generally constant rate.

3.2.2 Distribution Functions

We will not expand here on the mathematics of distribution functions; we will discuss briefly two continuous ones and one discrete.

The Gaussian (normal) Distribution is by far one of the most widely used stochastic 'model' and it describes many natural phenomena, such as human heights, weather patterns, and so on. However, there are limitations inherent in using this model in many engineering applications. This normal data distribution pattern occurs in many natural phenomena, and have the advantage that whenever several random variables are added together, the resulting sum tends to normal regardless of the distributions of the variables being added. This is known as the central limit theorem. The Gaussian function can be written as:

$$f(x) = \frac{1}{\sigma(2\pi)^{1/2}} \exp\left[-\frac{1}{2}\left(\frac{x-\mu}{\sigma}\right)^2\right], \tag{3.8}$$

where μ is the location parameter, equal to the mean. σ is the scale parameter, equal to the standard deviation (SD).

The Weibull distribution is the most popular statistical distribution used by reliability engineers. It has the great advantage in reliability work that by adjusting the distribution parameters it can be made to fit many life distributions. The Weibull probability density function (pdf) in terms of time t is:

$$f(t) = \begin{cases} \frac{\beta}{\eta^\beta} t^{\beta-1} \exp\left[-\left(\frac{t}{\eta}\right)^\beta\right] & (\text{for } t \geq 0) \\ 0 & (\text{for } t < 0) \end{cases}, \tag{3.9}$$

corresponding to the reliability function:

$$R(t) = \exp\left[-\left(\frac{t}{\eta}\right)^\beta\right], \tag{3.10}$$

with hazard rate:

$$\frac{\beta}{\eta^\beta} t^{\beta-1}, \tag{3.11}$$

and MTTF:

$$\mu = \eta\Gamma\left(\frac{1}{\beta} + 1\right) \tag{3.12}$$

with the Euler function Γ:

$$\Gamma(\alpha) = \int_0^\infty x^{\alpha-1} \exp(-x) dx \tag{3.13}$$

The value of β reflects the hazard function or the expected failure rate of the Weibull distribution and inferences can be drawn about a population's failure characteristics by considering whether the value of β is less than, equal to, or greater than one. Describing the classic bathtub curve:

- if $\beta < 1$ indicates a decreasing failure rate and is usually associated with infant mortality, or early failures,
- if $\beta \approx 1$ a constant failure rate and is usually associated with the useful life,
- if $\beta > 1$ indicates an increasing failure rate and is usually associated with wear out, corresponding to the end life of the product with closer inter-arrival failure times.

Figure 3.3 shows the bathtub curve along with typical β values.

The **binomial distribution** describes a situation in which there are only two outcomes, such as pass or fail, and the probability remains the same for all trials. The pdf for the binomial distribution is:

$$f(x) = \begin{pmatrix} n \\ x \end{pmatrix} p^x q^{(n-x)}. \tag{3.14}$$

with

$$\begin{pmatrix} n \\ x \end{pmatrix} = \frac{n!}{(n-x)!x!} \tag{3.15}$$

This is the probability of obtaining x good items and $(n - x)$ bad items, in a sample of n items, when the probability of selecting a good item is p and of selecting a bad item is q. Thus the expected value μ and the standard deviation σ are:

$$\mu = np, \qquad \sigma = (npq)^{\frac{1}{2}} \tag{3.16}$$

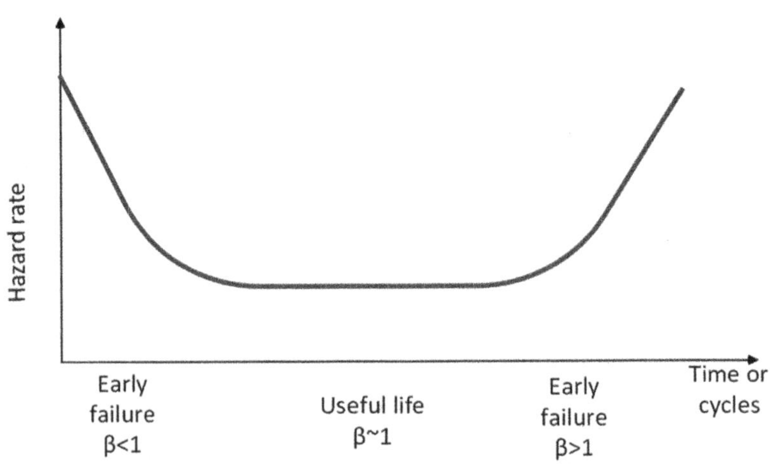

Figure 3.3 The bathtub curve, as described by the Weibull parameter β.

3.2.3 Confidence in Reliability and Prognosis

Confidence interval is a range of values that a parameter of a statistical distribution can have (e.g., mean or standard deviation), and includes the true value of this parameter with a given probability.

One can choose the confidence level. The higher this confidence level is (say 95%), the wider the confidence interval, that is, the range of values of the parameter of interest, that will include the correct value. If, instead, the confidence level is 75%, the confidence interval is narrow and will include the mean with 75% probability.

The confidence interval depends on the sample size; a larger sample will tend to produce a better estimate of the parameter, and hence a narrower confidence interval for the same confidence level.

The confidence interval or band is the interval between the upper and lower confidence limits. Confidence intervals are used in making an assertion about a population given data from a sample. Clearly, the larger the sample, the greater will be our intuitive confidence that the estimate of the population parameter will be close to, and the narrower the confidence band corresponding to an *a priori* set probability.

Similarly, a **prediction interval** or band is used to represent the uncertainty about the value of a new data point on the curve but subject to noise.

Let X be a random sample from a probability distribution with statistical parameter ϑ, which is a quantity to be estimated, and φ, representing quantities that are not of immediate interest. A confidence interval for the parameter ϑ, with confidence level or confidence coefficient γ, is an interval with random endpoints $(u(X), v(X))$, determined by the pair of random variables $u(X)$ and $v(X)$, with the property:

$$P_{r_{\theta,\varphi}}(u(X) < \theta < v(X)) = \gamma \quad \forall\,(\theta, \varphi) \tag{3.17}$$

The quantities φ, in which there is no immediate interest, are nuisance parameters. The number γ, with typical values close to but not greater than 1, is sometimes given in the form $1-\alpha$ (or as a percentage $100\%(1-\alpha)$), where α is a small non-negative number, in the range [0-1], close to 0.

A prediction interval for a random variable is defined similarly to a confidence interval for a statistical parameter. Consider an additional random variable Y, which may or may not be statistically dependent on the random sample X. Then $(u(X), v(X))$ provides a prediction interval for the as-yet-to-be observed value y of Y if

$$P_{r_{\theta,\varphi}}(u(X) < Y < v(X)) = \gamma \quad \forall\,(\theta, \varphi). \tag{3.18}$$

The prediction interval used in the fault failure prognosis and the determination of the management or mitigation is different than the confidence interval, and wider, see Simons (1972). It results in an earlier decision on management, for example, adaptation of the controller parameters, as compared to that resulting from the use of the confidence interval.

- prediction interval:

$$y \pm \frac{t_a}{2} \times S_{y \cdot x} \sqrt{\left(1 + \frac{1}{n} + \sum_{h}^{K} \sum_{1}^{K} (x_h - \hat{x}_h)(x_1 - \hat{x}_1)^{\mu^{h1}}\right)} \tag{3.19}$$

- confidence interval:

$$y \pm \frac{t_a}{2} \times S_{y \cdot x} \sqrt{\left(\frac{1}{n} + \sum_{h}^{K} \sum_{1}^{K} (x_h - \hat{x}_h)(x_1 - \hat{x}_1)^{\mu^{h1}}\right)}, \tag{3.20}$$

where $S_{y \cdot x}$ is the mean square error, MSE.

Rather than using sample statistics as estimators of population parameters and applying confidence intervals to these estimates, we consider the 'next sample', X_{n+1} as itself a statistic, and compute its sampling distribution.

In failure prognosis, we wish to interpret this as a prediction of the next sample, so we model 'the next sample' as a draw from this estimated population, using the (estimated) population distribution. By contrast, in predictive confidence intervals, we use the sampling distribution of (a statistic of) a sample of n or $n + 1$ observations from such a population, and the population distribution is not directly used. Instead, the assumption about its form, though not the values of its parameters, is used in computing the sampling distribution.

3.3 Component Reliability

As the reliability of a subsystem (inverter, controller, motor) depends primarily on the reliability of its components, we discuss first some of the more dominant ones.

Ma et al. (2018) discuss lifetime models in order to better predict the reliability of devices and systems, in that case, power electronics converters, based on different mission profiles and applications. **Mission Profile** is a term describing the expected performance of a drive, during an anticipated time interval or its lifetime, for example, torque, speed, power, voltage, and so on. A typical mission profile is a drive cycle of a vehicle, for example, the driving cycle FTP-72, Figure 3.4. In these 'lifetime prediction' approaches for the reliability assessment of power electronics, for example, the mission profiles are correlated with the reliability performances of the components. The accuracy of the predicted results strongly depends on the lifetime models of components that are used.

Some critical limitations in this "lifetime prediction" approach can be identified as follows:

1. for most of the existing lifetime models, the device is loaded passively or with short load pulses to emulate limited ranges of thermal or other stress intensities. The testing conditions are normally different from the real-field applications, where the characteristics of the thermal cycles are much more complex;
2. the technology and manufacturing of power semiconductors and other devices have been advanced continuously; some of the lifetime models, which were built decades ago, need to be updated; and
3. only constants of lifetime are provided in these lifetime models; the time-varying and probability-distributed characteristics of the reliability are not included.

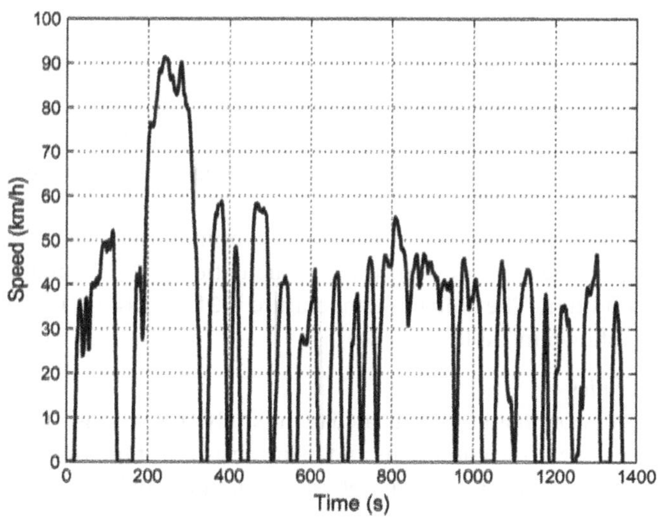

Figure 3.4 Driving cycle FTP-72.

Consequently, there is still little public literature showing that the accuracy of these predicted lifetimes of a converter has been experimentally verified.

Capacitors
Temperature and electrical stresses are the main stresses in DC capacitors. Ran et al. (2019) discuss the effects of temperature and electric stress on the aging of Metallized Film Capacitors (MFC); they estimate the lifetime of the electrical-thermal aging of the dielectric films. They present a reliability model of the MFC based on temperature and electrical stress:

$$L = L_0 \times \exp\left[-\alpha \times \Delta\left(\frac{1}{T}\right)\right] \times \left[\frac{E}{E_0}\right]^{-\left(\beta - \zeta \times \Delta\left(\frac{1}{T}\right]\right)} \tag{3.21}$$

$$\Delta\left(\frac{1}{T}\right) = \frac{1}{T_0} - \frac{1}{T}. \tag{3.22}$$

where L_0 is the lifetime when temperature is at T_0 and electric field is at E_0, L is the lifetime when temperature is at T and electric field is at E; α, β and γ are constant coefficients.

By including temperature and electric acceleration factors α_T and α_E, Ran et al. (2019) arrive at a reliability model:

$$L = L_0 \times \exp\left[-7321.98 \times \Delta\left(\frac{1}{T}\right)\right] \times \left[\frac{E}{E_0}\right]^{-\left(4.61 - 6230.26 \times \Delta\left(\frac{1}{T}\right)\right)}. \tag{3.23}$$

In the long-term operation, the aging degree of MFC is deepening, and the equivalent series resistance is increasing. Finally, under the same operating conditions, the temperature of MFC after aging is higher than that of non-aging MFC. Therefore, the aging rate and the failure rate of MFC are increasing under the same operating conditions. In order to establish a physical failure model of MFC considering aging degree, they establish an aging failure model of MFC from the aspect of equivalent series resistance.

Taking a practical MMC as an example, the minimum voltage of the MFC is $1587.5V$, and the peak to peak voltage fluctuation is 285 V under inverter operation condition. The ambient temperature is $313.73K(40^oC)$. The trend of aging rate is shown in Table 3.1.

Power Electronics
A large body of work has been dedicated to the study of power electronic devices under various operating conditions and using a variety of metrics. These studies have been based on modelling, physics of failure analysis, and extensive experiments.

Ma et al. (2018) estimated experimentally the effect on the cumulative distribution function (CDF) of IGBTs under different stress levels and cycles

Table 3.1 Reliability assessment of MFC under different aging degree; Ran et al. (2019).

Service time (year)	RESR (Ω)	Residual life (year)	Failure rate (occ/year)	reliability
0	2.10E-4	1122.434	0.00089	0.99911
4	2.51E-4	950.450	0.00105	0.99895
8	3.20E-4	717.597	0.00139	0.99861
12	4.22E-4	481.521	0.00208	0.99793
17	5.32E-3	187.938	0.00532	0.99469
20	7.90E-3	126.587	0.00790	0.99213
23	1.71E-2	58.462	0.01710	0.98304
25	6.56E-2	15.255	0.06555	0.93655
26	4.25E-1	2.354	0.42487	0.65385

Table 3.2 Operating Conditions of Testing Setup Under Different thermal cycling Frequencies. Ma et al. (2018).

Condition	f_{out}	f_s	I_{max}	V_{refmax}	T_H
Cl	0. 1 Hz	10 kHz	21 A	113 V	59°C
C2	0.2 Hz	10 kHz	22 A	115 V	57°C
C3	0.5 Hz	10 kHz	25 A	145 V	53°C
C4	1 Hz	10 kHz	30 A	140 V	48°C
C5	1.25 Hz	12 kHz	30 A	140 V	50°C
C6	1.7 Hz	15 kHz	30 A	144 V	47°C

(output and switching frequency, current, and voltage). They apply the developed model to the cases of one and multiple stress levels, that is, thermal cycles, as shown in Table 3.2 and Figure 3.5.

Wang et al. (2014) relate lifetime prediction results to specific failure mechanisms as in Table 3.3.

To improve reliability of IGBTs and to evaluate more accurately the thermal stresses and their effects, Ahsan et al. (2020) first developed a new driving strategy of IGBTs to mitigate wire-bond thermal stresses, and also developed machine-learning based-models to predict degradation from monitored accelerated aging due to thermal cycles.

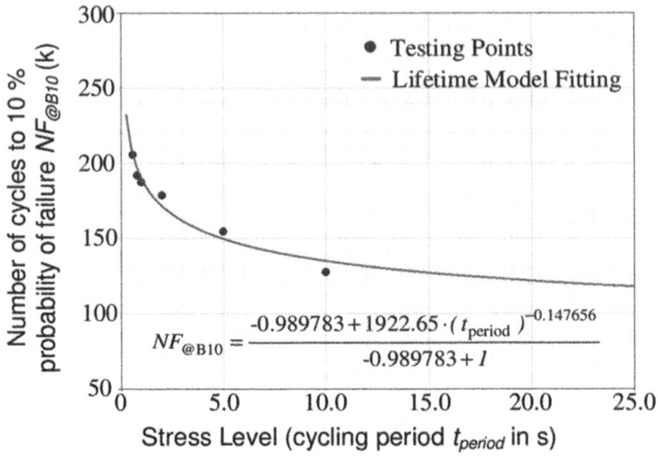

Figure 3.5 B10 Lifetime model of IGBT modules tested under different stress levels of cycling period with testing conditions shown in Table 3.2 Ma et al. (2018).

Table 3.3 Lifetime prediction results Wang et al. (2014).

Failure Mechanism	L_{10} lifetime	Year
Baseplate solder joints due to ΔT_c	358	24
IGBT chip solder joints (due to ΔT_{jIGBT})	438	22
Wire bonds due to ΔT_{jIGBT}	2633	74
Overall determined by the shortest one	358	22

Bearings

The reliability of bearings is greatly affected by their operating conditions. The dominant one and the one that is most probably included in the original design of a motor is the temperature of the lubricant (see Somes and Gerstenkorn (2015)). Manufacturers publish tables relating the type of bearing, temperature and life of grease or grease service, for example, Schatz, Sch (2020) give the L_{10} life as depending on the load and speed:

$$L_{10} = \frac{16.67}{S} \times \left(\frac{C_r}{F_r + ZF_a} \right)^3 \tag{3.24}$$

$$\text{Adjusted bearing Life} = L_{10} \cdot A1 \cdot A2 \cdot A3, \tag{3.25}$$

where F_r is applied radial load, F_a applied axial load, S bearing speed, L_{10} life in hours, and C_r dynamic load rating. L_{10} means $100 - 10 = 90\%$ of identical bearings, operating under the same conditions, can run before their end of life

Table 3.4 Illustrative Failure Rates of sensors Bazzi et al. (2012).

Failure Rate	failures/hour	Failure Rate	failures/hour
λ_{SEO}	7.4×10^{-7}	λ_{CSC}	1×10^{-7}
λ_{SEG}	1.9×10^{-7}	λ_{OC}	5×10^{-7}
λ_{SEC}	1.9×10^{-7}	λ_{SCG}	5×10^{-7}
λ_{CSO}	1×10^{-7}	λ_{SCDC}	5×10^{-7}
λ_{CSG}	1×10^{-7}	λ_{PP}	3.2×10^{-6}
λ_{CSB}	1×10^{-7}	λ_{BR}	3.2×10^{-6}

occurs. This bearing life is adjusted for material and lubrication (factors A2 and A3), while the reliability factor (A1), being 1 for 90% decreases to .62 for 95% (L_5) and .21 for 99% (L_1)

Other manufacturers, for example, Koyo, see Koy (2020), include total revolutions or running distance and a correction coefficient for operating temperature, varying from 1.0 at 125^oC to 0.75 at 250^oC. Correction for design reliability values varies between manufacturers but for increased reliability, for example, 99.95% the modification factor becomes 0.077.

Sensors

Position and current sensors reliability are vital to the health and operation of a drive. Bazzi et al. (2012) study the effects of the operating conditions and give illustrative failure rates for speed encoder constant (SEC), gain (SEG), and omission (no signal, SEC), as well as for current sensors, constant (CSC), bias (CSB), gain (CSG), and omission (CSO) in Table 3.4.

Insulation

Insulation defines the reliability of electrical machines. Their modes of failure and reliability have been the topic of extended laboratory and field studies and research.

Winding shorts due to insulation failure are of the most common causes of major failures in electrical machines and drive systems. Insulation reliability is a dominant determinant of the reliability of a drive. Failure mechanisms and factors have been studied extensively, and models have been developed.

Giangrande et al. (2020) discuss the reliability of the insulation systems in electrical machines. Of particular interest is the insulation lifetime for variable duty applications. Variable duty Electrical Machines (EM) are largely used in safety-critical systems, such as aerospace EMAs and automotive traction drives. Nonetheless, their insulation lifetime prediction is more complex than that for continuous duty electrical machines. Hence, an appropriate lifetime modeling approach can be reached by combining the Arrhenius law, effect of thermal aging, with the cumulative damage law (i.e., Miner's law). Taking into account a generic time-variable temperature profile $\theta(t)$, whose time duration is indicated with Δt_{cycle} (i.e., cycle's period), the insulation loss-of-life occurring during a single cycle (i.e., single temperature profile) at a specific Weibull Cumulative Distribution Function percentile is calculated as in the following equation:

$$\mathrm{LF}_{\text{cycle}-B\rho} = \int_0^{\Delta t_{\text{cycle}}} \frac{1}{L_{B\rho}[\theta_i(t)]} dt. \tag{3.26}$$

where $\mathrm{LF}_{\text{cycle}-B\rho}$ is the loss-of-life per cycle, $L_{B\rho}[\theta_i(t)]$ is the insulation life at the constant temperature $\theta_i(t)$, and dt is the infinitesimal interval of the temperature profile during which the temperature is supposed to be constant.

Therefore, the turn-to-turn insulation will experience a breakdown when the cumulative loss-of-life reaches unity. Considering the insulation exposed to the thermal cycling, its total life L_{tot} can be expressed by the following equation:

$$L_{\text{tot}} = K_{B\rho} \cdot \Delta t_{\text{cycle}}. \tag{3.27}$$

This is discussed, and experimental results and details are provided by Huang et al. (2016a), Huang et al. (2016b) and Maussion et al. (2015).

More complex reliability estimation by Penrose (2014) adds a number of attributes to the life estimate. It incorporates the additional stress based upon the initial electrical endurance of the winding insulation system (E_2), the voltage applied during the study (E_1), and the voltage applied to the winding (E_0):

$$L_{TE} = L \times \exp(-B \times cT) \times \left(\frac{E_0}{E_I}\right)^{-\left(\frac{\log(E_I)}{\log(E_2)} \times cT\right)}, \tag{3.28}$$

Thermal shock, defined as the low temperature at which a rapid increase towards operating temperature, will cause a degradation in the insulation system:

$$L_{TS} = \frac{T_{TI} - T_{SH}}{T_{TI}} \times L_{TE}, \tag{3.29}$$

where L_{TS} is the expected life of the system based upon the amount of time in the range of the thermal shock value (T_{SH}), usually within 10^oC, and T_{T1} is the log average life from severe thermal shock testing.

Vibration is one more stressor of insulation with a clear impact. Severe vibration is determined as the maximum shock, instant or continuous, that the system should see during operation.

$$L_{TV} = \frac{T_{VI} - T_{VS}}{T_{VI}} \times L_{TS}. \tag{3.30}$$

The L_{TV} is the expected life of the system based upon the amount of time in the range of the severe vibration value (T_{vs}) and T_{VI} is the log average life from severe vibration testing.

Operation from a fast switching inverter can exacerbate partial discharges and result in further loss of life. Seri and Montanari (2020) in a series of publications explore this in depth. They note that the aging mechanism based on electric field can change radically if the partial discharge inception voltage is exceeded. This reflects on much shorter life times and lower activation energy of the degradation process. Insulation life can be modelled in

$$L \propto \left(\frac{kT}{h}\right) exp\left[\frac{\Delta G - f(E)}{kT}\right], \tag{3.31}$$

where k and h are Bolzmann and Plank constants, ΔG the activation free energy, and $f(E)$ a function of the electric field.

Most of the insulation system design is based on the inverse power model, often adding a term related to supply voltage frequency:

$$L = t_0 \left(\frac{E}{E_{S0}}\right)^{-\frac{1}{n}} \cdot \frac{f_0}{f}, \tag{3.32}$$

where f_0 is the reference power frequency, parameters n, E_0, and t_0 depend on temperature and the presence of Partial Discharges, E_{S0}, the reference electrical field corresponding to life t_0.

3.4 Reliability of Subsystems and Systems

3.4.1 Analysis Tools

The goal of the safety assessment process is to identify all failures that cause hazardous situations and to demonstrate that their probabilities are sufficiently low. In the application domains of safety-critical systems, the safety assurance process is defined by means of safety standards (e.g., IEC 61508).

Reliability Block Diagram (RBD): O'Connor and Kleyner (2011) show the logical connections between components of the system (see Figure 3.2). The RBD is not necessarily the same as a block schematic diagram of the system's functional layout. For systems involving complex interactions, construction of the RBD can be quite difficult, and a different RBD will be necessary for different definitions of what constitutes a system failure. Block diagram analysis consists of reducing the overall RBD to a simple system which can then be analysed using the formulae for series and parallel arrangements. It is necessary to assume independence of block reliabilities.

Complex RBDs can be analysed using cut set or tie set methods. A cut set is produced by drawing a line through blocks in the system to show the minimum number of failed blocks which would lead to system failure. Tie sets (or path sets) are produced by drawing lines through blocks, which, if all were working, would allow the system to work.

Fault Tree Analysis (FTA) is a common top-down deductive approach to identify failure modes, their causes, and effects with impact on the system safety. With Component Fault Trees (CFTs) there is a model and component-based methodology for FTA, which supports a modular and compositional safety analysis strategy, Zeller and Montrone (2018). Figure 3.6 shows a typical fault tree.

A representative FTA shows all component failure modes. For a large system, such as a flight control system or a chemical process plant, it can be very complex and impracticable to draw out and evaluate manually. Computer programs are used for generating and evaluating FTAs. These perform cutset analysis and create the fault-tree graphics.

However, when using Fault Tree methodology, there is no possibility to express a temporal sequence of events or state-dependent behavior. But this is especially needed for modern, complex systems that are increasingly controlled by software components.

In contrast to this, *Markov Chains (MCs)* are a state-based analysis technique and enable safety and reliability analysis of systems with temporal or

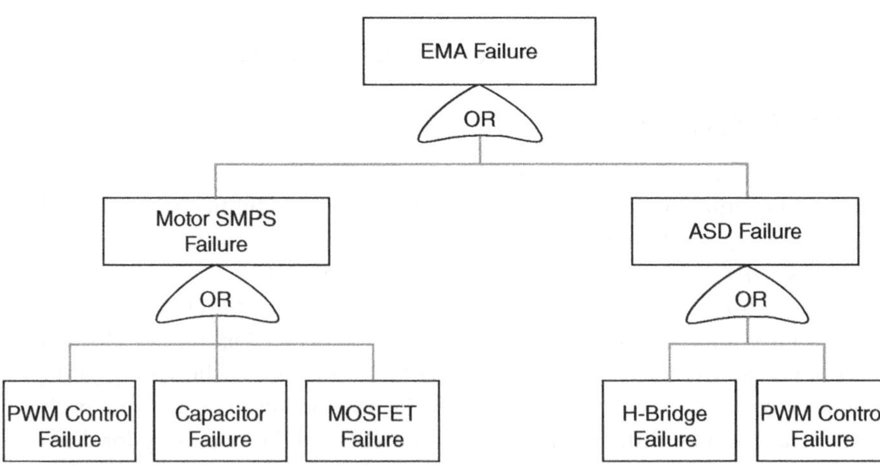

Figure 3.6 Example of Fault Tree graphical representation, Kim et al. (2014).

state-based behavior (e.g., fault tolerance capabilities). However, the use of MCs to analyze systems in industrial practice is limited due to exponential explosion of the size of the model when modeling large-scale systems.

Fault coverage can be included easily in a Markov reliability model, as discussed in Bazzi et al. (2012), thus providing a link between the system dynamic behavior for different operating conditions and system reliability. In this regard, there are two possible outcomes after a component fault triggers a transition. In the Markov reliability model, this is reflected by transitions to two different states: the one in which the system is still operational, and another in which it fails. The transition rate to the operational state is given by the product of the particular component failure rate and the fault coverage c. Similarly, the transition to a failed state is given by the product of the component failure rate and $1 - c$.

The Markov model is a convenient way to find dynamic transition probabilities of the drive, where the transition matrix $\mathbf{\Phi}$ is filled with the transition rates shown in eq. 3.33. These transition probabilities are used to find $R(t)$ and the MTTF. For the model shown in Figure 3.7, repair rates μ_1 and μ_2 are shown as an example of how to build $\mathbf{\Phi}$. The resulting $\mathbf{\Phi}$ for the three states i, j_1, and j_2 is thus :

$$\mathbf{\Phi} = \begin{bmatrix} -c\lambda_{ij_1} - (1-c)\lambda_{ij_2} & c\lambda_{ij_1} & (1-c)\lambda_{ij_2} \\ \mu_1 & -\mu_1 & 0 \\ \mu_2 & 0 & -\mu_2 \end{bmatrix}. \tag{3.33}$$

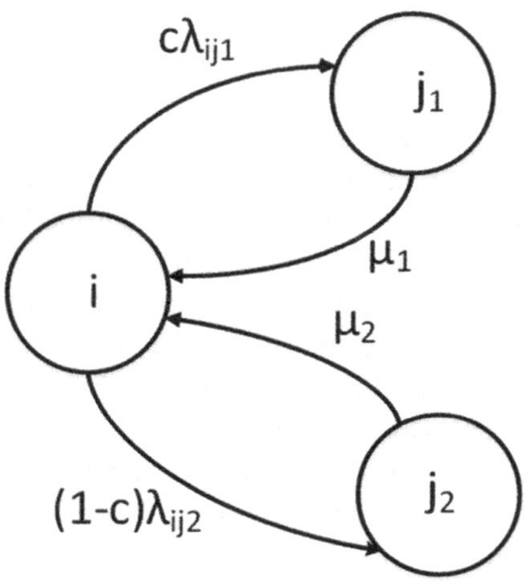

Figure 3.7 Part of a Markov model showing fault coverage Bazzi et al. (2012).

In the context of reliability modeling, the state-transition diagram associated with the Markov model describes the system status — failed or operational — for each system configuration reached after a unique sequence of component failures. The edges represent transitions between configurations triggered by component failures (or repair procedures). There are two types of nodes: absorbing nodes and nonabsorbing nodes. The system will fulfill its function whenever it is in a nonabsorbing node. The system will fail to deliver the functions for which it was designed whenever it transitions to an absorbing node. When a random fault occurs, the system transitions from one state to another. The transition rate is the failure rate of the component in which the fault occurred. A return to the previous state is referred to as recovery and could have a recovery rate associated with it.

The probability **P** of being in a given state at a given time is described by the Chapman–Kolmogorov equation:

$$\dot{\mathbf{P}}^T(t) = \frac{d\mathbf{P}^T}{dt} = \mathbf{\Phi}^T \mathbf{P}^T(t) \tag{3.34}$$

Solving this differential equation yields the probabilities of transitioning from every state to another as a function of time. The solution takes the form:

$$\mathbf{P}^T(t) = e^{\mathbf{\Phi}^T(t)} \mathbf{P}^T(0). \tag{3.35}$$

where $\mathbf{P}^T(0)$ is an $(m + 1) \times 1$ vector, except for $\mathbf{P}_0(0) = 1$.

In a reliability analysis context, system reliability is quantified by calculating the probability that the system will be in any of the nonabsorbing states at a given time. This can be accomplished by adding the entries of $\mathbf{P}^T(t)$ associated with nonabsorbing states.

As already discussed, at the first level, system reliability is computed from reliability of components, but there are some problems with this. Beyond including the mission profile and the changing operating conditions due to operating characteristics and demands, and environment, dramatic changes in topology and control based on faults, real, misidentified or predicted, modify the operating profile of individual components.

Kim et al. (2014) discuss the relationship between Fault Tree Flow Diagram and Bayesian Network. Unlike the tree-based structure of Fault Tree (FT), Bayesian Networks (BN) are directed acyclic graphs (DAGs) for representing the joint distribution and reasoning under uncertainty.

Reasoning under uncertainty refers to the capability of representing and drawing an inference of data with uncertainty. In practice, collected data may include noise, errors, or even missing values. Moreover, knowledge provided by experts may sometimes contain ambiguity or non-informative guesswork. Note that uncertainty, in this paper, refers to the randomness of the reliability and failure relationship among system elements. The reliability of a system element is treated as a random variable to quantify its subjective belief. The failure relationship is represented with conditional probabilities to provide a more generic and realistic representation. Uncertainty may influence both the accuracy and the precision of reliability estimation. High accuracy of the reliability estimation means that the expected reliability has small or no deviation from its true value. High precision means the variance of the estimated reliability is small.

Although in the FT every basic failure event is associated with a probability, the logic gates are deterministic and there is no belief-updating process as in the Bayesian inference scheme adopted by BNs. In BNs, each node is assigned a random variable with a certain probability distribution in describing a particular event. Probability distribution can fully capture the uncertain information of the data. Arcs direct from parent nodes to children nodes indicate the cause-and-effect relationships. Conditional probability tables (CPTs) are specified for each pair of connected nodes in quantifying the dependency strength. A CPT is rather flexible in handling logic relationships with uncertainty.

3.5 Lifetime, Reliability Prediction

There are several system-level reliability prediction tools (Relex, ItemSoft, PRISM, ReliaSoft, Isograph, etc.) commonly used in the commercial market. These tools enable the user to predict and possibly prevent failures by analyzing existing data. In other efforts to improve the education process, some of the researchers have developed tools to help students in the courses and class assignments, see Goel and Graves (2006).

Prediction values, however, are not a guarantee that the predicted reliability is actually realized in its operational application. Depending on the assumptions and methods used, actual reliability can vary significantly from the value obtained by a prediction.

Reliability prediction has also been criticized for hampering new design techniques. Codier (1968), raised a question as to the validity of the prediction models and the concept of assigning a failure rate to a component. There is no doubt that it is impossible to predict the field reliability, where any kind of mishandling can cause the product to fail. Loll (1998) introduced another term known as reliability budget. He suggested that instead of reliability prediction, project managers should use a reliability budget, which will guide them to the part of the design where early analysis and tests should be made. According to this theory, by adding the nonconstant failure rate in the budget, one can include wear out in the analysis. This allows the comparison of the reliability budget with the reliability target for the system. He also discussed how predictions hamper progress and argued that reliability prediction is nothing but a self-fulfilling prophecy.

In contrast to reliability prediction, lifetime prediction is a deterministic approach, see Hirschmann et al. (2007). The results are somehow comparable, but one has to keep in mind that the basic principle of the two calculation methods is fundamentally different. Lifetime prediction is based on physical models and makes the assumption that a component will withstand a certain amount of stress and then fail. Therefore, all identical components will fail at exactly the same time. As a result, the weakest component in a system will determine the lifetime of the whole system. Reliability prediction is based on empirical analysis, whereas lifetime prediction tries to develop formulas that describe the effects of the destructive processes on the lifetime. Hence, lifetime prediction does not account for random processes. It is comparable to the physics-of-failure model. In the following, two methods to determine lifetime are explained, see Hirschmann et al. (2007).

1. Analysis Based on Coffin–Manson

The Coffin–Manson relationship has been used and modified for many applications. Since all materials expand and contract due to temperature cycles, nowadays modified relationships are also used to predict the capability of semiconductors to withstand temperature cycles. One example formula is:

$$N = 10^7 \cdot e^{(-0.05\Delta T)}, \tag{3.36}$$

where N is the number of temperature cycles, the transistor will withstand if the junction temperature varies for ΔT. Note that this formula only considers temperature cycles. It appears as if the absolute component temperature does not have any influence on the predicted lifetime.

Due to the fact that the number of cycles N is not considered in lifetime predictions or in reliability predictions, Equation 3.36 had to be converted. If a component will withstand N cycles, it is reasonable to state that at each cycle a fraction, that is, $1/N$, of the above-mentioned device is damaged. Thus, this fraction represents a partial destruction of the device. If the fractions of all temperature cycles are summed up, this value can be interpreted as the probability of failure $F(t)$, which is defined as $1 - R(t)$. If the value reaches 1, the component will fail. This can be used to determine the number of driving cycles the component will work nominally.

2. Arrhenius Equation

Generally, chemical reactions can be accelerated by increasing the system temperature. The relation between the system temperature and the reaction rate is described quantitatively by the Arrhenius equation. Since it is a chemical process, the aging of, for example, organic film capacitors is accelerated by increasing its temperature. This is especially true for electrolytic capacitors. In the data sheets of the used capacitors, the lifespan is calculated by:

$$L = L_0 \left[\frac{E}{E_0}\right]^{-(n-bT)} exp(-BT), \tag{3.37}$$

where E_0 is the value of electrical stress below which electrical aging can be neglected; L the life time to be estimated at temperature Θ; L_0 base lifetime of capacitor at temperature Θ_0; $T = (1/\Theta_0 - 1/\Theta)$ and n, b, B model parameters.

This equation is called the Arrhenius law and has been applied, among others, to aluminum electrolytic capacitors.

Goel and Graves (2006) discuss the design aspects for reliability. A system designer needs reliability information during the design phase to determine the initial, maintenance, and total system costs, as well as system-level reliability and availability. Many organizations choose to meet the demand for their products by integrating the reliability considerations into the product design process right from the start. A reliability model presents a clear picture of functional interdependencies and provides a framework for developing quantitative system-level reliability estimates to guide the design tradeoff process.

An important element of the design for reliability process is reliability prediction, enabling some form of prediction for product failure rates. System-level predictions provide a quantitative means of evaluating design alternatives in terms of their reliability estimates.

Reliability prediction is therefore seen as:

- providing a quantitative forecast of the reliability;
- helping to achieve a reliable design and manufacturing process that meets end-user reliability;
- comparing competitive designs;
- identifying potential reliability problems, such as design imbalance, source of unreliability, etc.;
- helping in feasibility evaluation, which seeks to determine if design reliability goals can be met;
- predicting warranty cost and maintenance support requirements;
- assessing potential warranty risks and providing inputs to safety analysis.

Accurate and timely reliability predictions are an important part of a well-structured reliability program. If properly performed, they can provide insight into the design and maintenance of reliable systems. Many authors stress that reliability cannot be an afterthought. It must be a goal from the beginning of the design phase and implemented as a continuing activity throughout a project, with the predictions being regularly updated as more design, test, and evaluation data become available.

For series systems (i.e., systems without redundancy), reliability prediction methods make the following assumptions:

1. a system will fail if any of its components fails;
2. the failure rate of a system is calculated by summing the failure rates of individual components;
3. at the component level, the failure rate prediction methods assume that during the useful life of a system, the failure rates of the individual components are constant.

Some researchers have investigated new methods, which do not depend on the above assumptions and have proved the validity of their methods by comparing the predicted values with test results and/or field failure data.

3.6 Fault Management and Mitigation

Although it has been discussed earlier, it is useful at this point to relate faults to the appropriate actions as related to reliability.

Repair to 'as new' or less can be an option, but it is not the preferable approach, at least not as far as reliability is the measure. This can be inadequate when associated with mission profile, especially when interruption of service for repair is not feasible or too costly. The effect that the estimation of failure has on maintenance scheduling is discussed later. The alternatives to interruption and repair are either continuation of operation under the existing or modified operation, the change of the control, and the utilization of physical redundancies, either partial on full. The factors that are important in deciding mitigation are:

- the time it takes to activate the redundant subsystem, component, or alternative control;
- the increased load on components and subsystems after mitigation, the modified mission profile, and the reliability of the resulting system;
- the accuracy of the diagnosis and prognosis that precipitates the modification.

When it comes to power electronics in a drive, the major causes of failures, and hence for the mitigation that has to be considered first, the following options are possible:

1. In the simplest case, a three-phase electrical machine supplied from a two-level inverter, one of the switches can be disabled through a gate driver

malfunction. Similarly, one of the windings is opened, forcing the operation with only two phases. Operation can continue, although the control algorithm must have the option to do so. The resulting power rating will decrease, the loading of the power electronics will increase (assuming operation at the same torque), and torque pulsations may result. A different algorithm can be followed in the case of a neutral permanently connected to the midpoint of the DC supply (Kersten et al. (2019) Eickhoff et al. (2018)).

Detection of the fault can be very fast, within a fraction of the power cycle, as is the transfer to a new controller algorithm.

2. Mirafzal (2014) provides an extensive review of fault-tolerant topologies of two-level and multi-level inverters. With the activation of a redundant fourth leg of the inverter and possibly additional triacs a number of faults can be managed: phase loss (PL), switch open circuit (SOC), leg open (LOC), switch short circuit (SSC), Leg short circuit (LSC). When a split-bus capacitor is used, the post-fault PWM, pulses are between $\pm V_{dc}/2$. A topology using both triacs and fuses allows the continued operation under all these faults, but with a decrease of the post fault output power by $1/\sqrt{3}$. Other topologies are simpler but cannot mitigate all faults - this may be of lesser importance if some faults have a low probability of occurring.
3. An incipient fault, if caused by high temperature, can be detected in power electronics, and its effects can be mitigated by the modification of the PWM algorithm, decreasing the switching losses, for example, Song and Wang (2014), Ugur et al. (2019).
4. A multi-level inverter can be used, the operation of which will allow to mitigate an open form.
5. A multi-phase winding stator can operate with fewer phases, and with decreased or eliminated torque pulsation, as well as a relatively smaller increase of the switch and winding current.

A short circuit in the stator windings of an AC machine will cause increased currents and thermal problems, leading to subsequent faults. The problem is more pronounced in permanent magnet AC (PMAC) machines.

Short circuit faults in PMAC machines can quickly lead to decreased torque production, increased temperature in the shorted windings, and a failure that will migrate to neighboring slots and phases. The management of turn-turn short-circuit in PMAC machines has been reviewed extensively, for example, by Arumugam et al. (2015). It depends greatly on the machine design, and the available time to manage a fault before it expands is very limited. The mitigation techniques reviewed there are:

1. A method proposed for PM machines, where a separate H-bridge inverter for each phase is used to isolate the phase windings electrically and control

them under SC fault. The respective faulty phase is short-circuited through the H-bridge converter switches. To limit the magnitude of the current in the shorted phase, the machine has to be designed with more than 1.0 pu inductance.

2. A current injection method for limiting short-circuit (SC) current in an open slot, bar wound, large PM machine. The injected current must have the same phase angle as the terminal SC current, and its magnitude must be greater than the terminal SC current (complete phase short-circuited), whilst being considerably smaller than the single turn SC current. However, this method would not be effective in the case of a semi-open slot, since the injected current would be greater than the single turn SC current.
3. More involved techniques, which could both endanger reliability through increased complexity, as well as improve it by limiting flux, involve field windings and magnetic shunts. A magnetic shunt arrangement was proposed for a PM alternator. The shunts are placed between the pole pieces and the winding of each flux shunt; this generates a flux proportional to the supplied current. The flux through the stator winding is controlled by increasing or reducing the shunt flux and thus by controlling the amount of flux diverted from the magnets through the shunt. In the faulted case, the windings around the shunt are de-excited, thus, a large portion of the PM flux is shorted through the shunt. This method requires either the use of slip rings or rotating diodes, since the windings are in the rotor.
4. A stator winding arrangement with fuses would limit the short-circuit current, at increased cost and complexity.

Diao et al. (2013) proposed a fault detection and diagnosis structure of mechanical and current sensors faults of a Permanent Magnet Synchronous Machine (PMSM) drive. The method is based on two interconnected observers: an Extended Kalman Filter (EKF) and a Model Reference Adaptive System (MRAS) observer. The EKF, thanks to its optimality, is designed to estimate the position in case of mechanical sensor fault and despite sensor current fault. The MRAS estimates the phase currents using the actual position and speed. The computation and sort of the residuals (difference between measured and estimated values) allow the fault isolation.

The system can deal with simultaneous phase current sensor and mechanical sensor fault, using these measurements as well as stator voltage measurements.

In case of a mechanical rotor position sensor fault, the estimated values with the EKF are used to ensure continuity of the operation. For current sensor fault the principle of balanced machine is used and the faulted phase current sensor is isolated and the measured value is computed from the two other healthy phase current sensors. However, when reconfiguration can't be achieved, a safe stop is operated.

3.7 Design and Manufacturing

Increasing reliability of a system quickly translates to increasing reliability of components as they function within the system, as well as addressing issues of cascading faults and failures. To increase the reliability of drive, a number of actions are needed in the design of drives, the selection of components, the control, the operation, the diagnosis of possible faults, prognosis, and selection of mitigating actions, including redundancies.

To select these, the drive designer has to determine first the possible stresses that will lead to degradation and failures. This means that probable and anticipated modes of failure have to be considered and included in the design process. It is often a mistake, in this process, to over design components in the expectation that the overuse of materials will improve reliability. Besides the obvious problems (cost, weight, losses), over-design may have the opposite effect. It is a careful design that is needed, based on analysis and use of history of failures. To inform this design studies on materials hand software have to be available. There have been studies on how winding insulation materials fail under thermal, mechanical, and electrical stresses, the effects of duty cycles and temperature on power electronics switches, the causes of bearing currents and how they cause bearing degradation, and the detailed modeling of battery cells.

Interaction between components is important in the design for reliability. Among them, as already noticed, are the characteristics of inverters (e.g., speed of switching and voltage step size and their effect on bearing degradation and insulation stresses and aging). When the machine and power electronics are in close proximity, the temperature effects have to be studied together, while if they are distant, overvoltages due to traveling waves will stress insulation.

Of great importance is the increased testing of components, in intermediate and/or final stages, before integration and assembly. Among them, must be tested materials (e.g., magnets, steel), insulation (a variety of tests, high pot, partial discharges), power electronics, capacitors, batteries.

A second level of reliability analysis at the design stage is considering the effects of inclusion of additional, redundant elements to be used in the case of failure, providing a partial redundancy. They may include, as discussed earlier, a neutral connection of a multi-phase machine, additional inverter legs, additional batteries. Although these are expected to be used to mitigate faults, their mere presence may compromise the total reliability.

Based on these fundamentals, design starts with the selection of alternative topologies, where reliability is central. Topologies include:

- design of the machine, so that a fault does not expand. Typical is the case of single coils of phase windings isolated in slots, which are minimally

coupled with windings of other phases, magnetically, electrically and thermally,
- selection of slot design and the resulting value of inductance that will affect the short circuit current,
- selection of winding insulation based on the expected mission, so that aging will not cause premature deterioration,
- internal redundancies in the machine design can include:
  - more than three-phases,
  - access to the neutral, so that it can be connected to the DC mid-point or to an additional inverter leg,
  - completely separate terminals for each phase, so that each can be connected to a dedicated 1-phase inverter,
- design of the inverter supply that can be a standard three-phase two-level inverter, one with a fourth leg, DC midpoint, multilevel inverter, or
- design for redundancies, including:
  - Additional fuses or triacs,
  - multi-level inverters, from a plethora of alternatives, see for example Mirafzal (2014),
- controller that can allow for:
  - fault diagnosis and failure prognosis,
  - decision regarding redundancies, reconfiguration, etc,
  - operation with failure of sensors,
  - operation with reconfigured topologies.

Determination of the stresses of components during normal mission as well as after reconfiguration includes:

- electrical stresses in insulation (including partial discharges) and switching, including operation at full DC voltage after reconfiguration,
- thermal analysis of both inverter and electrical machine,
- stresses on bearings.

3.8 Applications and Case Studies

Although not directly addressing drive reliability, a number of studies address the reliability of the components by adaptive parameter estimation, using the degradation and failure mechanisms.

Si (2015) recognizes that the assumption that the degradation process has a linear path or can be linearized by logarithmic or time-scale transformations restricts the applications of these prognostic methods in several complex circumstances. An example is the lithium-ion batteries, which contain three different operational profiles (charge, discharge, and impedance). These environmental/operational conditions affect the degradation processes of systems and lead to the varying degradation rate of systems, that is, nonlinear stochastic degradation.

In the context of Prognostics and Health Monitoring (PHM), it is necessary and valuable to derive the PDF of the RUL with an explicit form to provide real-time RUL estimation for the subsequent maintenance scheduling. An adaptive and nonlinear prognostic model is presented for RUL estimation, in which the dynamics and nonlinearity of the degradation process are modeled by a time-dependent drift coefficient. This degradation model can cover conventionally linear models. In order to make the RUL estimation depending on the history of the observations, a state-space model is constructed, and Bayesian filtering is applied to update one parameter in the drifting function by treating this parameter as a hidden state variable. The PDF of the RUL is derived with an explicit form, and some commonly used linear model-based methods can be shown to be special cases of the proposed model. In addition, the Expectation Maximization (EM) algorithm in conjunction with the Kalman filter is utilized to update the drifting parameter and other parameters in the state-space model simultaneously and recursively. Finally, several numerical examples and an application to lithium-ion batteries data were provided to validate the proposed approach.

The author there uses a degradation model and defines RUL as the time that the degradation hits first the pre-established threshold. The essence here is to consider the observation history to estimate the parameters of a battery and adapt them to the observation data.

More explicitly referring to reliability, Wang et al. (2014) address reliability prediction (not based on constant failure rate λ) as an important tool to quantify the lifetime, failure level, and design robustness based on various sources of data and prediction models.

Figure 3.8 shows a generic prediction toolbox based on the Physics of Failure approach. The toolbox includes statistical models and lifetime models and various sources of available data (e.g., manufacturer testing data, simulation data, field data, etc.) for the reliability prediction of individual components and the overall system.

As not all stressors have equivalent effects, it is useful to analyze the most critical ones that will cause disabling damage to components. Based on the accumulated industrial experiences and future research needs, Table 3.5 shows the critical stressors for different components in power electronic systems. It can be noted that steady-state temperature, temperature swings, humidity, voltage, and vibrations have different levels of impact on semiconductor devices, capacitors, inductors, and low power control boards.

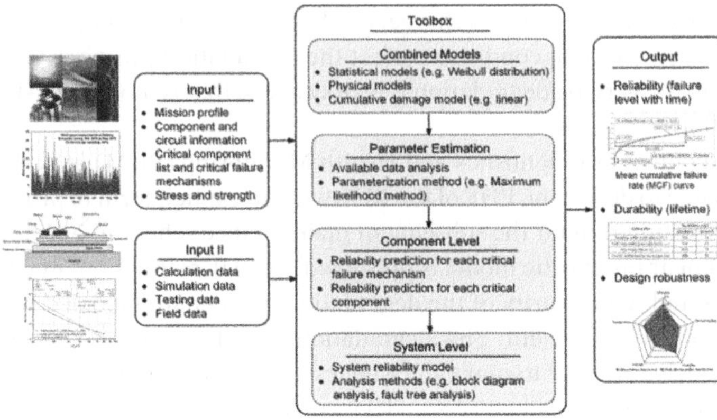

Figure 3.8 Reliability prediction toolbox for power electronic systems Wang et al. (2014).

Table 3.5 Focus Point Matrix for Power Electronics, Wang et al. (2014).

Load			Focus points									
Climate + Design => Stressor			Active power components			Passive power components		Control circuitry, IC, PCB, connectors…				
Ambient	Product design	Stressors	Die	LASJ	Wire-bond	Cap.	Ind.	Solder Joint	MLCC	IC	PCB	Connectors
Relative humidity -$RH(t)$ Temperature -$T(t)$	-thermal system -operation point -ON/OFF -power $P(t)$	Temperature swing ΔT	X	X	X			X				
		Average Temperature T	X	X	X	X		X	X	x	x	x
		dT/dt	x	x	x	x						
		Water								X	X	x
		Relative Humidity	x	x	x	X	x	x	x	X	X	x
Pollution	Tightness	Pollution						x			x	
Mains	Circuit	Voltage	x	x	x	X	X		x	x	x	x
Cosmic	Circuit	Voltage	x									
Mounting	Mechanical	Chock /vibration	x			x	x	x	x			x

LASJ - Large Area Solder Joint, MLCC - Multi-Layer Ceramic Capacitor, IC- Integrated Circuit, PCB – Printed Circuit Board, Cap. - Capacitor, Ind. - Inductor, Level of importance (from high to low): X-X-X-x.

They present two case studies on the application of IGBTs and capacitors in power converters, based on critical failure mechanisms related to thermal cycling and voltage stressing.

A case study on the lifetime prediction of IGBT modules in a 2.3 MW wind power converter has been studied and applied to a two-level back-to-back (2L-BTB) converter. The technical advantage of the 2L-BTB topology is the relatively simple structure and few components, which contributes to a well-proven robust and reliable performance. By following the prediction procedure from wind speed profile analysis, case temperature and junction temperature estimation, cycling counting of temperature swings to parameter estimation of lifetime models, the lifetime of two candidates of IGBT modules for the grid-side converter is predicted. The lifetime prediction is based on each of the three critical failure mechanisms related to thermal cycling.

Xu et al. (2016) present a design philosophy of a fault-tolerant ten-phase Fault Tolerant Permanent Magnet Synchronous Motor (FTPMSM). It has two stators and two rotors mounted on the same shaft. Each segment of the FTPMSM can be considered as an 8-pole/10-slot PMSM, in which the concentrated, single-layer, and alternative teeth wound winding is adopted, based on the reliability analysis.

First, they analyze the fault mode of the system. Then the Markov process is adopted to calculate the reliability of the phase fault, while the reliability prediction approach is used to calculate the reliability of the remaining fault modes. Finally, the reliability model of the proposed FTPMSM system is presented for the reliability analysis.

Five possible fault modes are identified:

1. Phase winding fault, including the short-circuit fault and the open-circuit fault;
2. Bearing fault;
3. H-bridge inverter fault, including the MOSFET short-circuit fault and the MOSFET open-circuit fault;
4. Micro control unit fault;
5. Position sensor fault in the signal acquisition circuit, where a rotary transformer is used.

For the Markov state space $E = \{s_0, \cdots, {}_8\}$, the Markov state transition diagram is represented as shown in 3.9. There, λ_0 denotes the failure rate of the phase open-circuit fault, λ_s, the failure rate of the phase short-circuit fault, s_0 the system without any phase fault, s_1 the system under one–phase open-circuit fault, s_2 the system under one–phase short-circuit fault, s_3 the system under two-phase open-circuit faults, s_4 the system under one–phase open-circuit fault and one–phase short-circuit fault, s_5 the system under two-phase short-circuit faults, s_6 the system under three-phase open-circuit faults, s_7 the system under two-phase open-circuit faults and one–phase short-circuit fault, s_8 the system failure, which means that the system cannot operate continuously without system performance degradation.

With the use of the Markov state transition diagram and the total probability formula, the Markov model for the phase fault can be represented by a system of equations below. λ_o is associated with the phase winding open-circuit fault and the H-bridge inverter fault, while λ_s is the failure rate of the phase winding short-circuit fault, for example:

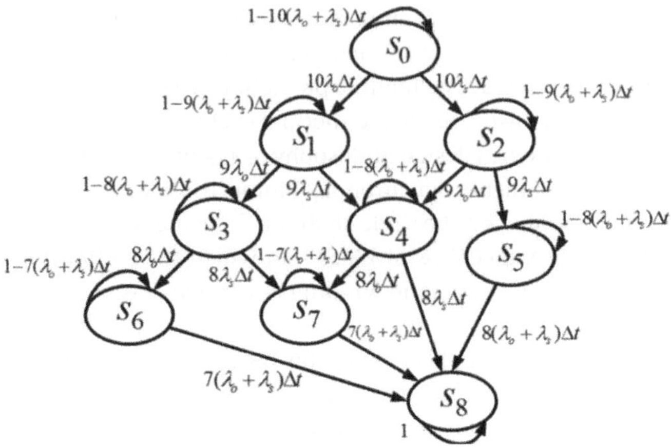

Figure 3.9 Markov state transition diagram for the phase fault Xu et al. (2016).

$$
\begin{aligned}
P_0(t+\Delta t) &= P_0(t)\,[1-10(\lambda_o+\lambda_s)\Delta t]\\
&\;\;\vdots\\
P_5(t+\Delta t) &= 9\lambda_s\Delta t P_2(t) + P_5(t)\,[1-8(\lambda_o+\lambda_s)\Delta t]\\
&\;\;\vdots\\
P_{10}(t+\Delta t) &= 7(\lambda_o+\lambda_s)\Delta t P_6 t) + 7(\lambda_o+\lambda_s)\Delta t P_7(t)\\
&\quad + 8\lambda_s\Delta t P_4(t) + 8(\lambda_o+\Lambda_s)\Delta t P_5(t) + P_8(t),
\end{aligned}
\tag{3.38}
$$

where $P_i(t)$ and $P_i(t+\Delta t)$ denote the probability of the system under the state s_i at time t and $t+\Delta t$, respectively. Then the differential form of the Markov model can be represented as follows:

$$
\begin{aligned}
P_0'(t) &= -10(\lambda_o+\lambda_s)P(t)\\
&\;\;\vdots\\
P_5'(t) &= 9\lambda_s P_2(t) - 8(\lambda_o+\lambda_s)P_5(t),\\
&\;\;\vdots\\
P_{10}'(t) &= 7(\lambda_o+\lambda_s)P_6(t) + 7(\lambda_o+\lambda_s)P_7(t)\\
&\quad + 8\lambda_s t P_4(t) + 8(\lambda_o+\Lambda_s)P_5(t).
\end{aligned}
\tag{3.39}
$$

Following this systematic approach, and including all previously identified faults, this results in the reliability $R_p(t)$ for the phase fault taking the form:

$$R_p(t) = \sum_{i=0}^{7} P_i(t) \tag{3.40}$$

with

$$\vdots$$

$$P_5(t) = \frac{45\lambda_s}{(\lambda_o + \lambda_s)^2}\left(e^{-8(\lambda_o+\lambda_s)t} - 2e^{-9(\lambda_o+\lambda_s)t} + e^{-10(\lambda_o+\lambda_s)t}\right), \tag{3.41}$$

$$\vdots$$

In Strangas et al. (2013), the authors take a similar approach for a limited number of faults in a PMAC machine, assuming that the first fault is not immediately catastrophic but can be mitigated. They include the effects of fault mitigation and subsequent changes in operation of the surviving system. They combine component reliability, diagnosis and prognosis, and the reliability of the faulted system components before and after mitigation to arrive at the reliability of a complete system that includes internal redundancies and the effects of the loading after mitigation. Figures 3.10 and 3.11 show the Markov chains in the two cases, without and with fault mitigation, as they can lead to a final failure, and from that, they estimate the reliability of the system in these two designs.

In Molaei et al. (2006), the authors start from the component models, develop block reliability characteristics by combining them, and use them in a complete model. In this, they do include replacement time in calculating inverter and rectifier reliability:

$$\lambda = \frac{-\ln\left[\sum_{i=r}^{n}\binom{n}{i}P^i(1-P)^{n-i}\right]}{TM} \tag{3.42}$$

$$P = e^{-\lambda TM} \tag{3.43}$$

$$P^i \text{ : Each Switch Reliability} \tag{3.44}$$

$$\lambda_0 \text{ : Each Switch Failure Rate} \tag{3.45}$$

$$TM \text{ : Replacement Time} \tag{3.46}$$

The authors include all components of the drive:

- Transformer
- Circuit Breaker
- Incoming and Outgoing Cable
- Incoming Filter Modeling
- Inverter

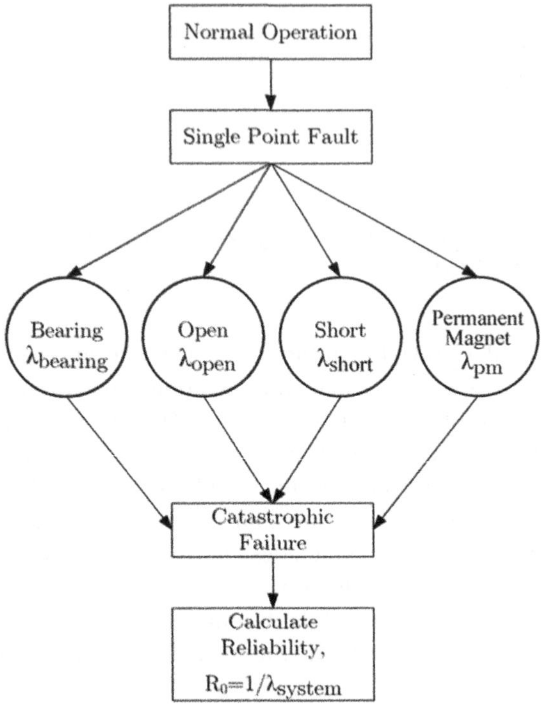

Figure 3.10 Failure rates of drive systems without fault mitigation, Strangas et al. (2013).

- DC Link Reactance
- DC Link Capacitor
- Output Filter
- Motor
- Control Circuit
- Cooling System

For determination of Markov modeling for the whole system reliability calculation, the equivalent component are combined with control circuit model as in Equations 3.47.

$$\lambda_s = \sum_{i=1}^{n} \lambda_i = \lambda_{\text{circuit breaker}} + \lambda_{\text{transformer}} + \lambda_{\text{input cable}}$$
$$+ \lambda_{\text{input filter}} + \lambda_{\text{rectifier}} + \lambda_{\text{inductor}}$$
$$+ \lambda_{\text{capacitor}} + \lambda_{\text{inverter}} + \lambda_{\text{output filter}}$$

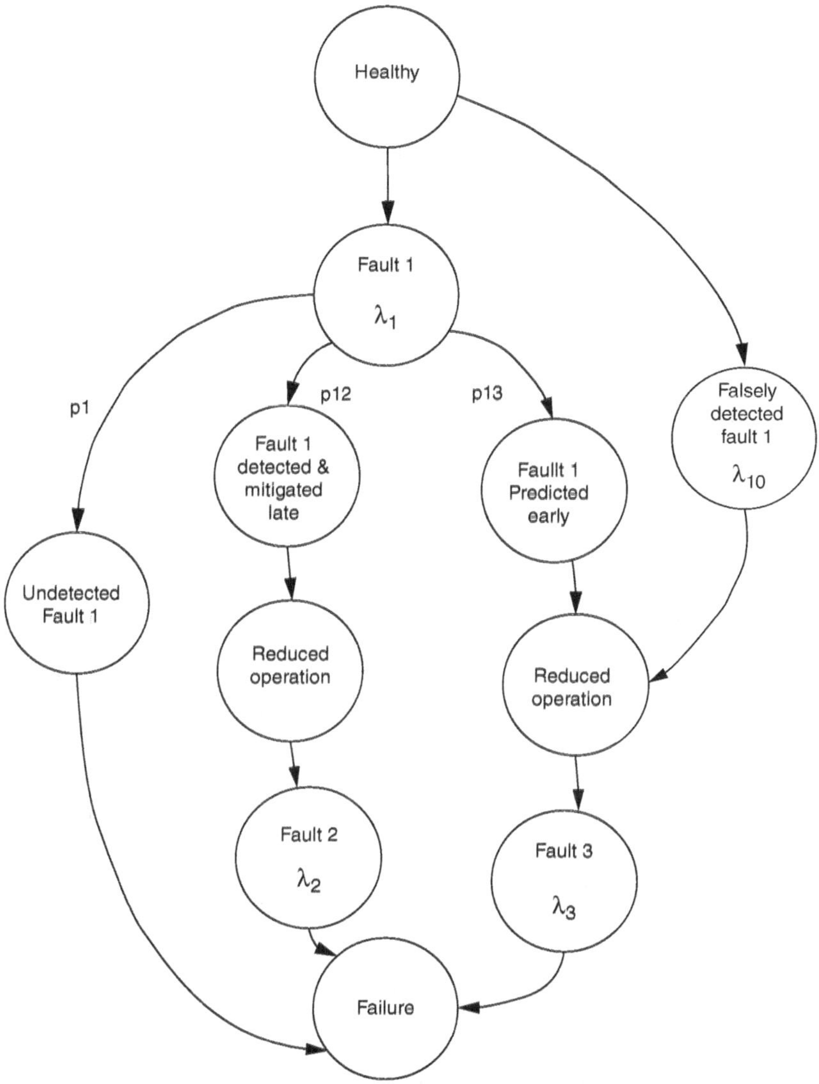

Figure 3.11 Paths to failure of a system with fault diagnosis and prognosis Strangas et al. (2013).

$$+ \lambda_{\text{output cable}} + \lambda_{\text{motor}} + \lambda_{\text{control circuit}}$$

$$r_s = \frac{1}{\mu_s} = \frac{\sum_{i=1}^{n} \lambda_i r_i}{\lambda_s}. \tag{3.47}$$

Each component has two possible states, Up and Down, except the control circuit model that has four states. Single-value liabilities are assigned to each component. The four-state control circuit is combined with other components, and applied to a drive with a three-phase transformer, a two level rectifier, a three-level VSI inverter, and an induction motor. The system reliability diagram versus time is computed, with both two stages and four stages control.

Bazzi et al. (2012) propose a framework that is distinct from general reliability modeling procedures as those in Rausand and Høyland (2004). It considers operational and reliability aspects of motor drives, including fault modes of drive performance requirements in different drive components, while being easy to follow by designers and researchers in the power electronics and drives community.

The proposed reliability model combines several reliability concepts such as the Markov reliability modeling and fault coverage with a dynamic model of the overall system that enables coupling the impacts of component faults with the overall system dynamic performance. The faults studied cover the power electronics, machine, and sensors, and the proposed framework provides the flexibility to go into a more detailed evaluation at the component level.

The system consists of an induction motor drive under IFOC (Indirect Field Oriented Control) requires measurements of the three-phase currents and speed and has two control inputs for torque, or speed, and rotor flux. Faults can occur in the following six possible subsystems or components as in Figure 3.12:

1. induction machine,
2. inverter,
3. current sensors,
4. speed encoder,
5. control and estimation platform,
6. connectors and wires.

The control and estimation platform, connectors, and wires are assumed to have very low failure rates compared to the rest of the system. This assumption is justified in part since the platform, which is usually a microcontroller, DSP, or field-programmable gate array (FPGA), is very reliable and does not involve high currents, high voltages, or mechanical movement. The wires and connectors are static and should have low failure rates, except in high-vibration or corrosive environments, if selected and installed correctly.

Although component redundancy is not considered when analyzing the faults, power electronics and current sensor faults are. System performance criteria are essential in determining whether performance is acceptable. After every fault, system performance is evaluated based on performance measures. These metrics can be functional, such as state overshoot or settling time, or

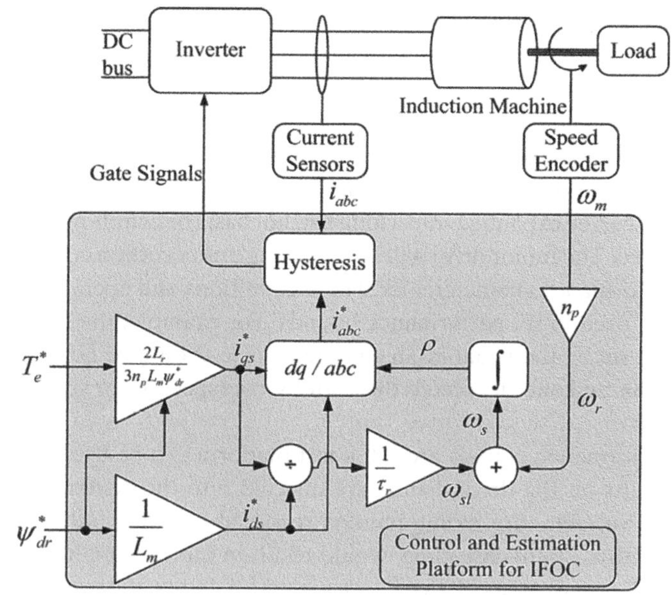

Figure 3.12 Induction motor drive under IFOC Bazzi et al. (2012).

nonfunctional, such as the total energy consumption or cost. In a motor drive, these measures should consider the safety of the operator, for example, passengers in an electric vehicle, and safety of the motor drive itself. Hence, if the performance is within predetermined bounds, then the system performance is acceptable; otherwise, the system has failed. In an IFOC motor drive, performance measures include the machine speed and torque, stator current peak, and settling time of all of these. Bounds are set based on desired operational limits.

Based on fault injection results and performance analysis, a system either survives or fails under the first or second fault. To limit complexity of the analysis, the system is assumed to be in a failure state following a third fault. In a procedure similar to that followed for Figure 3.7, the Markov model of the system is built but without considering recovery. The model includes 52 states: one initial state, seven survival states after 'Fault 1', one failed state after "Fault 1" that lumps five failures, 35 states that survived after "Fault 2" including states that survived only for certain inputs, seven failed states that occur after each survival under 'Fault 1', and one failed state at the third fault level. To solve for the probabilities, the rows and columns of the absorbing states, nine failure states, in this case, are eliminated from the transition matrix **Φ**. **Φ** is a sparse matrix, since not all states in the state-transition diagram are connected.

The reliability modeling methodology for a motor drive proceeds in simple steps. The first step is to define the essential system components or subsystems. These are the elements that can fail frequently and can affect system operation. Whether the whole system would later fail or not is part of the observations. After these components are determined, possible fault modes are analyzed, for example, Open Circuit or Short Circuit faults in an IGBT–diode module. The fault modes can be expanded into a long list, but basic or common faults should be considered. The motor drive will be operating under certain conditions with desired performance characteristics. These conditions and operational characteristics are used to set performance bounds. For example, the torque on the motor shaft set by the pedal in an electric vehicle should be bounded within certain limits, and once it exceeds these limits, overspeeding or sudden braking could occur.

Once performance bounds are set, a test platform can be used to study the effect of faults on the drive. Faults are injected into the system one by one, and critical variables are monitored and recorded. A fault could be followed by another; thus, 'fault cascading' would result in multiple fault levels. Given N vulnerable components in a system, each with K faults, the result is KN fault combinations. Therefore, fault combinations grow exponentially as component counts increase.

After the first level of faults is injected and critical variables are recorded, these variables are compared to the performance bounds. If any variable exceeds its allowed bounds, the system is considered in a "failure" state; otherwise, it has survived. If the system survives, another level of faults is injected after the first fault, and the analysis repeats. In this study, two fault levels are considered, and the rest are truncated into a failure state. The error between this truncation and the actual levels can be estimated. Varying one or more inputs and repeating the fault injection and evaluation analysis provides the fault coverage of the drive. Once all fault combinations are analyzed and fault coverage is found, a Markov model or a state-transition diagram can be built.

Hirschmann et al. (2007) use continuous monitoring to determine MTTF of power electronic devices in hybrid electric vehicles. They base their analysis on lifetime prediction and compare different tools. The RDF 2000 failure-rate catalog has been adopted on July 3, 2000, by the UTE (Union Technique de l'Électricité) board of directors in France. Several assumptions were made while drafting this standard. The wear-out period is disregarded because the majority of electronic components never reaches this period. For some electronic components, the wear-out period is not very far in the future. For these components the normal life period has to be determined. Also, the early-life failures are disregarded because they are caused by a wide set of reasons which relate to fabrication tolerances and process uncertainties (infant stage). In addition, these failures can be eliminated during subsequent improved design and production processes as well as burn-in tests.

This standard does contain data about IGBTs, and it also includes aging effects due to temperature cycles. The effects of temperature cycles are calculated on the basis of a typical year of use and had to be converted into load cycle-based calculations. Unlike in MIL-HDBK-217F, the failure rate λ_p here is the sum of partial failure rates, which in turn are calculated similarly to the failure rates λ_p in MIL-HDBK-217F.

As previously mentioned, in contrast to reliability prediction, *lifetime prediction* is a deterministic approach. These methods are based on distinct principles.

- Lifetime prediction is based on physical models and makes the assumption that a component will withstand a certain amount of stress and then fail. It tries to develop formulas that describe the effects of the destructive processes on the lifetime.
- Reliability prediction is based on empirical analysis.

The authors analyze the effect of the Coffin–Manson empirical relationship between the number N of temperature cycles a transistor can withstand if the junction temperature varies by ΔT. It appears as if the absolute component-temperature does not have any influence on the predicted lifetime

$$N = 10^7 \cdot e^{(-0.05\Delta T)} \tag{3.48}$$

Due to the fact that the number of cycles N is not considered in lifetime predictions or in reliability predictions, Equation 3.48 is converted. If a component will withstand N cycles, it is reasonable to state that at each cycle a fraction, that is, $1/N$, of the above mentioned device is damaged. Thus, this fraction represents a partial destruction of the device. If the fractions of all temperature cycles are summed up, this value can be interpreted as the probability of failure $F(t)$ which is defined as $1-R(t)$. If the value reaches 1, the component will fail. This can be used to determine the number of driving cycles the component will work nominally.

The authors then base their analysis on the detection of temperature cycles, using a driving cycle, for example, FTP-72. They developed a program that computes the component temperature over a whole driving cycle. The program is divided in four functional modules: the HEV and Motor Model, the Loss Calculation, the Thermal Simulation, and the Reliability Prediction modules They simulate the temperatures using 3.13, accounting for dormant cycles and neglecting small changes in junction temperatures and determine local extrema, for example, Figure 3.14.

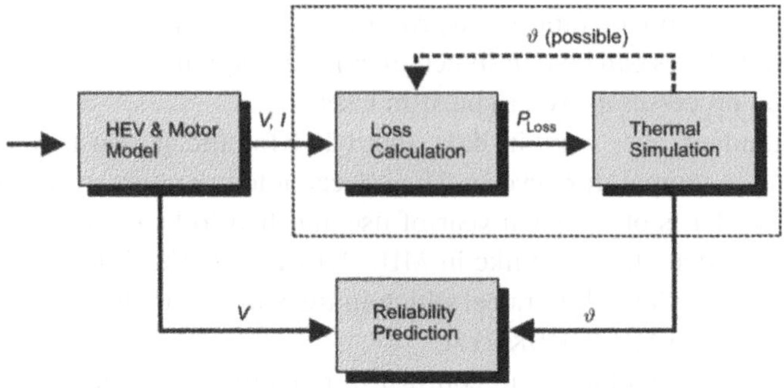

Figure 3.13 Block diagram showing basic structure of simulation tool Hirschmann et al. (2007).

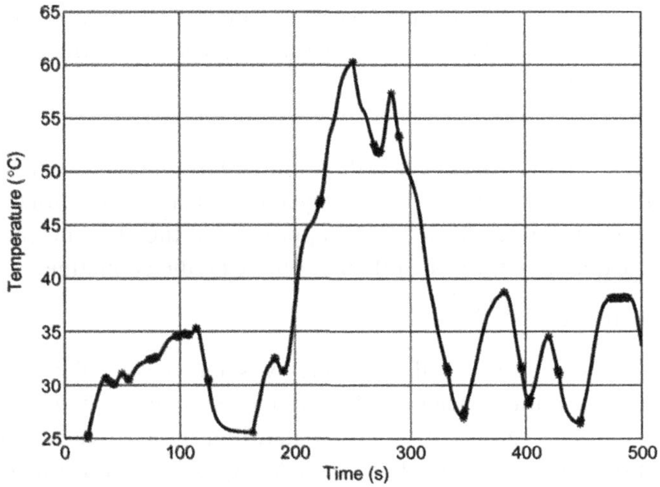

Figure 3.14 All detected extreme values of temperature curve (indicated with *) Hirschmann et al. (2007).

Results using the discussed techniques are shown in Table 3.6. These are quite puzzling, since different methods produce vastly different MTTF, and different least reliable components. The results establish, once more, that reliability prediction of complex mission profiles is not well established in catalogs, while lifetime prediction laws (i.e., Coffin-Manson and Arrhenius equations) require reliable measurement data.

El Murr et al. (2015) assess the mean-time-to-failure (MTTF) of a number of representative fault-tolerant PMAC drives, including dual converter open winding drives, and dual motor drive, where the payload is coupled to two

Table 3.6 Reliability assessment of MFC under different aging degree; Hirschmann et al. (2007).

MTTF in hours	IGBTs	Diodes	Capacitors	System
MIL-HDBK-217F	**$0.36 \cdot 10^6$**	$11.36 \cdot 10^6$	$3.44 - 10^6$	56,259.00
RDF 2000	43,291.00	43,233.00	$17.45 \cdot 10^6$	36,037.00
Coffin-MansonArrhenius	34,819.00	$0.61 - 10^6$	14,022.00	14,022.00

Table 3.7 Drive component failure rates, El Murr et al. (2015).

Components	Abbr	Failure rate (hrs)
Speed Encoder	SE	10.2×10^{-7}
Current sensor	CS	2×10^{-7}
IGBT + Gate drive Normal	SW	2.5×10^{-7}
IGBT + Gate drive Stressed	SW(Str)	4.5×10^{-7}
Capacitor Normal	Cap	2.5×10^{-7}
Capacitor (stressed)	Cap(Str)	10×10^{-7}
Windings	Wdgs	3.2×10^{-7}
hline Bearings	Brgs	6.4×10^{-7}

motors via a mechanical transmission, by employing Markov chain modeling technique.

They start from the drive component failure rates in Table 3.7, They then proceed to construct the Markov transition diagram (Figure 3.15) for the case of a single-motor, dual-converter drive, and the corresponding Chapman–Kolmogorov equations.

Considering the four different states, healthy (H), failure (F) and two remedial states, the state equations describing the Markov diagram in Figure 3.15 are:

$$\begin{bmatrix} \dot{H}(t) \\ \dot{R}_1(t) \\ \dot{R}_2(t) \\ \dot{F}(t) \end{bmatrix} = \begin{bmatrix} -\lambda_1 - \lambda_2 - \lambda_5 & 0 & 0 & 0 \\ \lambda_1 & -\lambda_3 & 0 & 0 \\ \lambda_2 & 0 & -\lambda_4 & 0 \\ \lambda_5 & \lambda_3 & \lambda_4 & 0 \end{bmatrix} \begin{bmatrix} H(t) \\ R_1(t) \\ R_2(t) \\ F(t) \end{bmatrix} \tag{3.49}$$

$$H(t) + R_1(t) + R_2(t) + F(t) = 1, \tag{3.50}$$

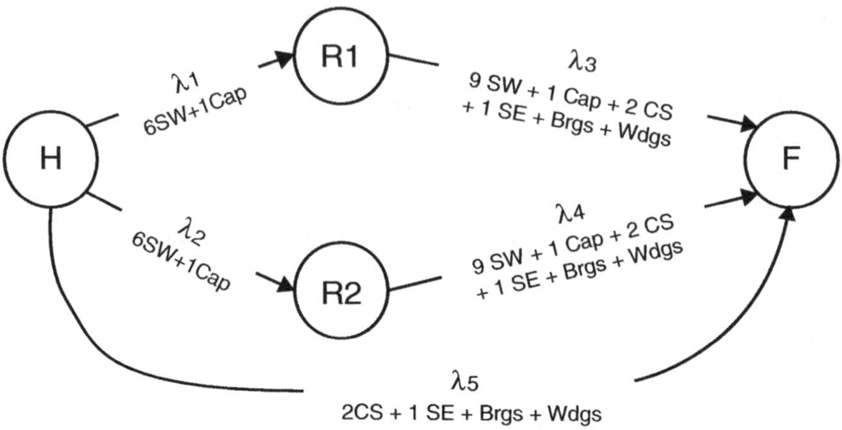

Figure 3.15 Markov chain state-transition diagram of open-winding dual converter drive El Murr et al. (2015).

where

$$\lambda_1 = \lambda_2 = 6\lambda_{SW} + \lambda_{Cap}, \tag{3.51}$$

$$\lambda_3 = \lambda_4 = 9\lambda_{SW} + \lambda_{Cap} + 2\lambda_{CS} + \lambda_{SE} + \lambda_{Brgs} + 4_{Wdgs}, \text{ and} \tag{3.52}$$

$$\lambda_5 = 2\lambda_{CS} + \lambda_{SE} + \lambda_{Brgs} + \lambda_{Wdgs}. \tag{3.53}$$

The reliability $R(t)$ of the drive is obtained by adding all the non-absorbing states as discussed in Section 3.4.1.

$$R(t) = \sum (H(t), R_1(t), R_2(t)) \tag{3.54}$$

They proceed in a similar fashion to study a dual-converters drive, a 9-phase fault-tolerant PMAC drive supplied by three 3-phase inverters dual motor drive with and without repair and a fully redundant system with two motors, each with an inverter and controller.

Given that bearing reliability is critical in both 9-phase drive and dual motor drive, bearing condition monitoring can potentially provide a cost effective method for increased availability by early detection of degradation or incipient fault. To assess the effect of bearing condition monitoring and maintenance on drive reliability, it is assumed that the bearing is replaced at its half lifetime and the period to carry the replacement/maintenance is denoted by μ. The Markov chain modelling is intrinsically stochastic, and the maintenance based on online monitoring by a transition probability is valid in rigorous sense since half lifetime is of stochastic nature.

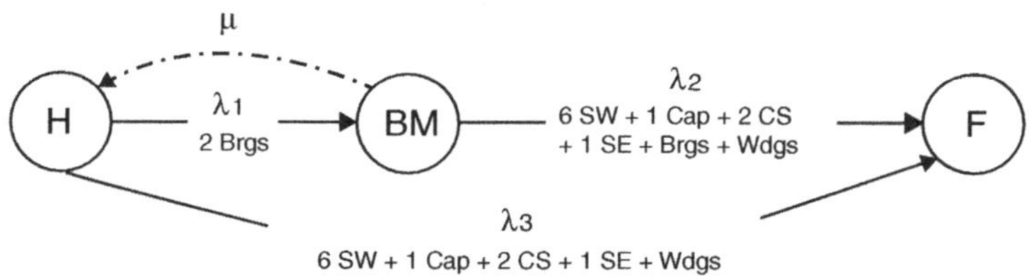

Figure 3.16 Markov chain state-transition diagram of standard 3-phase drive with bearing monitoring and replacement, El Murr et al. (2015)

By introducing a bearing remedial state BM, which represents a fatigued bearing at its half lifetime before complete failure resulting in the Markov chain state-transition diagram shown in Figure 3.16 for the standard 3-phase drive. Since failure rate is inversely proportional to lifetime, the rate that triggers the state transition from H (Healthy state) to BM (Bearing remedial state) is twice the bearing failure rate, that is, $2\times\lambda_{\text{bearings}}$. The standard 3-phase drive reliability can increase by 100% by applying bearing monitoring and performing replacement before failure. The same methodology can be applied to the 9-phase drive. Speed sensor fault can still lead to direct failure of the drive system, but the drive reliability will be governed mainly by the speed sensor failure rate which is relatively small. The resultant reliability is improved by order of magnitude reaching 800,000 hours, when repair and bearing replacement are performed. Further reliability improvement can be achieved by applying sensorless control as a back-up in case of sensor fault.

To evaluate the reliability of the converter, Zhou et al. (2016) estimate the lifetime of the power modules by using multi-time scales of the mission profile. They analyze the thermal behavior of IGBT module, developing the loss profiles and thermal profiles for estimating lifetime.

In a thermal network of an IGBT module, R_{jc} is thermal resistance from the case to junction of the IGBT chip, R_{ch} is the thermal resistance from sink to case, R_{ha} is thermal resistance from air-cooling system to heat sink, T_a is the ambient temperature, in which subscripts d and t denote the diode and IGBT. The average junction temperature T_j and the fluctuation of the junction temperature dT_j are usually selected as two foremost indexes for evaluation of power semiconductors reliability, and the T_j and the dT_j of IGBT can be analytically solved by:

$$T_{tj} = P_t \cdot \left(\sum_{i}^{4} R_{tjc(\mathrm{i})} + R_{tch} + R_{ha} \right) + P_d \cdot R_{ha} + T_a \tag{3.55}$$

$$dT_{tj} = 2P_t \cdot \sum_{i}^{4} R_{tjc(i)} \cdot \frac{\left(1 - e^{-\frac{t_{on}}{m_{tjc(i)}}}\right)^2}{1 - e^{-\frac{t_p}{m_{tjc(i)}}}}, \tag{3.56}$$

where P_t denotes the power loss of IGBT chip, P_d the loss of diode chip, t_p and m denote the fundamental cycle of the current and the thermal time constant of each Forster layer, and t_{on} is the on state time in each fundamental cycle.

The relationship between the junction temperature and the reliability can be analyzed by different models. Here, the Coffin–Manson model is adopted to evaluate the power cycling capabilities of power devices:

$$N_f(T_j, dT_j) = A_1 \cdot dT_j^{-n_1} \cdot \exp\left(\frac{E_a}{kT_j}\right). \tag{3.57}$$

In the expression of eq. 3.57, N_f denotes the number of cycles to failure, A_1 and n_1 are constants based on experiments, E_a and k are the activation energy and the Boltzmann constant. The lifetime consumption of power devices under various operation modes is based on Miner's rule:

$$CL_{IGBT} = \sum_{i=1}^{N} \frac{N_i}{N_{f,i}}, \tag{3.58}$$

where N_t is the cycle lifetime, $N_{f,i}$ the number of power cycles to failure, in which subscripts i denote the ith thermal cycle, N_i represents the number of corresponding thermal cycles.

In the analysis of the device average junction temperature and its fluctuation, the following are considered: the impact of wind speed, ambient temperature, and thermal stress of the power semiconductors. Figure 3.17 defines these time scales, as switching time scale, fundamental frequency time scale, and low-frequency time scale. The switching frequency of power devices is usually several thousand Hz, and the junction temperature fluctuation is very little. It is shows that very small junction temperature fluctuation affects very little the lifetime of devices. Therefore, the junction temperature fluctuation in switching scale is ignored for estimating lifetime.

By this method for reliability evaluation due to the fundamental frequency junction temperature fluctuation, the consumed lifetime and wind speed distribution per year of IGBT modules is calculated. The percentage of consumed lifetime of IGBT module considering different thresholds of wind speed is shown in Figure 3.18, while the percentage of consumed lifetime of power semiconductors can be calculated in 3.59.

$$p(CL_{v_{th}}) = CL_{IGBT}(v > v_{th})/CL_{IGBT}, \tag{3.59}$$

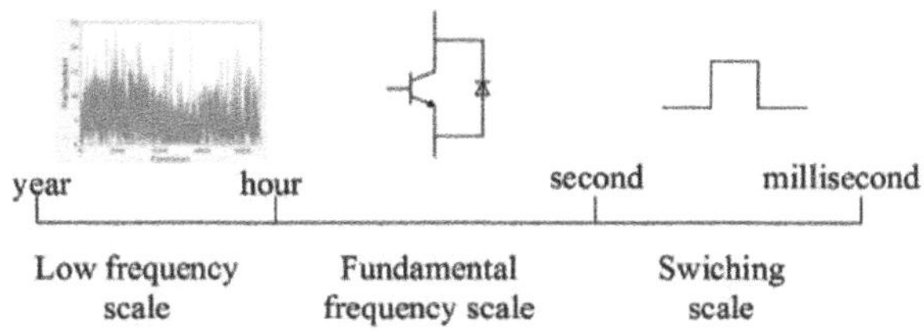

Figure 3.17 Thermal stress of power devices in a power converter with multi-time scale, Zhou et al. (2016).

where υ is wind speed and υ_{th} is wind speed threshold; CL_{IGBT} is calculated by 3.58

As shown in Figure 3.18, the wind speed is mainly concentrated in low speed area, but the lifetime of IGBT module consumed intensively at $9 - 12m/s$.

Bolvashenkov et al. (2016) presented a systematic method of estimation the reliability and fault tolerance of a multi-phase permanent magnet synchronous motors based on a multi-level Markov model of the lifecycles for the multi-state system. The impact on the fault tolerance degree regarding the number of phases of permanent magnet synchronous motors was investigated as well as the practical application of the proposed technique for the multi-phase traction motor.

As a basis for assessing the level of reliability and fault tolerance of the system and its compliance with the requirements of the project, the authors considered the processes occurring in the system as discrete-states and continuous time (DSCT) Markov processes. Hence, a hierarchical Markov model of the vehicle lifecycle with embedded Markov models of the reliability and fault tolerance analysis was applied. A block diagram is shown in Figure 3.19, where the following notation was used:

- EPS — electric power source,
- PE — power electronics,
- EM — electric traction motor,
- CU — control unit.

The four levels analyzed were:

A. Element Level

At the level of elements, either analytically (e.g., via Reliability Block Diagram-RBD, Failure Tree Analysis-FTA), or using Markov models of binary-states, reliability characteristics of components of the units were determined. For the power electronics unit, it is failure rates of diodes, IGBTs, capacitors, etc. For the Permanent Magnet Synchronous motor, the reliability is based on the

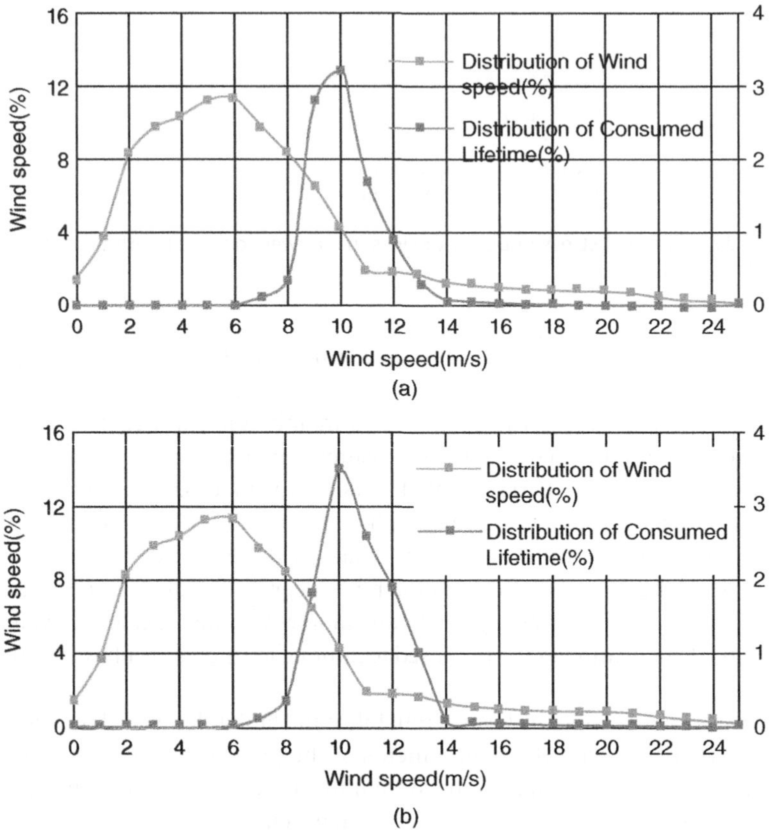

Figure 3.18 Distribution of lifetime consumption and wind speed of IGBT modules in power converter. (a) rotor side. (b) grid side, Zhou et al. (2016).

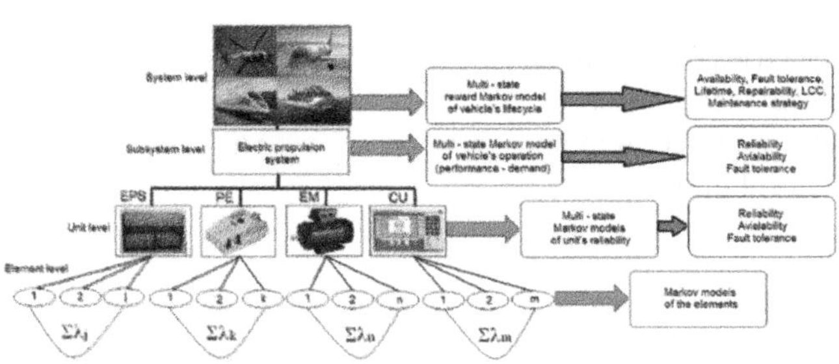

Figure 3.19 Block diagram of a hierarchical Markov model of electric vehicles lifecycle. Bolvashenkov et al. (2016).

characteristics of the stator, the rotor, and the bearings:

$$\lambda_{PSM}(t) = \lambda_S(t) + \lambda_R(t) + \lambda_B(t), \tag{3.60}$$

where λ_S, λ_R, and λ_B are the failure rates of parts of the electric motor, respectively of stator, rotor and bearing. The probability of failure-free operation of a system, which consists of these elements is determined as:

$$P_{PSM}(t) = \exp[-(\lambda_S + \lambda_R + \lambda_B)]. \tag{3.61}$$

B. Unit Level

For the accurate assessment of the reliability and fault tolerance of each unit of the electric propulsion system and compliance (or non-compliance) with project requirements, based on the structural model of multi-phase PMSMs, an MMS reliability Markov model with N degradation levels was built. This model is available in two versions — with recovery (repairable), and without the restoration (not repairable). Regarding fault tolerance, a Markov model is most interesting with absorbing states and without the possibility of repair within operational conditions. The transition parameters are determined based on the calculation results on the element level of the model.

C. Subsystem Level

In order to construct a combined Markov model of the traction electric drive, including electric power source, the concepts of demand W and performance P were defined. The model of the electric propulsion system operation was represented as a stochastic model of multi-state system reliability with the change of discrete operating modes:

1. start (takeoff),
2. acceleration (climb),
3. constant speed (cruise),
4. deceleration (reduction of altitude),
5. stop (landing).

Each of these modes complies with specific power characteristics. On one hand we have the requirements for safe vehicle operation, which form a model of demand: W; on the other hand, we have the guaranteed generated electric power, which forms the model of performance, P.

Considering the multi-state system, its output performance can be represented by a stochastic process $P(t)$ and described as a continuous-time Markov chain with five different possible states $P_1, \ldots, P_5$, and corresponding transition intensities matrix $\alpha = [\alpha_{ij}]$. Respectively, the demand process $W(t)$ is also modeled as a continuous-time Markov chain with five different states $W1, \ldots, W5$ and corresponding transition intensities with the matrix $\beta = [\beta_{xi}]$.

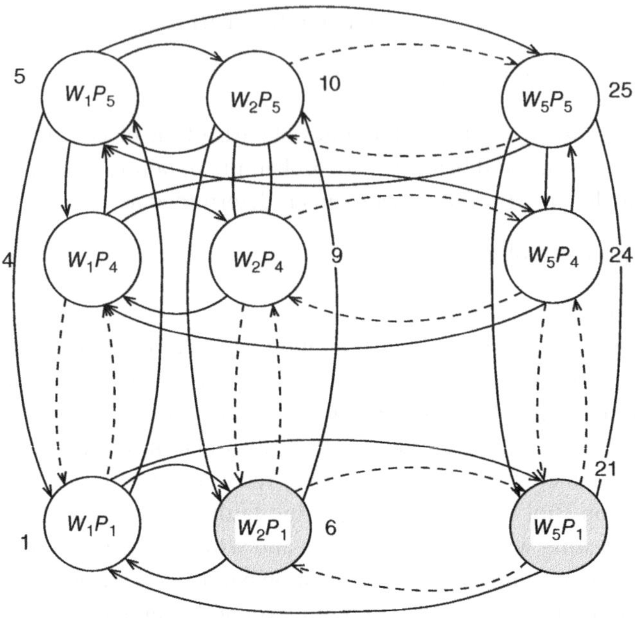

Figure 3.20 Combined performance-demand Markov model of electric propulsion system, realized as 5-state demand model and 5-state output performance model including the demand level W and the performance rate P, Bolvashenkov et al. (2016).

The performance and demand models can be combined, based on the independence of events in these two models. The probabilities of transitions in each model are not affected by the events that occur in another one. The state-space diagram for the combined 5-state demand model and 5-state output performance model is shown in Figure 3.20. Each state in the diagram is designated by two indices indicating the demand level $W \in \{w_1 \dots w_5\}$ and the output performance rate $P \in \{P_1 \dots P_5\}$. Some states (gray colored) are unacceptable considering limitations and safety.

D. System Level

The model of the electric vehicle's lifecycle allows to increase the reliability and efficiency restored (repairable) systems by optimizing maintenance strategies according to intensity of the scheduled and unscheduled repairs, the use of functional systems of monitoring, forecasting reliability, and diagnostics. Current and forecasted values of reliability indices, are determined considering the external and internal operating conditions of the vehicle, as well as taking into account the availability (or non-availability) of structural or functional redundancy. Thus, the study and optimization task of the so-called reliability associated costs (RAC) is used here, see Lisnianski et al..

To build the model the process of the vehicle operations is represented by a chain of the lifecycle:

- to — operational,
- tn — non-operational,
- tw — working,
- ts — standing.

The data on duration of each cycle are obtained based on the analysis of statistical operational data of a particular type of vehicle on certain routes and areas.

For a preliminary assessment of RAC the following formula was used:

$$RAC = t_0 k_d (1 - A) C_{UR}, \tag{3.62}$$

where t_0 is the operational time duration in hours, k_d the driving time rate, A the average value of operational availability, C_{UR} the average specific cost of vehicle downtime due to the unreliability. It was proposed to determine the value of the availability based on the above Markov model "performance demand".

More profound study of RAC can be carried out on the basis of Markov reward models, which are useful for Monitoring Service System (MSS) lifecycle cost analysis and reliability-associated cost computation. The Markov reward model considers the continuous-time Markov chain with a set of states $\{1, \dots, K\}$, and the transition intensity matrix α. It is suggested that if the process stays in any state i during the time unit, a certain amount of money r_{ii} should be paid. It is also suggested that each time that the process transits from state i to state j a certain amount of money r_{ij} should be paid. This amount is called reward, and it may also be negative when it characterizes losses or penalties. The method is discussed in more detail in Section 3.9.1.

Based on the combined performance-demand model, the theory of the Markov reward processes can be applied for evaluation of important reliability indices for Markov MSS, such as MSS average availability, the mean reward $V_i(T)$ accumulated during the interval $[0, T]$ the mean number of MSS failures during the time interval $[0, T]$, expected accumulated performance deficiency, mean time to failure, expected costs for various maintenance strategy considering reliability prediction, monitoring and diagnostics.

This analysis was applied to a 9-phase traction drive, taking into account that the electric power drive, besides the PSM, includes an electric power source, a power electronic unit, and a control unit. Fault tolerance performance each of them should have a significant reserve, see Figure 3.21.

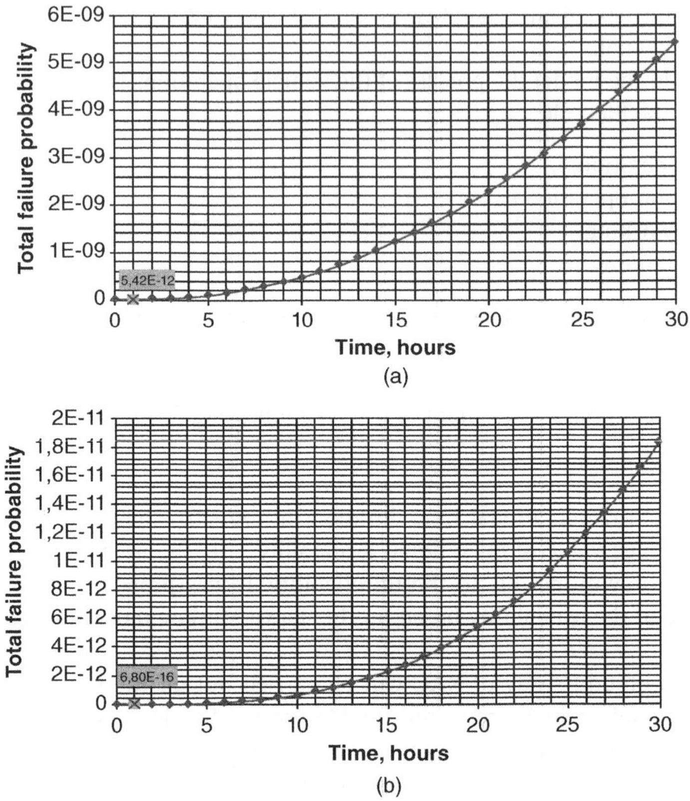

Figure 3.21 Total failure probability of 9-phase PSM at 85 % load (a) and 65 % load (b), Bolvashenkov et al. (2016).

Acharya et al. (2018) present one application of a mission-profile based analysis, in order to determine the temperature rise in the devices and consequently, its effects on the lifetime of the devices. It is applied to a multi-megawatt permanent magnet synchronous generator-based wind energy conversion system (PMSG-WECS) used for medium voltage (MV) grid integration.

It was considered necessary to investigate the reliability of the system, since maintenance of the systems leads to additional incurred costs. Numerous factors like thermal stress, mechanical stress, etc., can cause failure in the power devices. However, the most frequently observed failure mechanism is attributed to the thermal stresses caused by the temperature swings and the increased mean junction temperature in the device. Thermal and reliability analysis has been carried out on the power semiconductor devices using the procedure

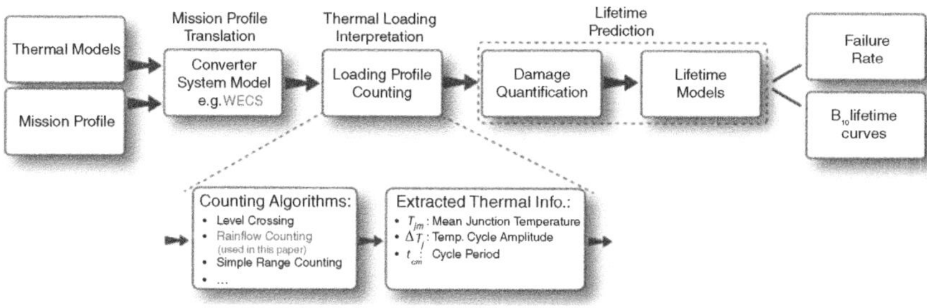

Figure 3.22 Flowchart for the reliability analysis of power devices caused by long term thermal cycles, Acharya et al. (2018).

shown in Figure 3.22. Rainflow analysis method is used to identify the corresponding temperature swings of each thermal cycle. An analytical lifetime model was used, and the damage is quantified by Palgrem Miner rule. The reliability of the DC-link capacitors and the high-frequency transformers have not been taken into account.

To estimate lifetime, a rainflow counting method is employed on the long-term thermal loading profile. This helps in the conversion of randomly changing thermal loading profiles to regulated thermal cycles. The rainflow counting method which extracts the change in junction temperature (ΔT_j), mean junction temperature (T_{jm}), and the number of cycles (t_{cycle}) was used. In an example a total of 563 thermal cycles have been identified.

However, the state-of-the-art lifetime model provides a good estimation for the lifetime, since it is based on the packaging of the device, Sintamarean et al. (2014):

$$N_f = A \times (\Delta T_j)^{\alpha} \times (ar)^{\beta_1 \Delta T_j + \beta_0} \times \left[\frac{C + (t_{ON})^{\gamma}}{C+1}\right] \tag{3.63}$$

$$\times \exp\left(\frac{E_a}{k_b \times T_{\text{im}}}\right) \times f_{\text{diode}}, \tag{3.64}$$

where $A, \alpha, \beta_0, \beta_1, \gamma$ and C are the model parameter. ar is the bond wire aspect ratio, f_{diode} is the diode chip-thickness and t_{ON} is the cycling period.

The damage modeling is done using Palgrem-Miner's rule , which states that the accumulated life consumption is linearly dependent on the contribution of different thermal cycles.

$$LC = \sum_i \frac{n_i}{N_{fi}}, \tag{3.65}$$

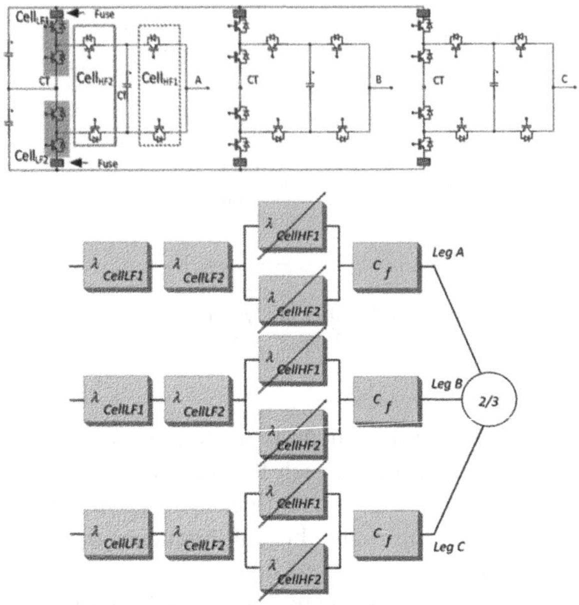

Figure 3.23 Structure and reliability block diagram of a three-phase five-level ANPC, Richardeau and Pham (2013).

and the expected lifetime was calculated from the damage caused by the identified cycles in t_{MP}, which is the mission profile period, as:

$$LF = \frac{t_{\mathrm{MP}}}{LC}. \tag{3.66}$$

Multilevel converters can be considered vulnerable to a decrease of the direct reliability based on first failure, as they have many power devices and drivers. Richardeau and Pham (2013), taking into account symmetrical multi-level converters such as the X -level active neutral point clamped (ANPC) family, show that several redundancies clearly can be identified, which can be managed to increase the global reliability. A general and theoretical methodology is used to calculate reliability laws and failure rates and is applied to compare two-, three-, and five-level topologies. Figure 3.23 shows the structure and reliability block diagram for a five-phase inverter, Results in Figure 3.24 show that the fault handling of three- and five-level three-phase topologies permits a great increase in reliability over a "relatively" short time duration, in addition to other benefits. The failure mode considered is a permanent fully on-state mode of the devices that is modeled by an ideal short-circuit state. They base their analysis on a reliability block diagram.

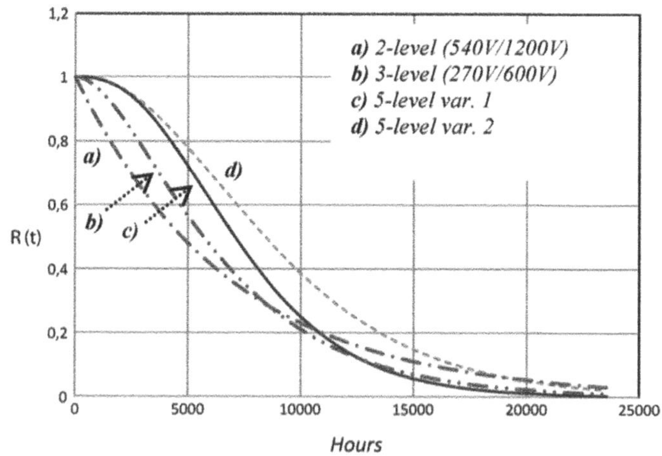

Figure 3.24 Reliability curve comparison concerning three-phase two-, three-, and five-level ANPC (variants 1 and 2) inverters including multiple fault-tolerant capabilities, Richardeau and Pham (2013).

3.9 Scheduled Maintenance, Condition-Based Maintenance and Prognosis-Enhanced Reliability

Scheduled maintenance has been the standard way to increase reliability in operating systems. Like pre-commissioning maintenance, it involves often visual inspection and testing using off-line tools and instruments not often available during operation. In that, it has major advantages and disadvantages: The first is that it can be extensive, detailed, accurate, and independent of the history of other, similar components and devices. But it will not happen when at the moment when it is needed, hence either wasting effort and time on a healthy system, or not applied before it is too late to repair or replace.

Condition-based maintenance (CBM) is based primarily on detection of faults during operation, although this detection could be quasi-off-line, during short periods of inactivity. Although the condition is estimated from statistics or data-based tools, it can be fairly accurate. Some of the weaknesses result from the similarity of manifestation of dissimilar faults. Implicit in CBM is a calculation of the Remaining Useful Life, which determines the urgency and extent of the maintenance. The effect of CBM on reliability is not always clear, and it is based on the tentative relation between perceived condition and anticipated failure.

The prognosis tools developed for different faults and devices produce usually two outputs: an estimate of the remaining useful life of a device, and the confidence in this estimate. When multiple faults are possible, each fault is identified and associated with the RUL and the confidence in it.

Based on these, a possible strategy can be followed: continue operation, activate internal or external redundancy, mitigate in other ways, modify mission profile, schedule off-line maintenance, emergency shut off. Since the RUL estimate is not exact, it is important that a decision to act be early to avoid a catastrophic failure. On the other hand, a decision taken unnecessarily early may lead to the loss of operating time, decrease performance, or stress other components.

Let us assume that the failure prognosis scheme at the moment of test, when a decision is considered, t_{test} gives an RUL estimate of t_{RUL} and that the probability of failure after t_{test} is $P(t_{test})$. Let us also assume the cost of failure is C_f. Then, the cost associated with the decision to not take action at t_{test}, to avoid failure is $C_f P(t_{test})$, increasing with t_{test}.

At the same time, taking action has a cost associated with it. This cost is more difficult to assess, since the action itself is costly, but also most probably will result in a system less robust, reliable and efficient than the one before action. Possible reasons are: a resulting heavier load, increased stress on parts or components, and decreased ability to handle a subsequent fault. The total cost to be considered at the time of evaluation, t_{test}, will have two components: one associated with the decision to act, C_f, and one decreasing with the time to act, $C_1 \cdot (t - t_{\text{test}})$, since the continued operation of the system without considering the specific fault is less costly before action.

Looking a the sum of these costs, it is clear that early action may be preferable, as delay will give higher probability of fault, while unnecessary too early action will lead to a significant cost in the operation of the modified system. This is discussed in more detail in subsection 3.9.1.

As the fault behavior is usually a more dynamic one, how to control the fault growth is important for the faulty system, Pei et al. (2010):

1. A fault will usually cause a component failure. Improved or safer reconfiguration (by lowing the performance, for example) of the control system may slow down the fault growth and prevent the total system failure.
2. A fault brings degraded performance, the fault growing will cause the performance further degrading. Suitable prognosis-enhanced reconfiguration may manipulate the predictable/controllable system performance degradation of the near future.
3. Under some critical circumstances, when the assigned mission needs to be completed somehow within the given time duration, we must design an

algorithm to reconfigure the faulty system controller to maintain acceptable performance and keep the system not running down for the whole mission duration.
4. With such algorithm determined, the prognosis-based controller design could also be used in the integrated product lifecyle/performance/cost consideration.

While this is costly, the prognosis-enhanced reconfiguration of a faulty system needs the relaxation on system performance or system constraints. Prognosis-enhanced reconfiguration strategy reconfigures the balance between the performance degradation and the fault growth before total system failure. Suitable controlled performance degradation methods are the key to solve reconfiguration problems of this kind.

Besides the conceptual formulation, workable prognosis-enhanced reconfiguration strategies as well as some illustration examples are discussed by Pei et al. (2010). There are two possible methods to implement the reconfiguration procedure:

1. direct performance relaxing, for example, a pre-filter could be designed to degrade the performance rigidity;
2. manipulating the control magnitude constraints to implement the performance adaptation.

Usually, a fault will start growing within a safe region $L \in W$, where $L(t)$ is the fault dimension, W is the safe region of fault propagation. When the fault reaches a given threshold, $L \in L^*$ (named hazard zone) at time $t = t_f$, the fault progresses into a failure. It is supposed that the fault dynamics (fault propagation) is known directly or indirectly via the existing fault diagnosis/prognosis algorithms. In practice, the fault dimension is formulated in probability distribution function (pdf). One important character of the fault dimension is the monotony with all the variables. Pei et al. (2010).

The purpose of the control reconfiguration is to change the operating condition (mainly through the control variable) to avoid the fault dimension reaching the given hazard zone before the time of mission completion $t = t_{\text{mission}}$. The reconfiguration of the control mechanism will ensure the mission completion within the given time, though it might cause some performance degradation.

The prognosis-enhanced reconfiguration is defined as:

For a faulty system, with fault propagation, and the baseline controller $u = k(y, r, \vartheta)$ of the nominal system, and given a reference input r, reconfiguration, by adapting the parameter ϑ of the controller $u = k(y, r, \vartheta)$, can ensure that the faulty system will run within the safe region $L(t) \in W$, $t = t_{\text{mission}}$, before the time of mission completion. Meanwhile, it will track the reference r with suitable performance. $J(y, r, u, \vartheta)$, $\vartheta \in \Theta$ is the reconfiguration parameter, which may parameterize the controller as well as the performance index. Detection and management is the weakest point of fault prognosis in condition-based management systems.

For nonlinear control systems:

$$\dot{x} = f(x) + g(x)u, \; u \in U, \; x \in X, \tag{3.67}$$

$$y = h(x), \tag{3.68}$$

where $x \in R^m$, $y \in R^n$ and $u \in R^p$ represent the system state, output, and control variables. U and X represent available regions of u and x, $f(x)$ and $g(x)$ are smooth functions with suitable dimensions. The baseline controller $u = k(y, r, \theta)$ will ensure the system output tracking $y \rightarrow r$ of a given reference r, under the engaged performance index $J(y, r, u, \theta)$, where $\theta \in \Theta$ is considered as some parameters of the controller, representing the control bounds, performance index structures, etc.

Long-term prediction of the time to failure of critical systems entails large-grain uncertainty that must be represented effectively and managed efficiently, that is, as more data becomes available, means must be devised to narrow or "shrink" the uncertainty bounds. Prediction accuracy and precision are typical performance metrics employed to access the performance of prognostic algorithms. That is, we would like the predicted time to failure to be as close as possible to the real one. Also, the bounds or limits of uncertainty must be as 'narrow' as possible (Khawaja et al. (2005)).

The fundamental problem of prognosis-enhanced reconfiguration is to design the suitable performance degradation algorithm to ensure the fault growth is being controlled.

Pei et al. (2010) provide an example of three motor drives connected in parallel mechanically to the drive single load through an common gear and elastic shaft (Figure 3.25). For this three motors case, if there is a fault in one motor, often the existing management strategy is simply to turn off this motor and re-distribute the pre-shared load (plus the inertial as well as frictions of this faulty motor) to the other two motors. The abnormal heavy load might raise the fault probability of the rest of the motors. For the maximal redundancy consideration of the mission completion, a better mechanism will be to let the faulty motor still share some load, which means to reduce the possibility of new fault of the rest of the motors by keeping the moderate load for the rest two motors.

This reconfiguration mechanism is meaningful, for a fault usually will not cause a sudden failure in practice, and there exists a time slot to reconfigure the system while the fault grows. By estimating the remaining useful life of the

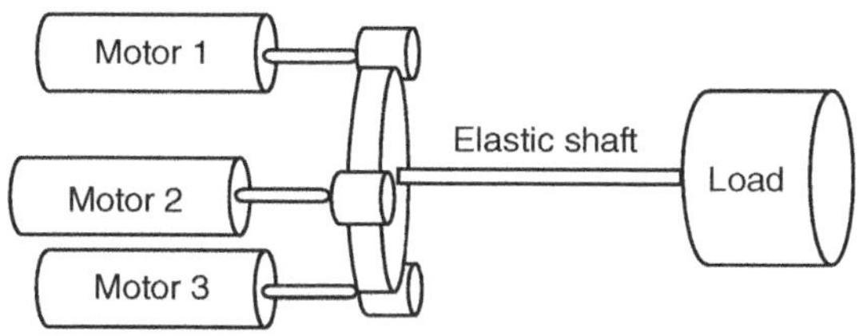

Figure 3.25 Three geared motors servomechanism, Pei et al. (2010).

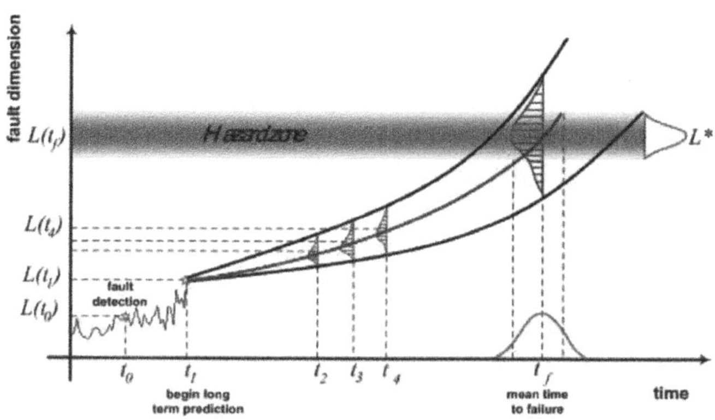

Figure 3.26 Fault dimension shown in pdf, Pei et al. (2010).

faulty motor via the prognosis scheme, if the RUL of the faulty motor doesn't meet the mission demand, the prognosis enhanced reconfiguration design will adapt the load share of the faulty motor to a suitable level, such that the faulty motor can still work till the completion of the mission.

Neither the hazard zone nor the fault dimension are uniquely defined, and that is illustrated in Figure 3.26. The uncertainties of the prognosis are further discussed by Tang et al. (2010). The focus of prognosis uncertainty management is to identify and manage the reducible uncertainties by applying available data using appropriate uncertainty management algorithms.

Particularly for dynamic model-based systems, opportunities exist to apply nonlinear filtering to provide a systematic way of dealing with the propagation of system damage at some future time, whenever imprecise diagnostic information is obtained. To quantify and reduce the prognostic uncertainty, the evolution of fault indicators and critical state variables should be treated as stochastic processes such that probability distribution functions can be predicted for a future time, and other important attributes, such as confidence intervals, may be computed. It is imperative to represent the prediction step as accurately as possible by correctly accounting for uncertainties in the damage state and measurements as well as in the model parameters and structure.

Moreover, when health information (diagnostics) is present, it is essential to update the current state based on the availability of this data in a statistically meaningful manner. The dynamics of all damage processes are inherently nonlinear, with uncertainties that are often non-Gaussian, therefore propagating these effects through time becomes a challenging and error-prone task.

Regression-based techniques are applicable to many systems, but the form of the regression curves are quite general and are not guaranteed to extrapolate well, especially when the failure significantly changes system dynamics. These techniques are commonly considered trending instead of prognosis. Data-driven prognostic methods utilize monitored operational/usage data and relate them to system health prediction. A principal disadvantage is that data-driven approaches require a substantial amount of data for training, whereas obtaining run-to-failure data, in particular for new systems, can be a lengthy and costly process. Limited by the absence of online health monitoring data, most usage-based prognostic methods rely on fleet reliability data and statistical models, and usually generate RUL predictions with large uncertainties.

Dynamic models, on the other hand, allow a number of systematic tools to be applied for prediction in the presence of noisy diagnostic information, loading profiles, and modeling uncertainties. The key elements of dynamic model-based prognostic problem are illustrated in Figure 3.28. Each time measured data is received, the data is filtered to form a posterior distribution based on the likelihood formed by the incoming data. Then, a prediction is performed to project the state distribution to next time step (or an arbitrary future time step). The recursive computation of the state's probability density from a measurement update step and a prediction step is known as the filtering problem. The prediction step can be repeated at several points in time until a critical threshold is reached at which logistical or maintenance actions are ordered; this is a long-term prediction (prognosis) process.

A potential issue with any non-deterministic prediction technique is that any errors or approximations in the initial probability density function (pdf), however small, can accumulate and grow over certain time horizon, and can severely distort the predicted probability distribution over a long time frame. Uncertainty representation and uncertainty management become important problems. The uncertainty representation problem focuses on how well an algorithm can propagate the true probability distribution of (RUL) through time. In contrast, the uncertainty management problem is focused on how can be extracted the most information from available measurements to reduce the system uncertainties.

The authors of Tang et al. (2010) set up a dynamic model-based system prognosis system for a case where such a model is available, the problem of crack growth. The added noise brings additive uncertainties to the model. As the model is applied, the parameters are learned based on incoming data. In this case, the hazard zone is set to 2.9 units and the true end of life is 840 cycles. Figure 3.27 shows how the parameters adapt as the fault progresses.

A few additional considerations:

- Long-term prediction, as discussed earlier and shown in Figure 3.28 increases the uncertainty of the RUL determination, and hence accelerate the need to adapt the control algorithm;
- The range of the parameters of the controller ϑ available for adaptation of the control algorithm and the possible relaxation of system performance may not be adequate to bring the fault dimension within the safe region $L(t) \in W$, before t_{mission}. In addition, certain faults, for example, bearing degradation, do not lend themselves to adaptation and possibly to use of redundancies;
- In that case, safety and/or cost considerations may warrant changes in mission, as discussed below.

The prognosis interval of a process that results in a failure, is specific to this process, in contrast to the confidence interval of a parameter, for example, mean value, standard deviation etc., which depends on a class of group of processes, see Geisser (1999), Simons (1972). A decision to manage an impending failure based on RUL alone, after time t_{test}, 3.29a, entails the risk that the failure may occur before the end of the mission of the process. In Figure 3.29b, the prognosis interval is shown around the RUL estimation, while the RUL estimate reaches the hazard zone late, the prediction interval does so too early, and does not allow for safe fault management within the necessary action time. The opposite is shown in 3.29c, where the lower bound of prognosis reaches the hazard zone late enough, after the mission is completed, allowing the safe completion of the drive mission. Finally, in Figure 3.29d, a narrower prediction interval allows on-time management action.

Figure 3.30 shows the resulting timing, including the estimate of RUL and the prediction interval associated with that, the time to act and the decision time, considering time to implement fault management action. The figure aims to make it clear that once a specific prediction interval is established, a decision has to be made at least adequately before the lower bound of this interval, and action has to be taken. The failure probability that defines the prediction interval is a parameter set at the beginning of the monitoring process.

Figure 3.31 shows the performance of the drive before and after mitigation. Assuming that fault management results in the drive components being more heavily loaded, for example, increased machine currents, power electronics switching losses, degraded performance, it is to be expected that the resulting drive degrades faster than the original one, in addition to the already accumulated degradation.

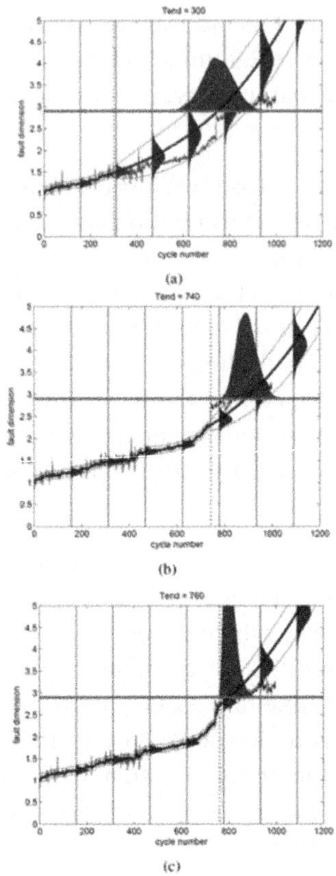

Figure 3.27 Slow parameter adaptation scenario using filter, Tang et al. (2010).

A few simple examples, some already discussed in a different context, illustrate the basic ideas:

- If a switch of a two-level, three-phase inverter supplying a three-phase machine is faulty, and the fault escalates to open switch, a standard management technique that is used is to operate with two phases, Mirafzal (2014), Eickhoff et al. (2018). To maintain the same torque, the current in the remaining two healthy phases has to be increased, typically by a factor of $\sqrt{3}$, with increased losses in the corresponding winding increasing by a factor if 3. This leads to stresses in the winding insulation and increases the possible subsequent fault in the regaining active inverter legs and the windings.

  If the fault is detected early, the losses in the faulty switch can be decreased by switching to discontinuous three-phase PWM, Song and Wang (2014), Ugur et al. (2019), to around 72% of those in the sinusoidal PWM. The switch

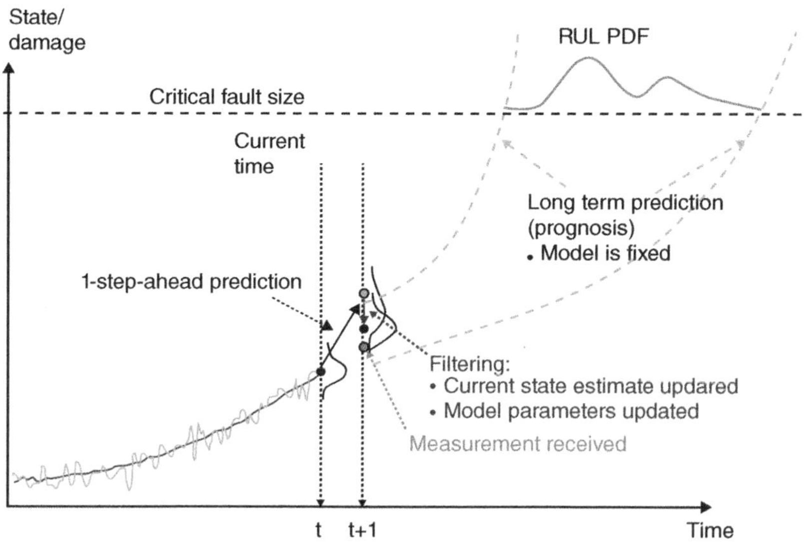

Figure 3.28 Concept of dynamic model-based prognostics, Tang et al. (2010).

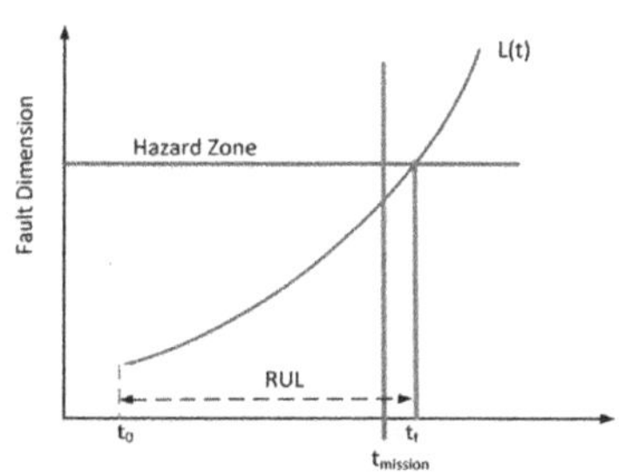

(a) Decision based on RUL v. T_{mission}.

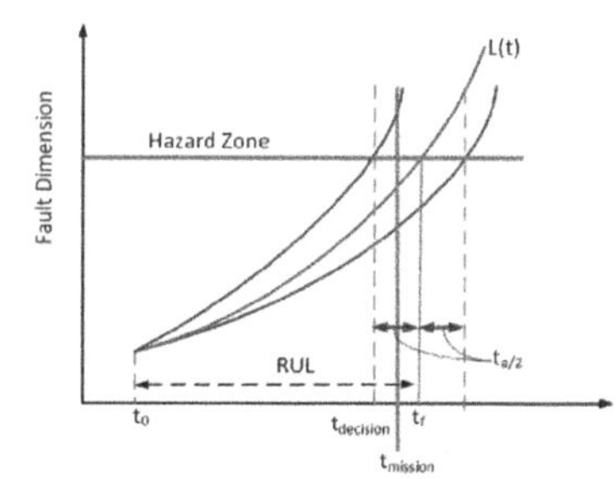

(b) Fault prognosis: RUL minus prognosis interval is less than T_{mission}.

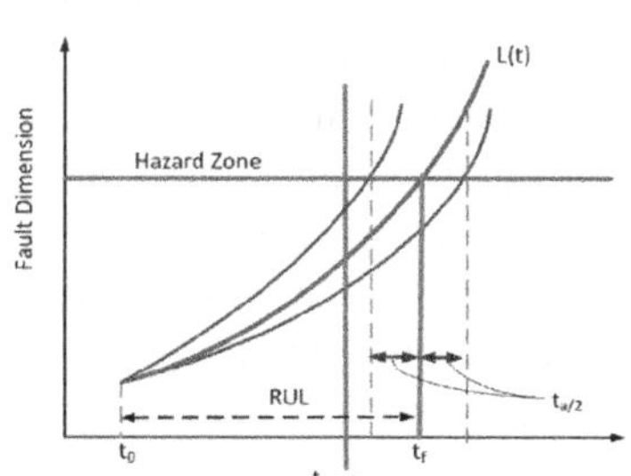

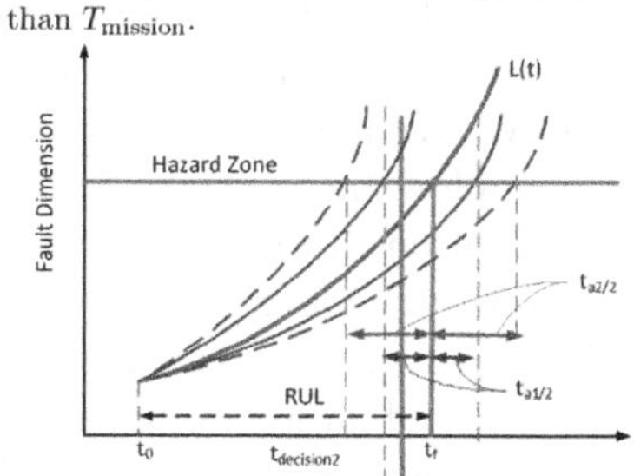

Figure 3.29 Effect of prognosis interval on reconfiguration decision.

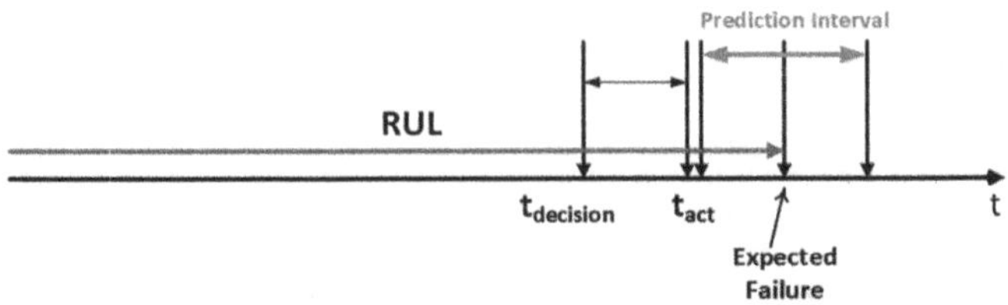

Figure 3.30 Time sequence: Estimate of RUL and prediction interval, decision and action timing.

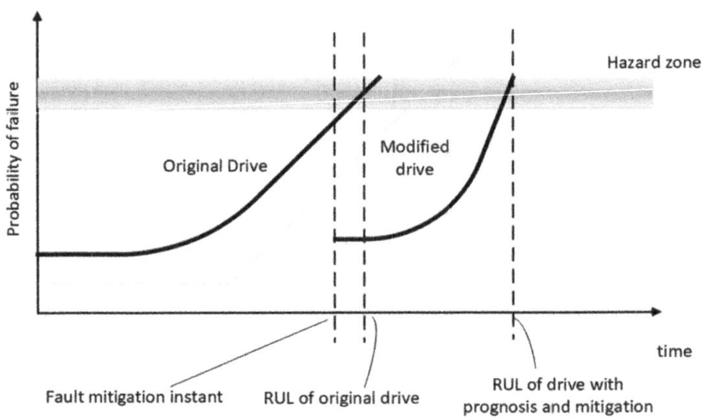

Figure 3.31 Fault prediction, mitigation, and RUL of the system with and without mitigation.

to DPWM will increase the harmonic distortion factor up to 0.6. Decreasing the switching losses, this way may allow continuing operation, and nevertheless keep the faulty switch operating, thus still allowing the subsequent solution of disengaging that switch and reverting to the classic solution.

- In a PMAC machine, early detection and/or prognosis of an internal short circuit, may allow the rapid injection of demagnetization current, Welchko et al. (2004), Cintron-Rivera et al. (2015), allowing continuous operation, albeit at a lower torque level. This will avoid a winding fault that will spread to adjoining phases, increased current in the windings and inverter, and negative developed torque. A major concern is the accuracy of the detection of the fault and the prediction interval, so that first adequate time is given for mitigation, and secondly avoiding unnecessary and possibly damaging and expensive modification of the operation.
- As already seen, Pei et al. (2010) discuss the case of three motors servomechanism, where a fault detected in one motor allows either the operation with

the remaining healthy two at increased load and losses, or, by decreasing the torque of the faulty one before it reaches the end of its RUL, allows for lighter load in the two healthy ones, and extension of the servomechanism life.

- Bearings are not conducive to most online mitigation techniques. Still, failure prognosis when the mission profile does not allow for repairs will result in timely modification of the mission profile and avoidance of a catastrophic failure. This is particularly important if the long-term prediction, as discussed earlier by Tang et al. (2010), can correctly lead to a decision to cancel a specific mission.

3.9.1 Reliability and Costs

The prognosis tools developed for different topologies, faults, and devices produce usually two outputs: an estimate of the remaining useful life (RUL) of a device, and the confidence in this estimate. When multiple faults are possible, each fault is identified and associated with the RUL and confidence in it.

Lisnianski et al. propose Markov Reward Models, useful in life cycle cost analysis and computation of the cost associated with reliability. This model considers the continuous-time Markov chain with a set of states $\{1, \dots, K\}$ and transition intensity matrix $\mathbf{a} = [a_{ij}, i, j = 1 \dots K]$. If the process stays in any state i during the time unit, a certain amount of money r_{ii} should be paid. It is also suggested that each time that the process transits from state i to state j a certain amount of money or other quantity r_{ij} should be paid. These amounts of money r_{ii} and r_{ij} are called rewards. Rewards may be, for example, the energy of a power generating system, information quantity of a communications system, the productivity of a production line, etc. Such a Markov process with rewards associated with its states and transitions is called a "Markov process with rewards". For these processes, an additional matrix $\mathbf{r} = [r_{ij}, i, \dots, K]$ of rewards is determined. If all rewards are zeroes, the process reduces to the ordinary continuous-time discrete state Markov process.

r_{ij} is measured in cost units, while reward r_{ii} is measured in cost units per time unit. The value that is of interest is the total expected reward accumulated up to time instant t under specified initial conditions.

At any time t, a total reward $V_i(t)$ is accumulated at state i. Then a system of differential equations must be solved to find the total expected rewards:

$$\frac{dV_i(t)}{dt} = r_{ii} + \sum_{j=1 j \neq i}^{K} a_{ij} r_{ij} + \sum_{j=1}^{K} a_{ij} V_j(t), \quad i = 1, \dots K \tag{3.69}$$

Assume that at time instant $t = 0$ the process is at state i. During the time increment Δt, the process can remain in this state or transit to some other state

j. If it remains in state i during time Δt, the expected reward accumulated during this time is $r_{ii}\Delta t$. Since at the beginning of the time interval $[t, \Delta t + t]$ the process is still in state i, the expected reward during this interval rate is $V_i(t)$ and the expected reward during the interval $[t, \Delta t + t]$ is $V_i(\Delta t + t) = r_{ii}\Delta t + V_i(t)$. The probability that the process will remain in state i during the time interval Δt equals 1 minus the probability that it will transit to any other state $j \neq i$ during this interval.

$$\pi_{ii}(0, \Delta t) = 1 - a_{ii}\Delta t. \tag{3.70}$$

But instead, during Δt the process may transit to some other state, $j \neq i$, with probability $\pi_{ij}(0, \Delta t) = a_{ij}\Delta t$. At the beginning of the new time interval Δt, the process is in state j. So the expected reward during this interval is V_j and the expected total reward by the end of this interval is $V_i(t + \Delta t) = r_{ij} + V_j(t)$. This leads to the system 3.69. Defining the vector column of total expected rewards $\mathbf{V}(t)$ with components $V_1(t), V_2(t), \ldots, V_K(t)$ and a vector $\mathbf{u}$:

$$u_i = r_{ii} + \sum_{j\neq i\ j=1}^{K} a_{ij}r_{ij},\ i = 1, \ldots, K, \tag{3.71}$$

we obtain a system of equations for $\frac{dV(t)}{dt}$, and a long run (stead-state solution):

$$\mathbf{0} = \mathbf{u} + \mathbf{a}\mathbf{V}(t) \tag{3.72}$$

This method is valuable in analyzing the costs associated with probabilities of transitions and the overall reliability of a multi-dimensional system, but it omits the costs associated with the accuracy of the determination of the probabilities π_{ii} and π_{ij}.

Combining cost analysis of Lisnianski et al. and the reconfiguration analysis of Pei et al. (2010), the decision to reconfigure could be based on a cost analysis.

Let us assume that at any point after t_0, at time t_{test}, the RUL, and the prediction interval t_a are determined for a predetermined level of fault probability P (i.e., there is an accepted value of $P(t_a)$). This means that the probability that the failure will occur within RUL$\pm t_a/2$, and a decision whether to manage a probable failure has to be taken. For mission time t_{mission}, determined at the commencement of the process, that is, at t_0, the act to manage the predicted failure is taken at time RUL-$t_a/2$ in order to avoid the predicted failure. It is also assumed that the time of reconfiguration of the system is negligible, or can be included in $t_a/2$. Note that the acceptable probability of failure, $P(RUL - t_a/2)$, is determined as a design parameter, independently of RUL and t_a, and that the associated prediction interval $t_a/2$ are determined as part of the RUL estimation process.

Associated with a decision to mitigate or not, there are the following costs:

- The cost of unmitigated failure after the decision time, t_{test}, is C_f. This cost can have a single value, while the cost of a decision to modify the system through control or redundancies, or through modification of the mission profile, will result in lower performance, and cost of a probable secondary fault. This cost will be proportional to the remaining time of healthy operation that was initially planned, that is, the time originally expected as time left in the process, as well as the cost of a probable secondary fault, associated with decreased reliability of the modified system, $C_1(t_{\text{mission}} - t_1)$. The prediction of failure at t_{test} will lead to a predicted probability of failure monotonically increasing with time, and will reach the predetermined value $P(t_a)$, at $RUL - t_a/2$ with a cost $C_f(t_{\text{mission}} - RUL - t_a/2)$. Of course, a failure can occur before that moment of decision but with lower probability.
- At the moment of test and decision, two costs are then considered: One, due to inaction, with a probability of possible failure $P(t_{\text{test}})$, $K_{\text{inaction}} = C_f \cdot P(t_{\text{test}})$; and another due to action, with a probability of no failure $(1 - P(t_{\text{test}})$ and cost of missed healthy operation, $K_{\text{action}} = (1 - P(t_{\text{test}}))\,[C_1(t_{\text{mission}} - t_{\text{test}})]$.

At each a decision point, t_{test}, the failure mitigation algorithm has to determine the total costs of action and inaction, based on the cost of the decision:

$$K_{\text{inaction}} = C_f \cdot P(t_{\text{test}}), \tag{3.73}$$

$$K_{\text{action}} = (1 - P_{(\text{test})}) \cdot [C_1(t_{\text{mission}} - t_{\text{test}})]\,. \tag{3.74}$$

It is reasonable, then, to select and to base action on the prediction interval, t_a, including the necessary time to act, so that the two cost values are equal, $K_{\text{inaction}} = K_{\text{action}}$.

$$K_{\text{inaction}} = C_f P(RUL - t_a/2), \tag{3.75}$$

$$K_{\text{action}} = (1 - P(RUL - t_a/2))C_1 \tag{3.76}$$

If it is possible to interrupt the mission and repair the system, then Equations 3.75–3.76 should be appropriately augmented.

3.10 Conclusions

Reliability of a drive design, as discussed to some extent earlier, is based on the probability of failure, and is determined by the characteristics and operating conditions of its components and subsystems, as well as its architecture. It is a property of a specific design class, and determined, explicitly or implicitly, by its design. Diagnosis of faults and failure prognosis of an electrical drive rather than a class of drives results in decisions about its architecture and control during its operation. They are "personalized" and they cannot be determined or estimated at the design stage. In that sense, they produce failure rates of the drive they are used are conceptually different than those predicted by the reliability prediction of a design.

Decisions to manage a fault made during operation, based on diagnosis and prognosis, can improve significantly the quality of a drive. As discussed in the section on costs, quality of operation is not only the narrow definition of reliability but includes operation at a reduced performance, and even appropriate modification and cancellation of a mission, thus decreasing the total costs.

Combining the two, by incorporating diagnosis and prognosis along with the resulting decisions making considerations and mechanisms in the operation of a drive design, increases the reliability of this design, especially through the expanded cost measures that go beyond failure rates.

In order to apply diagnosis and prognosis tools to the improvement of drive performance (including reliability) a lot of progress has been made in the last decades. Three aspects remain if not elusive, at least topics of rigorous research: estimation of the Remaining Useful Life through minimal supervised learning, minimization of the prediction interval, and increase of the meaningful and useful prediction time.

Bibliography

KOYO: Calculation of service life, 2020. URL https://koyo.jtekt.co.jp/en/support/bearing-knowledge/5-2000.html.

SCHATZ Bearing Life Calculation, 2020. URL http://https://schatzbearing.com/ball-bearing-technical-info/engineering-calculations/.

S. Acharya, A. Anurag, G. Gohil, S. Hazra, and S. Bhattacharya. Mission profile based reliability analysis of a medium voltage power conversion architecture for PMSG based wind energy conversion system. In *2018 IEEE Industry Applications Society Annual Meeting (IAS)*, pages 1–6, 2018. doi: 10.1109/IAS.2018.8544579.

M Ahsan, S T Hon, C Batunlu, and A Albarbar. Reliability Assessment of IGBT Through Modelling and Experimental Testing. *IEEE Access*, 8:39561–39573, 2020.

P Arumugam, T Hamiti, and C Gerada. Turn–turn short circuit fault management in permanent magnet machines. *IET Electric Power Applications*, 9(9):634–641, 2015.

A. M. Bazzi, A. Dominguez-Garcia, and P. T. Krein. Markov reliability modeling for induction motor drives under field-oriented control. *IEEE Transactions on Power Electronics*, 27(2):534–546, 2012. doi: 10.1109/TPEL.2011.2168543.

I. Bolvashenkov, J. Kammermann, and H. Herzog. Research on reliability and fault tolerance of multi-phase traction electric motors based on Markov models for multi-state systems. In *2016 International Symposium on Power Electronics, Electrical Drives, Automation and Motion (SPEEDAM)*, pages 1166–1171, 2016. doi: 10.1109/SPEEDAM.2016.7525928.

J. G. Cintron-Rivera, S. N. Foster, and E. G. Strangas. Mitigation of turn-to-turn faults in fault tolerant permanent magnet synchronous motors. *IEEE Transactions on Energy Conversion*, 30(2):465–475, 2015.

Ernest O. Codier. Reliability growth in real life. In *Proceedings 1968 Annual Symposium on Reliability*, pages 458–469, January 1968.

S. Diao, Z. Makni, J. Bisson, D. Diallo, and C. Marchand. Sensor fault diagnosis for improving the availability of electrical drives. In *IECON 2013 - 39th Annual Conference of the IEEE Industrial Electronics Society*, pages 3108–3113, 2013. doi: 10.1109/IECON.2013.6699625.

Patrick D. T. O'Connor and Andre Kleyner. *Practical Reliability Engineering*. J. Wiley, 2011.

H. T. Eickhoff, R. Seebacher, A. Muetze, and E. G. Strangas. Post-fault operation strategy for single switch open-circuit faults in electric drives. *IEEE Transactions on Industry Applications*, 54(3):2381–2391, 2018.

G. El Murr, A. Griffo, J. Wang, Z. Q. Zhu, and B. Mecrow. Reliability assessment of fault tolerant permanent magnet AC drives. In *IECON 2015 - 41st Annual Conference of the IEEE Industrial Electronics Society*, pages 002777–002782, 2015. doi: 10.1109/IECON.2015.7392522.

Seymour Geisser. *Diagnosis and Prediction*. Springer, 1999.

P. Giangrande, V. Madonna, S. Nuzzo, and M. Galea. Moving toward a reliability-oriented design approach of low-voltage electrical machines by including insulation thermal aging considerations. *IEEE Transactions on Transportation Electrification*, 6(1):16–27, 2020. doi: 10.1109/TTE.2020.2971191.

A. Goel and R. J. Graves. Electronic system reliability: collating prediction models. *IEEE Transactions on Device and Materials Reliability*, 6(2):258–265, 2006. doi: 10.1109/TDMR.2006.876570.

D. Hirschmann, D. Tissen, S. Schroder, and R. W. De Doncker. Reliability prediction for inverters in hybrid electrical vehicles. *IEEE Transactions on Power Electronics*, 22(6):2511–2517, 2007. doi: 10.1109/TPEL.2007.909236.

Z Huang, A Reinap, and M Alaküla. Degradation and fatigue of epoxy impregnated traction motors due to thermal and thermal induced mechanical stress - part I: Thermal mechanical simulation of single wire due to evenly distributed temperature. In *8th IET International Conference on Power Electronics, Machines and Drives (PEMD 2016)*, pages 1–6, 2016a.

Z Huang, A Reinap, and M Alaküla. Degradation and fatigue of epoxy impregnated traction motors due to thermal and thermal induced mechanical stress - part II: Thermal mechanical simulation of multiple wires due to evenly and unevenly distributed temperature. In *8th IET International Conference on Power Electronics, Machines and Drives (PEMD 2016)*, pages 1–5, 2016b.

A. Kersten, K. Oberdieck, A. Bubert, M. Neubert, E. A. Grunditz, T. Thiringer, and R. W. De Doncker. Fault detection and localization for limp home functionality of three-level NPC inverters with connected neutral point for electric vehicles. *IEEE Transactions on Transportation Electrification*, 5(2):416–432, 2019. doi: 10.1109/TTE.2019.2899722.

T. Khawaja, G. Vachtsevanos, and B. Wu. Reasoning about uncertainty in prognosis: a confidence prediction neural network approach. In *NAFIPS 2005 - 2005 Annual Meeting of the North American Fuzzy Information Processing Society*, pages 7–12, 2005. doi: 10.1109/NAFIPS.2005.1548498.

B U Kim, D Goodman, M Li, J Liu, and J Li. Improved reliability-based decision support methodology applicable in system-level failure diagnosis and prognosis. *IEEE Transactions on Aerospace and Electronic Systems*, 50(4):2630–2641, 2014.

A. Lisnianski, I. Frenkel, and Y. Ding. *Multi-state System Reliability Analysis and Optimization for Engineers and Industrial Managers*.

V. Loll. From reliability-prediction to a reliability-budget. In *Annual Reliability and Maintainability Symposium. 1998 Proceedings. International Symposium on Product Quality and Integrity*, pages 421–427, 1998. doi: 10.1109/RAMS.1998.653814.

K. Ma, U. Choi, and F. Blaabjerg. Prediction and validation of wear-out reliability metrics for power semiconductor devices with mission profiles in motor drive application. *IEEE Transactions on Power Electronics*, 33(11):9843–9853, 2018.

P Maussion, A Picot, M Chabert, and D Malec. Lifespan and aging modeling methods for insulation systems in electrical machines: A survey. In *2015 IEEE Workshop on Electrical Machines Design, Control and Diagnosis (WEMDCD)*, pages 279–288, 2015.

B. Mirafzal. Survey of fault-tolerance techniques for three-phase voltage source inverters. *IEEE Transactions on Industrial Electronics*, 61(10):5192–5202, 2014.

M. Molaei, H. Oraee, and M. Fotuhi-Firuzabad. Markov model of drive-motor systems for reliability calculation. In *2006 IEEE International Symposium on Industrial Electronics*, volume 3, pages 2286–2291, 2006. doi: 10.1109/ISIE.2006.295929.

H. Pei, D. Brown, and G. Vachtsevanos. Prognosis-enhanced reconfiguration control. In *2010 International Conference on Networking, Sensing and Control (ICNSC)*, pages 71–75, 2010.

H. W. Penrose. Evaluating reliability of insulation systems for electric machines. In *2014 IEEE Electrical Insulation Conference (EIC)*, pages 421–424, 2014. doi: 10.1109/EIC.2014.6869422.

Y Ran, Z Meimei, L Hui, L Wei, W Xiao, and L Haiyang. Reliability Modeling and Analysis on Metallized Film Capacitors for MMC. In *2019 10th International Conference on Power Electronics and ECCE Asia (ICPE 2019 - ECCE Asia)*, pages 1854–1860, 2019.

M. Rausand and A. Høyland. *SYSTEM RELIABILITY THEORY Models, Statistical MMethods, and Applications*. Wiley Interscience, 2004.

F. Richardeau and T. T. L. Pham. Reliability calculation of multilevel converters: Theory and applications. *IEEE Transactions on Industrial Electronics*, 60(10): 4225–4233, 2013. doi: 10.1109/TIE.2012.2211315.

Paolo Seri and Gian Carlo Montanari. A voltage threshold in operating condition of PWM inverters and its impact on reliability of insulation systems in electrification transport applications. *IEEE Transactions on Transportation Electrification*, 2020.

X Si. An Adaptive Prognostic Approach via Nonlinear Degradation Modeling: Application to Battery Data. *IEEE Transactions on Industrial Electronics*, 62(8): 5082–5096, 2015.

K.K Simons. Confidence and prediction interval determination. *IBM Technical Disclosure Bulletin*, 15(1):123–124, June 1972.

N. Sintamarean, F. Blaabjerg, H. Wang, and Y. Yang. Real field mission profile oriented design of a SiC-based PV-inverter application. *IEEE Transactions on Industry Applications*, 50(6):4082–4089, 2014. doi: 10.1109/TIA.2014.2312545.

T. Somes and R. Gerstenkorn. Impacts of reduced motor cooling on reliability. In *2015 IEEE Petroleum and Chemical Industry Committee Conference (PCIC)*, pages 1–6, 2015. doi: 10.1109/PCICON.2015.7435131.

Y. Song and B. Wang. Survey on reliability of power electronic systems. *IEEE Transactions on Power Electronics*, 28(1):591–604, 2013. doi: 10.1109/TPEL.2012.2192503.

Y. Song and B. Wang. Evaluation methodology and control strategies for improving reliability of HEV power electronic system. *IEEE Transactions on Vehicular Technology*, 63(8):3661–3676, 2014. doi: 10.1109/TVT.2014.2306093.

E G Strangas, S Aviyente, J D Neely, and S S H Zaidi. The Effect of Failure Prognosis and Mitigation on the Reliability of Permanent-Magnet AC Motor Drives. *IEEE Transactions on Industrial Electronics*, 60(8):3519–3528, 2013.

L. Tang, J. DeCastro, G. Kacprzynski, K. Goebel, and G. Vachtsevanos. Filtering and prediction techniques for model-based prognosis and uncertainty management. In *2010 Prognostics and System Health Management Conference*, pages 1–10, 2010.

E. Ugur, S. Dusmez, and B. Akin. An investigation on diagnosis-based power switch lifetime extension strategies for three-phase inverters. *IEEE Transactions on Industry Applications*, 55(2):2064–2075, 2019.

H Wang, M Liserre, F Blaabjerg, P de Place Rimmen, J B Jacobsen, T Kvisgaard, and J Landkildehus. Transitioning to Physics-of-Failure as a Reliability Driver in Power Electronics. *IEEE Journal of Emerging and Selected Topics in Power Electronics*, 2(1):97–114, 2014.

B. A. Welchko, T. M. Jahns, and T. A. Lipo. Short-circuit fault mitigation methods for interior PM synchronous machine drives using six-leg inverters. In *2004 IEEE 35th Annual Power Electronics Specialists Conference (IEEE Cat. No.04CH37551)*, volume 3, pages 2133–2139 Vol.3, 2004.

Jinquan Xu, Hong Guo, Xiaolin Kuang, and Tong Zhou. Reliability Analysis Approach to Fault Tolerant Permanent Magnet Synchronous Motor System. In *2016 IEEE Vehicle Power and Propulsion Conference, VPPC 2016 - Proceedings*. Institute of Electrical and Electronics Engineers Inc., dec 2016.

M. Zeller and F. Montrone. Combination of Component Fault Trees and Markov Chains to Analyze Complex, Software-Controlled Systems. In *2018 3rd International Conference on System Reliability and Safety (ICSRS)*, pages 13–20, 2018. doi: 10.1109/ICSRS.2018.8688854.

Q. Zhou, S. Xue, J. Li, C. Xiang, and S. Chen. Evaluation of wind power converter reliability considering multi-time scale and mission profiles. In *2016 IEEE International Conference on High Voltage Engineering and Application (ICHVE)*, pages 1–4, 2016. doi: 10.1109/ICHVE.2016.7800597.

Index

Fault Diagnosis, Prognosis, and Reliability for Electrical Machines and Drives. First Edition.
Elias G. Strangas, Guy Clerc, Hubert Razik, and Abdenour Soualhi.